FUNDAMENTALS OF MOLECULAR SPECTROSCOPY

T0200681

Fourth Edition

Colin N. Banwell

Tessé France

Elaine M. McCash

Department of Chemistry
University of York

The McGraw-Hill Companies

London · New York · St Louis · San Francisco · Auckland
Bogotá · Caracas · Lisbon · Madrid · Mexico · Milan
Montreal · New Delhi · Panama · Paris · San Juan
São Paulo · Singapore · Sydney · Tokyo · Toronto

Published by
McGraw-Hill Publishing Company
Shoppenhangers Road, Maidenhead, Berkshire, SL6 2QL, England
Telephone 01628 23432
Fax 01628 770224

British Library Cataloguing in Publication Data
Banwell, C. N.
  Fundamentals of Molecular Spectroscopy.
  4Rev.ed
  I. Title II. McCash, Elaine
  535.84

  ISBN 0–07–707976–0

Library of Congress Cataloging-in-Publication Data
Banwell, C. N.
  Fundamentals of molecular spectroscopy / Colin Banwell, Elaine
McCash. — 4th ed.
    p. cm.
  Includes bibliographical references and index.
  ISBN 0–07–707976–0
  1. Molecular spectroscopy.   I. McCash, Elaine.
  II. Title.
  QD96.M65B36 1994
  543′.0858—dc20   94–869
                                                          CIP

McGraw-Hill
A Division of The McGraw-Hill Companies

45 CUP 10

Typeset by Keyword Publishing Services, London
Printed and bound by CPI Group (UK) Ltd, Croydon, CR0 4YY

To
Professor Norman Sheppard, FRS
for bringing together the partnership which produced this volume,
and for his great influence on the teaching and practice of chemical spectroscopy
over a period of some forty years.

CONTENTS

Having finally retired to the peace and tranquility of rural south-west France, it came as something of a shock to be firmly reminded by my publisher that the apparently statutory eight years had passed since the last edition of this text. Even worse, for several years my only connection with spectroscopy has been the fact that France now uses nuclear magnetic resonance spectroscopy routinely in the quality control of its wines—a tenuous, but nonetheless very pleasant, connection. With some misgivings, therefore, I accepted McGraw-Hill's suggestion that I find a co-author to do the necessary work for me. My fears were groundless—not only was the experience of again working with a young and lively mind a great pleasure, but I feel sure we have also produced a much-improved book. A bonus was that my co-author's spouse, being a physicist, was able to contribute a different viewpoint to our discussions (which almost never became arguments), and greatly helped us to clarify our writing in several areas.

The most obvious change in this revision is the addition of a totally new chapter on the spectroscopy of surfaces and solids. These techniques require very sensitive instrumentation, and it is only in recent years that some of them have become routine; they are included here because the topic has most certainly come of age and will become increasingly important in the future. Many rely on the use of high-intensity sources, such as the synchrotron, or on the application of Fourier transform principles. The latter were introduced in the last edition, but, as their use has become much more widespread, they receive an enhanced coverage in this, some techniques sections having been completely revised to take account of the new methods.

Many of the example spectra used in the book have been re-recorded on modern instruments, illustrating their better sensitivity or resolution. Also many of the diagrams have been redrawn, often taking advantage of the larger format for the book to include more illustrative detail, and of suggestions from my co-author, to whom students perhaps found it easier to offer criticisms of the book, rather than directly to me. The basics of spectroscopy do not change, however, and little, except for some very historical and outdated material, has been omitted from the previous edition. A few sections have been moved as more logical groupings became available to us with the introduction of the new chapter, and much rewriting appears.

Previous users of the book will find that there has been no change in our overall philosophy; we still aim to offer a book which, with the minimum of mathematics, will enable a student to grasp the basics of spectroscopy and to gain a clear mental picture of the principles and processes which underlie the subject.

As always the thoughts and suggestions of a great many people have been incorporated into this edition, and it would be impossible to thank them all individually. But again, as always, paramount among these is Professor Sheppard of the University of East Anglia, whose comments have been legion, kind, and invariably totally to the point. Without his influence when I was a student I would not have even been able to contemplate writing the first edition of this book so many years ago, and his guiding hand has ever been within reach since. May it long so continue.

C. N. Banwell
Tessé

One of the very first undergraduate textbooks I purchased, as a young student, was the second edition of a book entitled *Fundamentals of Molecular Spectroscopy* by Colin Banwell. It is a book I have used continuously ever since and I have always regarded it as a key reference when teaching spectroscopy. One can imagine my surprise, therefore, when Professor Sheppard telephoned and asked if I had ever considered writing a book, and would I consider co-writing the fourth edition. He can take pleasure in knowing that, for the first time in my life, I was reduced to total silence!

It is very difficult to alter a book which one looks upon as an old friend, but, after a week of well disguised relaxation in the depths of the French countryside Colin and I were well under way. Our shared philosophy and working approach, combined with my co-authors' magnanimity, cemented an excellent working relationship—as effective both when working at the same desk as when we were 600 miles apart.

I must express my gratitude to all the friends and colleagues who have provided spectra, ideas, and moral support during the production of the fourth edition. I am particularly indebted to Professor Michael Chesters (University of Nottingham) and Dr Elaine Seddon (SERC Daresbury Laboratory) for a number of enlightening conversations and to my research students who have carried a heavier burden than would normally be expected at this stage of their careers. I am also grateful to all the undergraduates in York and Cambridge who have taught me so much about molecular spectroscopy. Finally, I would like to thank my co-author's spouse for providing a soothing hand when it was needed.

E. M. McCash
York
June, 1993

INTRODUCTION

1.1 CHARACTERIZATION OF ELECTROMAGNETIC RADIATION

Molecular spectroscopy may be defined as the study of the interaction of electromagnetic waves and matter. Throughout this book we shall be concerned with what spectroscopy can tell us of the structure of matter, so it is essential in this first chapter to discuss briefly the nature of electromagnetic radiation and the sort of interactions which may occur; we shall also consider, in outline, the experimental methods of spectroscopy.

Electromagnetic radiation, of which visible light forms an obvious but very small part, may be considered as a simple harmonic wave propagated from a source and travelling in straight lines except when refracted or reflected. The properties which undulate—corresponding to the physical displacement of a stretched string vibrating, or the alternate compressions and rarefactions of the atmosphere during the passage of a sound wave—are interconnected electric and magnetic fields. We shall see later that it is these undulatory fields which interact with matter giving rise to a spectrum.

It is trivial to show that any simple harmonic wave has properties of the sine wave, defined by $y = A \sin \theta$, which is plotted in Fig. 1.1. Here y is the displacement with a maximum value A,

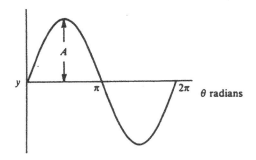

Figure 1.1 The curve of $y = A \sin \theta$.

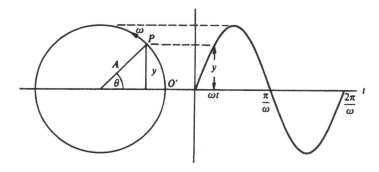

Figure 1.2 The description of a sine curve in terms of the circular motion of a point P at a uniform angular velocity of ω rad s^{-1}

and θ is an angle varying between 0 and 360° (or 0 and 2π radians). The relevance of this representation to a travelling wave is best seen by considering the left-hand side of Fig. 1.2. A point P travels with uniform angular velocity ω rad s^{-1} in a circular path of radius A; we measure the time from the instant when P passes O' and then, after a time t seconds, we imagine P to have described an angle $\theta = \omega t$ radians. Its vertical displacement is then $y = A \sin \theta = A \sin \omega t$, and we can plot this displacement against time as on the right-hand side of Fig. 1.2. After a time of $2\pi/\omega$ seconds, P will return to O', completing a 'cycle'. Further cycles of P will repeat the pattern and we can describe the displacement as a continuous function of time by the graph of Fig. 1.2.

In one second the pattern will repeat itself $\omega/2\pi$ times, and this is referred to as the *frequency* (v) of the wave. The SI unit of frequency is called the hertz (abbreviated to Hz) and has the dimensions of reciprocal seconds (abbreviated s^{-1}). We may then write:

$$y = A \sin \omega t = A \sin 2\pi v t \tag{1.1}$$

as a basic equation of wave motion.

So far we have discussed the variation of displacement with time, but in order to consider the nature of a *travelling* wave, we are more interested in the distance variation of the displacement. For this we need the fundamental distance–time relationship:

$$x = ct \tag{1.2}$$

where x is the distance covered in time t at a speed c. Combining (1.1) and (1.2) we have:

$$y = A \sin 2\pi v t = A \sin \frac{2\pi v x}{c}$$

and the wave is shown in Fig. 1.3. Besides the frequency v, we now have another property by which we can characterize the wave—its *wavelength* λ, which is the distance travelled during a complete cycle. When the velocity is c metres per second and there are v cycles per second, there are evidently v waves in c metres, or

$$v\lambda = c \qquad \lambda = c/v \quad \text{metres} \tag{1.3}$$

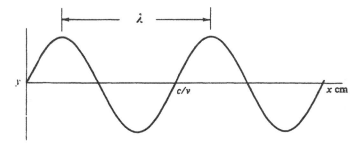

Figure 1.3 The concept of a travelling wave of wavelength λ.

so we have:

$$y = A \sin \frac{2\pi x}{\lambda} \qquad (1.4)$$

In spectroscopy wavelengths are expressed in a variety of units, chosen so that in any particular range (see Fig. 1.4) the wavelength does not involve large powers of ten. Thus, in the microwave region, λ is measured in centimetres or millimetres, while in the infra-red it is usually given in micrometres (μm)—formerly called the *micron*—where:

$$1\,\mu\text{m} = 10^{-6}\,\text{m} \qquad (1.5)$$

In the visible and ultra-violet region λ is usually expressed in nanometres (10^{-9} m), although occasional use is still made of the non-SI ångström unit where:

$$1\,\text{Å} = 10^{-10}\,\text{m} \qquad \text{or} \qquad 1\,\text{nm} = 10^{-9}\,\text{m} = 10\,\text{Å} \qquad (1.6)$$

There is yet a third way in which electromagnetic radiation can be usefully characterized, and this is in terms of the *wavenumber* $\bar{\nu}$. Formally this is defined as the reciprocal of the wavelength expressed in *centimetres*:

$$\bar{\nu} = 1/\lambda \;\text{cm}^{-1} \qquad (1.7)$$

and hence

$$y = A \sin 2\pi\bar{\nu}x \qquad (1.8)$$

It is more useful to think of the wavenumber, however, as the number of complete waves or cycles contained in each centimetre length of radiation.

It is unfortunate that the conventional symbols of wavenumber ($\bar{\nu}$) and frequency (ν) are similar; confusion should not arise, however, if the units of any expression are kept in mind, since wavenumber is invariably expressed in reciprocal centimetres (cm^{-1}) and frequency in cycles per second (s^{-1} or Hz). The two are, in fact, proportional: $\nu = c\bar{\nu}$, where the proportionality constant is the velocity of radiation expressed in *centimetres* per second (that is 3×10^{10} cm s^{-1}).

1.2 THE QUANTIZATION OF ENERGY

Towards the end of the last century experimental data were observed which were quite incompatible with the previously accepted view that matter could take up energy continuously. In 1900 Max Planck published the revolutionary idea that the energy of an oscillator is discontinuous and that any change in its energy content can occur only by means of a jump between two distinct energy states. The idea was later extended to cover many other forms of the energy of matter.

A molecule in space can have many sorts of energy; e.g. it may possess rotational energy by virtue of bodily rotation about its centre of gravity; it will have vibrational energy due to the periodic displacement of its atoms from their equilibrium positions; it will have electronic energy since the electrons associated with each atom or bond are in unceasing motion, etc. The chemist or physicist is early familiar with the electronic energy states of an atom or molecule and accepts the idea that an electron can exist in one of several discrete energy levels: he learns to speak of the energy as being *quantized*. In much the same way the rotational, vibrational, and other energies of a molecule are also quantized—a particular molecule can exist in a variety of rotational, vibrational, etc., energy levels and can move from one level to another only by a sudden jump involving a finite amount of energy.

Consider two possible energy states of a system—two rotational energy levels of a molecule, for example—labelled E_1 and E_2 in the following diagram:

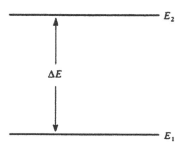

The suffixes 1 and 2 used to distinguish these levels are, in fact, *quantum numbers*. The actual significance of quantum numbers goes far deeper than their use as a convenient label—in particular, we shall later see that analytical expressions for energy levels usually involve an algebraic function of one or more quantum numbers. Transitions can take place between the levels E_1 and E_2 provided the appropriate amount of energy, $\Delta E = E_2 - E_1$, can be either absorbed or emitted by the system. Planck suggested that such absorbed or emitted energy can take the form of electromagnetic radiation and that the frequency of the radiation has the simple form:

$$\nu = \Delta E/h \quad \text{Hz}$$

i.e.

$$\Delta E = h\nu \quad \text{joules} \tag{1.9}$$

where we express our energies E in terms of the joule, and h is a universal constant—Planck's constant. This suggestion has been more than amply confirmed by experiment.

The significance of this is that if we take a molecule in state 1 and direct on to it a beam of radiation of a single frequency ν (monochromatic radiation), where $\nu = \Delta E/h$, energy will be absorbed from the beam and the molecule will jump to state 2. A detector placed to collect the radiation after its interaction with the molecule will show that its intensity has decreased. Also if we use a beam containing a wide range of frequencies ('white' radiation), the detector will show that energy has been absorbed *only* from that frequency $\nu = \Delta E/h$, all other frequencies being undiminished in intensity. In this way we have produced a spectrum—an *absorption* spectrum.

Alternatively the molecule may already be in state 2 and may revert to state 1 with the consequent emission of radiation. A detector would show this radiation to have frequency

$v = \Delta E/h$ only, and the *emission* spectrum so found is plainly complementary to the absorption spectrum of the previous paragraph.

The actual energy differences between the rotational, vibrational, and electronic energy levels are very small and may be measured in joules per molecule (or atom). In these units Planck's constant has the value:

$$h = 6.63 \times 10^{-34} \text{ joules s}$$

Often we are interested in the total energy involved when a gram-molecule of a substance changes its energy state: for this we multiply by the Avogadro number $N = 6.02 \times 10^{23}$.

However, the spectroscopist measures the various characteristics of the absorbed or emitted radiation during transitions between energy states and often, rather loosely, uses frequency, wavelength, and wavenumber as if they were energy units. Thus in referring to 'an energy of 10 cm^{-1}', the spectroscopist means 'a separation between two energy states such that the associated radiation has a wavenumber value of 10 cm^{-1}'. The first expression is so simple and convenient that it is essential to become familiar with wavenumber and frequency energy units if one is to understand the spectroscopist's language. Throughout this book we shall use the symbol ε to represent energy in cm^{-1}.

It cannot be too firmly stressed at this point that the frequency of radiation associated with an energy change does *not* imply that the transition between energy levels occurs a certain number of times each second. Thus an electronic transition in an atom or molecule may absorb or emit radiation of frequency some 10^{15} Hz, but the electronic transition does not itself *occur* 10^{15} times per second. It may occur once or many times and on each occurrence it will absorb or emit an energy quantum of the appropriate frequency.

1.3 REGIONS OF THE SPECTRUM

Figure 1.4 illustrates in pictorial fashion the various, rather arbitrary, regions into which electromagnetic radiation has been divided. The boundaries between the regions are by no means precise, although the molecular processes associated with each region are quite different. Each succeeding chapter in this book deals essentially with one of these processes.

In increasing frequency the regions are:

1. Radiofrequency region: 3×10^6–3×10^{10} Hz; 10 m–1 cm wavelength. Nuclear magnetic resonance (n.m.r.) and electron spin resonance (e.s.r.) spectroscopy. The energy change involved is that arising from the reversal of spin of a nucleus or electron, and is of the order 0.001–10 joules/mole (Chapter 7).
2. Microwave region: 3×10^{10}–3×10^{12} Hz; 1 cm–100 μm wavelength. Rotational spectroscopy. Separations between the rotational levels of molecules are of the order of hundreds of joules per mole (Chapter 2).
3. Infra-red region: 3×10^{12}–3×10^{14} Hz; 100 μm–1 μm wavelength. Vibrational spectroscopy. One of the most valuable spectroscopic regions for the chemist. Separations between levels are some 10^4 joules/mole (Chapter 3).
4. Visible and ultra-violet regions: 3×10^{14}–3×10^{16} Hz; 1 μm–10 nm wavelength. Electronic spectroscopy. The separations between the energies of valence electrons are some hundreds of kilojoules per mole (Chapters 5 and 6).
5. X-ray region: 3×10^{16}–3×10^{18} Hz; 10 nm–100 pm wavelength. Energy changes involving the inner electrons of an atom or a molecule, which may be of order ten thousand kilojoules (Chapter 5).

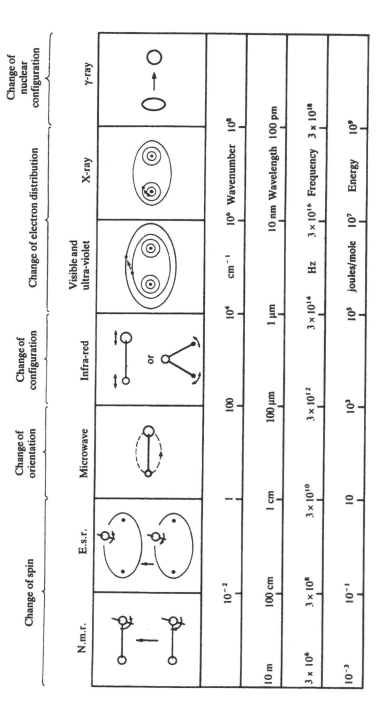

Figure 1.4 The regions of the electromagnetic spectrum.

6. γ-ray region: 3×10^{18}–3×10^{20} Hz; 100 pm–1 pm wavelength. Energy changes involve the rearrangement of nuclear particles, having energies of 10^9–10^{11} joules per gram atom (Chapter 9).

One other type of spectroscopy, that discovered by Raman and bearing his name, is discussed in Chapter 4. This, it will be seen, yields information similar to that obtained in the microwave and infra-red regions, although the experimental method is such that observations are made in the visible region.

In order that there shall be some mechanism for interaction between the incident radiation and the nuclear, molecular, or electronic changes depicted in Fig. 1.4, there must be some electric or magnetic effect produced by the change which can be influenced by the electric or magnetic fields associated with the radiation. There are several possibilities:

1. The radiofrequency region. We may consider the nucleus and electron to be tiny charged particles, and it follows that their spin is associated with a tiny magnetic dipole. The reversal of this dipole consequent upon the spin reversal can interact with the magnetic field of electromagnetic radiation at the appropriate frequency. Consequently all such spin reversals produce an absorption or emission spectrum.
2. The microwave region. A molecule such as hydrogen chloride, HCl, in which one atom (the hydrogen) carries a permanent net positive charge and the other a net negative charge, is said to have a permanent electric dipole moment. H_2 or Cl_2, on the other hand, in which there is no such charge separation, have a zero dipole. If we consider the rotation of HCl (Fig. 1.5, where we notice that if only a pure rotation takes place, the centre of gravity of the molecule must not move), we see that the plus and minus charges change places periodically, and the component dipole moment in a given direction (say upwards in the plane of the paper) fluctuates regularly. This fluctuation is plotted in the lower half of Fig. 1.5, and it is seen to be exactly similar in form to the fluctuating electric field of radiation (cf. Fig. 1.2). Thus interaction can occur, energy can be absorbed or emitted, and the rotation gives rise to a spectrum. All molecules having a permanent moment are said to be 'microwave active'. If there is no dipole, as in H_2 or Cl_2, no interaction can take place and the molecule is 'microwave inactive'. This imposes a limitation on the applicability of microwave spectroscopy.

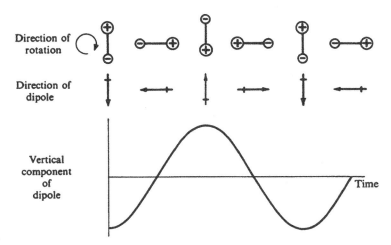

Figure 1.5 The rotation of a polar diatomic molecule, showing the fluctuation in the dipole moment measured in a particular direction.

Figure 1.6 The symmetric stretching vibration of the carbon dioxide molecule with amplitude much exaggerated.

3. The infra-red region. Here it is a vibration, rather than a rotation, which must give rise to a dipole change. Consider the carbon dioxide molecule as an example, in which the three atoms are arranged linearly with a small net positive charge on the carbon and small negative charges on the oxygens:

$$O \overset{\delta-}{\text{—}} C \overset{2\delta+}{\text{—}} O \overset{\delta-}{}$$

During the mode of vibration known as the 'symmetric stretch', the molecule is alternately stretched and compressed, both C—O bonds changing simultaneously, as in Fig. 1.6. Plainly the dipole moment remains zero throughout the whole of this motion, and this particular vibration is thus 'infra-red inactive'.

However, there is another stretching vibration called the anti-symmetrical stretch, depicted in Fig. 1.7. Here one bond stretches while the other is compressed, and vice versa. As the

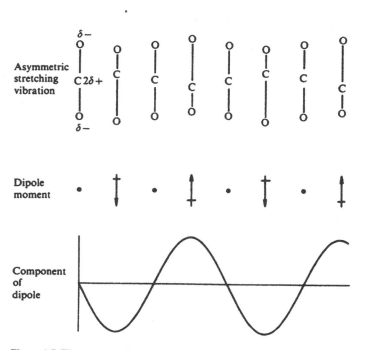

Figure 1.7 The asymmetric stretching vibration of the carbon dioxide molecule, showing the fluctuation in the dipole moment.

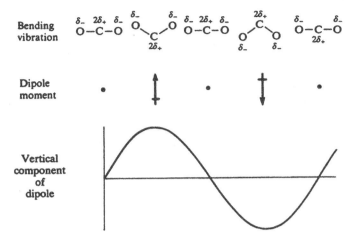

Figure 1.8 The bending motion of the carbon dioxide molecule and its associated dipole fluctuation.

figure shows, there is a periodic alteration in the dipole moment, and the vibration is thus 'infra-red active'. One further vibration is allowed to this molecule (see Chapter 3 for a more detailed discussion), known as the bending mode. This, as shown in Fig. 1.8, is also infra-red active. In neither of these motions does the centre of gravity move. Note particularly that the relative motions of the atoms are very much exaggerated in Figs 1.6, 1.7, and 1.8; in real molecules the displacement of atoms during vibrations is seldom more than about 10 per cent of the bond length.

Although dipole change requirements do impose some limitation on the application of infra-red spectroscopy, the appearance or non-appearance of certain vibration frequencies can give valuable information about the structure of a particular molecule (see Chapter 3).

4. The visible and ultra-violet region. The excitation of a valence electron involves the moving of electronic charges in the molecule. The consequent change in the electric dipole gives rise to a spectrum by its interaction with the undulatory electric field of radiation.
5. There is a rather special requirement for a molecular motion to be 'Raman active'; this is that the electrical *polarizability* of the molecule must change during the motion. This will be discussed fully in Chapter 4.

1.4 REPRESENTATION OF SPECTRA

We show in Fig. 1.9 a highly schematic diagram of a recording spectrometer suitable for use in the ultra-violet, visible and infra-red regions of the spectrum; since it uses a grating (a block of reflective material with a grid of parallel lines ruled on its surface) to select the frequencies which pass through, it is usually called a 'grating spectrometer'; another term frequently used is 'dispersive spectrometer' since the grating 'disperses' the radiation into its frequency components.

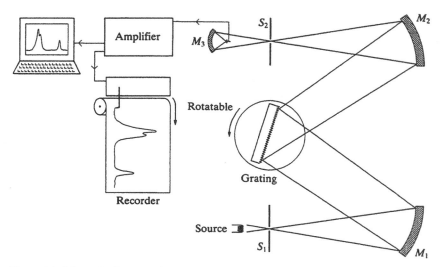

Figure 1.9 Schematic diagram of a grating spectrometer.

Radiation from the source, shown here as a hot or incandescent filament, passes through the vertical entrance slit S_1 (seen from above in the diagram) on to the spherical mirror M_1, from which a parallel beam is reflected on to the grating. The source is 'white', i.e. it radiates energy over a wide range of frequencies and, because of interference at the grid of the grating, different frequencies reflect from the grating at different angles. Thus when the grating is rotated as shown, a succession of frequencies is swept across the mirror M_2, from where it is focused on to the exit slit, S_2. So we see that the frequency of the radiation arriving at S_2 depends on the angle of the grating.

From S_2 the radiation falls on M_3, by which it is focused on to the detector. The latter, perhaps a thermocouple for infra-red radiation or a photomultiplier for visible and ultra-violet, responds with an electrical output proportional to the intensity of the radiation falling on it. This signal is amplified electronically, and then used either to drive a pen, which records the spectrum immediately, or is collected and stored on a computer for later processing and display.

The purpose of the slits is threefold. Firstly, they provide a sharply defined image which can be focused on the detector; secondly, the narrower they are the smaller is the frequency range which passes through, and consequently the resolving power (see Sec. 1.6) of the instrument is increased; and finally they help to prevent stray radiation, perhaps scattered from various components of the spectrometer or arriving from outside, getting to the detector and thus being falsely recorded as part of the spectrum.

The sample (not shown in the diagram) can be placed almost anywhere in the radiation beam. In order to be able to minimize its size, it is often placed close to one of the slits, where the beam is smallest; also it is preferable to put it near the exit slit, S_2, where it is bathed in only a narrow range of frequencies at any one time, rather than near S_1 where it is subjected to the full output of the source and may suffer some degradation from the intense heat or light.

Let us consider now what happens when a spectrum is recorded. If there is no sample present the detector output will, ideally, be constant over the range of frequencies covered by the instrument. On the other hand, if we imagine putting into the beam a substance having just two possible energy levels, E_1 and E_2, the detector output will show a sudden fall at a frequency given by $\nu = (E_2 - E_1)/h$, since some energy at this frequency will be absorbed by the sample and will no longer reach the detector. The resulting trace on the chart paper is illustrated in Fig. 1.10(a), where we show, on the left, the detector output (with 100 per cent at the bottom) and, on the right, the percentage of the energy absorbed by the sample. We would say that the spectrum has been *scanned* between the beginning and ending frequencies, and such a picture is referred to, rather grandly, as a spectrum in the 'frequency domain', to indicate that it records detector output against frequency. In Section 1.8 we shall discuss 'time domain' spectroscopy, where the detector output is recorded as a function of time.

The ideal situation of Fig. 1.10(*a*) is seldom attained. Not only does the source emissivity vary with frequency, but often the sensitivity of the detector is also frequency-dependent. Thus the baseline—the no-sample condition—is seldom horizontal, although matters can usually be arranged so that it is approximately linear. Further, since it is impossible to make the slits infinitely narrow, a *range* of frequencies, rather than just a single frequency, falls on the detector at any given position of the grating. Additionally, we shall see in Sec. 1.7 that no energy transitions in atoms and molecules are absolutely sharp, but they always occur over a range of frequencies. These factors cause a broadening of the spectral absorption and Fig. 1.10(*b*) illustrates a more typical spectrum showing their effects. Also shown are the small peaks and troughs arising from randomly generated electrical signals in the detector, the amplifier, or the pen recorder, usually collectively referred to as 'noise' on the spectrum.

Figure 1.10 The spectrum of a molecule undergoing a single transition: (*a*) idealized and (*b*) usual appearance.

At this point it is helpful to pause briefly in our discussion of spectra and consider what happens to the energy absorbed in the sort of process discussed above. In the ultra-violet, visible, and infra-red regions it is an experimental fact that a given sample continues to show an absorption spectrum for as long as we care to irradiate it—in other words a finite number of sample molecules appear to be capable of absorbing an infinite amount of energy. Plainly the molecules must be able to rid themselves of the absorbed energy. The most common mechanism for this is thermal collisions. An energized molecule collides with its neighbours and gradually loses its excess energy to them as kinetic energy—the sample as a whole becomes warm.

Another mechanism is that energy gained from radiation is lost as radiation once more. A molecule in the ground state absorbs energy at frequency ν and its energy is raised an amount $\Delta E = h\nu$ above the ground state. It is thus in an excited, unstable, condition, but by emitting radiation of frequency ν again, it can revert to the ground state and is able to reabsorb radiation from the source beam once more. In such a case it may be asked how an absorption spectrum can arise at all, since the absorbed energy is re-emitted by the sample. The answer is simply that the radiation is re-emitted in a random direction and the proportion of such radiation reaching the detector is tiny—in fact re-emitted radiation has the same chance of reaching the source as the detector. The net effect, then, is an absorption from the directed beam and, when re-emission occurs, a scattering into the surroundings. The scattered radiation can, of course, be collected and observed as an emission spectrum which will be—with important reservations to be discussed in Chapter 4—the complement of the absorption spectrum. Under the right conditions much of the radiation emitted from a sample can be in a very coherent beam—the so-called laser radiation. We discuss this in Sec. 1.10.

On the other hand, the mechanism for reradiating *radiofrequency* radiation is not particularly efficient. In this region it is quite possible for samples to become 'saturated' and be unable to continue to absorb radiation.

Returning now to our general description of spectra: if there are several energy levels available to the sample, it is very unlikely that there is the same probability of transition between the various levels. This question will be discussed more fully in Sec. 1.7, but we may here note that differences in transition probability will mean that the absorbance (or transmittance) at each absorbing frequency will differ—the spectrum will show several peaks of varying height.

Figure 1.10(*b*) shows the sort of record which is produced by most modern spectrometers, whatever the region in which they operate. However, another form of presentation is sometimes adopted—that showing the *derivative* of the spectral trace. A prime example of a technique using this form is electron paramagnetic resonance which shows relatively broad lines in the microwave region. The derivative of a curve is simply its slope at a given point; in calculus notation, the derivative of the spectral trace is $dA/d\nu$, where A is the energy absorbed or emitted. The derivative record is thus a plot of the slope $dA/d\nu$ against ν. In Fig. 1.11(*a*) we show a typical broad absorption peak and in Fig. 1.11(*b*) its derivative curve. Imagine moving from left to right across the spectrum in Fig. 1.11(*a*); we note that its slope is initially positive and small, that it increases until we reach a point midway up the peak, where it begins to decrease, becoming zero at the top of the peak, where the trace is horizontal. From then on the slope is always negative, increasingly so until half-way down the peak, and then steadily decreasing until it eventually becomes zero. The curve of Fig. 1.11(*b*) traces out this pattern.

Although at first sight more complex, the derivative trace has advantages over the direct record in some circumstances. Firstly, it indicates rather more precisely the centre of a broad absorbance peak; at this point the A curve is horizontal, hence $dA/d\nu$ is zero, and the centre is marked by the intersection of the derivative curve with the axis. Secondly, the width of a peak is often more accurately measured from the derivative curve. The width is normally defined as the 'half-width', i.e. the width of the absorbance peak at half its height, and this, as we can see from

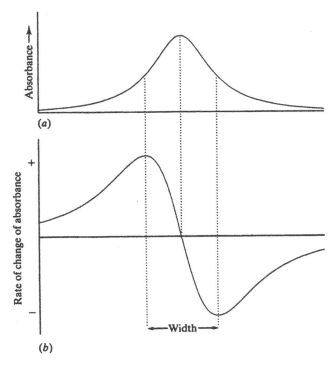

Figure 1.11 The relationship between absorption and derivative spectra: (*a*) a broad absorption band and (*b*) its derivative.

Fig. 1.11(*b*), is shown very clearly by the frequency difference between the maximum and the minimum of the derivative. Finally, for instrumental reasons, it is often better to measure the relative intensities of absorptions from the derivative curve rather than the direct trace.

1.5 BASIC ELEMENTS OF PRACTICAL SPECTROSCOPY

Spectrometers used in various regions of the spectrum naturally differ widely from each other in construction. These differences will be discussed in more detail in the following chapters, but here it will probably be helpful to indicate the basic features which are common to all types of spectrometer. We may, for this purpose, consider absorption and emission spectrometers separately.

1. *Absorption instruments.* Figure 1.12(*a*) shows, in block diagram form, the components of an absorption spectrometer which might be used in the infra-red, visible, and ultra-violet regions. The radiation from a white source is directed by some guiding device (e.g. the mirror of Fig. 1.9) on to the sample, from which it passes through an analyser (e.g. the grating of Fig. 1.9), which selects the frequency reaching the detector at any given time. The signal from the latter passes to a recorder which is synchronized with the analyser so as to produce a trace of the absorbance as the frequency varies.

Placed, often, between the sample and the analyser is a *modulator*; this mechanical or electronic device interrupts the radiation beam a certain number of times per second, usually fixed somewhere between 10 and 1000 times, and its effect is to cause the detector to send an alternating current signal to the recorder, with a fixed frequency of 10–1000 Hz, rather than

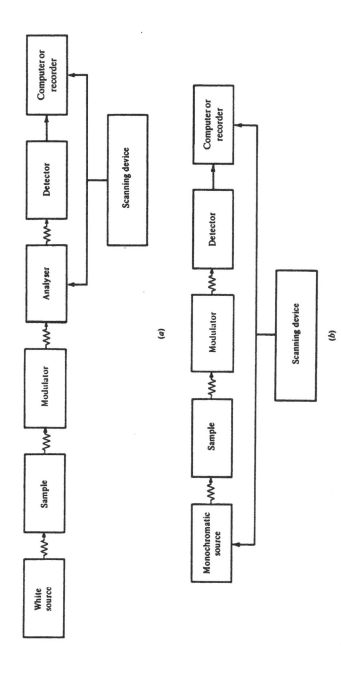

Figure 1.12 Block diagram of a typical scanning absorption spectrometer for use in (*a*) the infra-red, visible and ultra-violet regions, where a 'white' source is available, and (*b*) the microwave and radiofrequency regions where the source can be tuned over a range of frequencies.

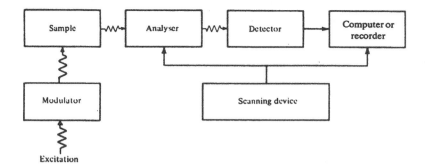

Figure 1.13 Block diagram of a typical emission spectrometer.

the direct current signal which would result from a steady, uninterrupted beam. This has two main advantages: (*a*) the amplifier in the recorder can be of a.c. type which is, in general, simpler to construct and more reliable in operation than a d.c. amplifier, and (*b*) the amplifier can be tuned to select only that frequency which the modulator imposes on the signal, thus ignoring all other signals. In this way stray radiation and other extraneous signals are removed from the spectral trace and a better, cleaner spectrum results.

In the microwave and radiofrequency regions it is possible to construct monochromatic sources whose emission frequency can be varied over a range. In this case, as Fig. 1.12(*b*) shows, no analyser is necessary, the source being, in a sense, its own analyser. Now it is necessary for the recorder to be synchronized with the source-scanning device in order that a spectral trace be obtained.

2. *Emission instruments.* The layout now differs in that the sample, after excitation, is its own source, and it is necessary only to collect the emitted radiation, analyse, and record it in the usual way. Figure 1.13 shows, schematically, a typical spectrometer. The excitation can be thermal or electrical, but often takes the form of electromagnetic radiation. In the latter case it is essential that the detector does not collect radiation directly from the exciting beam, and the two are placed at right angles as shown. A modulator placed between the source of excitation and the sample, together with a tuned detector-amplifier, ensures that the only emission recorded from the sample arises directly from excitation; any other spontaneous emission is ignored.

1.6 SIGNAL-TO-NOISE: RESOLVING POWER

Two other spectroscopic terms may be conveniently discussed at this point since they will recur in succeeding chapters.

1.6.1 Signal-to-Noise Ratio

Since almost all modern spectrometers use some form of electronic amplification to magnify the signal produced by the detector, every recorded spectrum has a background of random fluctuations caused by spurious electronic signals produced by the source or detector, or generated in the amplifying equipment. These fluctuations are usually referred to as 'noise'. In order that a real spectral peak should show itself as such and be sufficiently distinguished from the noise, it must have an intensity some three or four times that of the noise fluctuations (a signal-to-noise ratio of three or four). This requirement places a lower limit on the intensity of observable signals. In Sec. 1.9 we refer briefly to a computer-averaging technique by which it is possible to improve the effective signal-to-noise ratio.

1.6.2 Resolving Power

This is a somewhat imprecise concept which can, however, be defined rather arbitrarily and is often used as a measure of the performance of a spectrometer. We shall here consider it in general terms only.

No molecular absorption takes place at a single frequency only, but always over a spread of frequencies, usually very narrow but sometimes quite large (see Sec. 1.7); it is for this reason that we have up to now drawn spectra with broadened line shapes.

Let us consider two such lines close together, as on the right of Fig. 1.14(*a*): the dotted curve represents the absorption due to each line separately, the full line their combined absorption. We shall first take the exit slit width to be larger than the separation between the lines. Scanning the spectrum plainly involves moving the twin absorbance peaks steadily to the left so that they pass across the exit slit and into the detector; the situation at successive stages is shown in (*b*), (*c*), and (*d*) of Fig. 1.14, the shaded area showing the amount of absorbance which the detector

Figure 1.14 Illustrating the relationship between slit width and resolving power.

would register. At (*e*) of this figure, the absorbance is plotted against frequency, together with the approximate positions of stages (*a*), (*b*), (*c*), and (*d*).

It is quite evident that the separation between the lines has disappeared under these conditions—the lines are not *resolved*. It is equally evident that the use of a much narrower slit would result in their resolution—the *resolving power* would be increased. In fact, provided the slit width is less than the separation between the lines, the detector output will show a minimum between them.

However, it must be remembered that a narrower slit allows less total energy from the beam to reach the detector and consequently the intrinsic signal strength will be less. There comes a point when decreasing the slit width results in such weak signals that they become indistinguishable from the background noise mentioned in the previous paragraph. Thus spectroscopy is a continual battle to find the minimum slit width consistent with acceptable signal-to-noise values. Improvements in resolving power may arise not only as a result of obtaining better dispersion of the radiation by the analyser but also by using a more sensitive detector.

1.7 THE WIDTH AND INTENSITY OF SPECTRAL TRANSITIONS

In the preceding sections we have seen that a spectral transition has the important property of *position*, measured in terms of its frequency, wavelength, or wavenumber; there are two other important properties, its *width* and its *intensity*, and we shall consider these briefly here.

1.7.1 The Width of Spectral Lines

Throughout this chapter we have drawn spectral absorptions and emissions not as infinitely sharp lines but as more or less broad peaks; we have seen that one reason for this is that the mechanical slits in spectrometers are not infinitely narrow and thus allow a *range* of frequencies, rather than a single frequency, to fall on the detector, hence blurring the pattern. While improvements in spectrometer design can improve the resolving power of an instrument, however, there is nonetheless a minimum width inherent in any atomic or molecular transition—the *natural line width*—beyond which no instrument, however superior, will show a sharpening. This width arises essentially because the energy levels of atomic and molecular systems are not precisely determined, but have a certain fuzziness or imprecision. Several factors contribute to this.

1. *Collision broadening.* Atoms or molecules in liquid and gaseous phases are in continual motion and collide frequently with each other. These collisions inevitably cause some deformation of the particles and hence perturb, to some extent, the energies of at least the outer electrons in each. This immediately gives a possible explanation for the width of visible and ultra-violet spectral lines, since these deal largely with transitions between outer electronic shells. Equally vibrational and rotational spectra are broadened since collisions interfere with these motions too. In general, molecular interactions are more severe in liquids than in gases, and gas-phase spectra usually exhibit sharper lines than those of the corresponding liquid.

    In the case of solids, the motions of the particles are more limited in extent and less random in direction, so that solid-phase spectra are often sharp but show evidence of interactions by the splitting of lines into two or more components.

2. *Doppler broadening.* Again in liquids and gases the motion of the particles causes their absorption and emission frequencies to show a Doppler shift; since the motion is random in a given sample, shifts to both high and low frequencies occur and hence the spectral line is broadened. In general, for liquids collision broadening is the most important factor, whereas

for gases, where collision broadening is less pronounced, the Doppler effect often determines the natural line width.

3. *Heisenberg uncertainty principle.* Even in an isolated, stationary molecule or atom the energy levels are not infinitely sharp, due to the operation of a fundamental and very important principle, the uncertainty principle of Heisenberg. In effect this says that, if a system exists in an energy state for a limited time δt seconds, then the energy of that state will be uncertain (fuzzy) to an extent δE where

$$\delta E \times \delta t \approx h/2\pi \approx 10^{-34} \text{ J s} \tag{1.10}$$

where h is again Planck's constant. Thus we see that the lowest energy state of a system *is* sharply defined since, left to itself, the system will remain in that state for an infinite time; thus $\delta t = \infty$ and $\delta E = 0$. However, the lifetime of an excited electronic state, for example, is usually only about 10^{-8} s, which gives a value for δE of about $10^{-34}/10^{-8} = 10^{-26}$ J. A transition between this state and the ground state will thus have an energy uncertainty of δE and a corresponding uncertainty in the associated radiation frequency of $\delta E/h$, which we can write as:

$$\delta v = \frac{\delta E}{h} \approx \frac{h}{2\pi h \delta t} \approx \frac{1}{2\pi \delta t} \tag{1.11}$$

Thus for our example of an excited electronic state lifetime of 10^{-8} s, $\delta v \approx 10^8$ Hz. This apparently large uncertainty is, in fact, small compared with the usual radiation frequency of such transitions, 10^{14}–10^{16} Hz, and so the natural line width is said to be small; in fact, the apparent widths of electronic transitions are far more dependent on collision and Doppler broadening than on energy uncertainties.

On the other hand an excited electron *spin* state may exist for some 10^{-7} s which, from Eq. (1.11), leads to a frequency uncertainty of some 10^7 Hz for a transition. This, compared with the usual frequency of such transitions, 10^8–10^9 Hz, represents a very broad transition indeed, and here the Heisenberg uncertainty relation is by far the most important effect.

Further examples of the application of Heisenberg's principle will be given in later chapters.

1.7.2 The Intensity of Spectral Lines

When discussing spectral intensities there are three main factors to be considered: the likelihood of a system in one state changing to another state—the *transition probability*; the number of atoms or molecules initially in the state from which the transition occurs—the *population*; and the amount of material present giving rise to the spectrum—the *concentration* or *path length* of the sample.

1. *Transition probability.* The detailed calculation of absolute transition probabilities is basically a straightforward matter, but as it involves a knowledge of the precise quantum mechanical wave functions of the two states between which the transition occurs, it can seldom be done with accuracy and is, in any case, beyond the scope of this book. We shall generally content ourselves with qualitative statements about relative transition probabilities without attempting any detailed calculations.

At a much lower level of sophistication, however, it is often possible to decide whether a particular transition is forbidden or allowed (i.e. whether the transition probability is zero or non-zero). This process is essentially the deduction of *selection rules*, which allow us to decide between which levels transitions will give rise to spectral lines, and it can often be carried out through pictorial arguments very like those we have already used in discussing the activity or otherwise of processes in Sec. 1.3.

2. *Population of states.* If we have two levels from which transitions to a third are equally probable, then obviously the most intense spectral line will arise from the level which initially has the greater population. There is a simple statistical rule governing the population of a set of energy levels.

For example, if we have a total of N molecules distributed between two different energy states, a lower and an upper with energies E_{lower} and E_{upper}, respectively, we would intuitively expect most of the molecules to occupy the lower state. Proper statistical analysis bears this out and shows that, *at equilibrium*:

$$\frac{N_{\text{upper}}}{N_{\text{lower}}} = \exp(-\Delta E/kT) \tag{1.12}$$

where $\Delta E = E_{\text{upper}} - E_{\text{lower}}$, T is the temperature in K, and k is a universal constant. The expression is known as the Boltzmann distribution, after its originator, and k, which has a value of $1.38 \times 10^{-23}\,\text{J K}^{-1}$, as Boltzmann's constant. Examples showing the use of this very important expression will recur throughout the remaining chapters.

3. *Path length of sample.* Clearly if a sample is absorbing energy from a beam of radiation, the more sample the beam traverses the more energy will be absorbed from it. We might expect that twice as much sample would give twice the absorption, but a very simple argument shows that this is not so. Consider two identical samples of the same material, S_1 and S_2, and assume that S_1 or S_2 alone absorb 50 per cent of the energy falling on them, allowing the remaining 50 per cent to pass through. If we pass a beam of initial intensity I_0 through S_1, 50 per cent of I_0 will be absorbed and the intensity of the beam leaving S_1 will be $\frac{1}{2}I_0$; if we then pass this beam through S_2 a further 50 per cent will be absorbed, and $\frac{1}{2} \times \frac{1}{2} I_0 = \frac{1}{4} I_0$ will leave S_2. Thus two 50 per cent absorptions in succession do not add up to 100 per cent but only to 75 per cent absorption. An exactly similar relationship exists between the *concentration* of a sample and the amount of energy absorption—a doubling of the concentration produces something less than a doubling of the absorption.

The relationship between concentration (c), path length (l), and the incident and transmitted intensities of radiation, (I_0 and I, respectively) can be expressed in many ways, all based on the *Beer–Lambert law*, which is often written:

$$I/I_0 = \exp(-\kappa cl) \tag{1.13a}$$

where κ is a constant for the particular spectroscopic transition under investigation. To remove the inconvenience of using an exponential function, Eq. (1.13a) may be recast as:

$$I/I_0 = 10^{-\epsilon cl} = T \tag{1.13b}$$

where we have introduced the symbol T for *transmittance*, defined simply as the ratio I/I_0; ϵ is called the *molar absorption coefficient*, which is the term now favoured by the International Union of Pure and Applied Chemistry (IUPAC), although other terms still common are the *decadic extinction coefficient*, or *decadic absorptivity* (*decadic* meaning related to base 10 rather than base e).

When spectroscopy is used to measure concentration of material it is convenient to have a relationship which, unlike Eqs (1.13a) or (1.13b), is linear in concentration. Inverting Eq. (1.13b) and taking logarithms we have:

$$I_0/I = 10^{\epsilon cl} \quad \text{or} \quad \log(I_0/I) = \epsilon cl = A \tag{1.13c}$$

where we define the important quantity A, called the *absorbance*, or *optical density*. A is directly proportional to concentration and, because of its convenience in this respect, some spectrometers are constructed to record spectra directly in absorbance units.

1.8 FOURIER TRANSFORM SPECTROSCOPY

One of the major disadvantages of the conventional method of producing a spectrum such as that of Fig. 1.10(b) is its inherent slowness. Each point of the spectrum has to be recorded separately—the spectrometer is set to start reading at one end, the frequency is swept smoothly across the whole span of the spectrum, and the detector signal is monitored and recorded. The inefficiency of such a method is clear when one considers taking a spectrum with only one or two peaks in it; we have to sweep from one end to the other in order to find the peaks, but most of the time is spent recording nothing but background noise. Initially it was only in the visible and ultra-violet regions that the whole of a spectrum could be recorded simultaneously (on a photographic plate), but the development of Fourier transform (FT) spectroscopy now provides simultaneous and almost instantaneous recording of the whole spectrum in the magnetic resonance, microwave and infra-red regions. In this section we shall briefly discuss the basic ideas of the technique, leaving to later chapters more detailed consideration of its methods and applications.

Although equally applicable to both emission and absorption spectroscopy, it is easier to visualize Fourier transform spectroscopy in terms of emission. For the moment, too, we shall ignore the line-broadening discussed in the previous section and think of the emitted radiation as a pure cosine wave at some quite precise frequency, ν. If a detector capable of responding sufficiently rapidly receives this emitted radiation, its output will be an oscillating signal, again of frequency ν. Note carefully that here we think of the detector output as a function of *time* ('time domain spectroscopy') rather than as the function of frequency ('frequency domain') previously considered.

Now imagine a sample emitting radiation at two different frequencies; a detector receiving the total radiation will 'see' the *sum* of the two cosine waves. We illustrate, diagrammatically, two separate but superimposed waves in Fig. 1.15(a), where the lower frequency wave is shown dashed, and their sum in Fig. 1.15(b); clearly the detector output, plotted against time, shows both an oscillation due to the frequency of the two waves, and a slow increase and decrease in overall amplitude. The frequency of this latter oscillation is often called the 'beat' frequency, by analogy with a similar phenomenon for musical tones, and it arises because the two component waves are sometimes in step, where they totally reinforce each other (points labelled A in the figure), and sometimes out of step, where they cancel (points B).

If we *decrease* the difference in frequency between the component waves, they will get out of step more slowly and it takes longer for them to get back totally in step again. This is shown in Fig. 1.15(c) and (d), with the points marked A and B indicating, as before, where the waves are in step and totally out of step, respectively. The separate waves are drawn in Fig. 1.15(c) and their sum in Fig. 1.15(d), and we note that the beat frequency is lower than in Fig. 1.15(b). In general, it is easily shown that the beat frequency is equal to the difference in frequency of the two component waves.

Mathematically it is simple, although tedious, to resolve a combined wave such as Fig. 1.15(b) or (d) into its components. Essentially each component wave has its own frequency and maximum amplitude, so two components require the evaluation of four unknowns from the composite curve. In principle, then, observations of the time domain signal at four points and solution of four simultaneous equations will yield the information we seek.

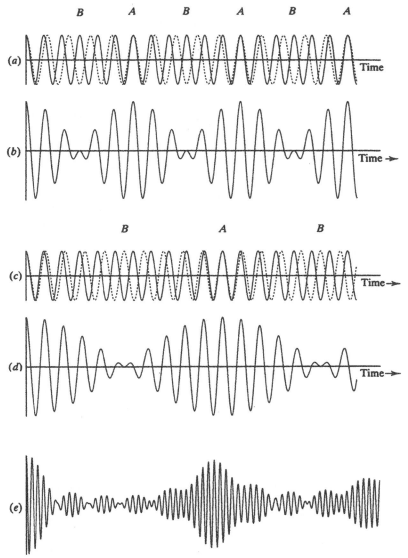

Figure 1.15 Adding cosine waves: (*a*) and (*c*) show the superposition of two cosine waves with slightly different frequencies; (*b*) and (*d*) show their sums. The summation of six cosine waves with different frequencies is shown in (*e*).

Adding more than two cosine waves complicates the resultant combined wave and makes the resolution into components even more tedious, but does not change the principle. Figure 1.15(*e*) shows the result of superimposing six cosine waves of different frequencies. It would need 12 measured points and the solution of 12 simultaneous equations to determine the frequency and relative amplitude of each component. Fortunately there is a simple and quite general way to resolve a complex wave into its frequency components; this is the mathematical process known as Fourier transformation, named after the French mathematician Jean Baptiste Fourier, who developed the method in the early 1800s. Even more fortunately we do not need to know how the process works; it suffices to say that it is essentially a matter of integration of the complex waveform and that it is now carried out very conveniently by computer.

As an example of its operation let us imagine that a suitable detector is responding to the complex waveform of Fig. 1.15(b). A computer receiving the detector output might typically be set to sample it once every millisecond and to store, say, 2000 samplings in separate memory locations; it would thus need to collect the signal for just two seconds. The computer would then apply the Fourier transform process to the stored data, taking a further second or so, and the component cosine wave frequencies and amplitudes could be displayed. Conventionally the display would not take the form of Fig. 1.15(a), where the actual periodic variation of the waves is shown, but would instead be the *spectrum* of the waves—i.e. two very sharp peaks of equal height plotted on a suitable frequency scale to show where the two frequencies occur. This is shown in Fig. 1.16, where the complex wave in Fig. 1.16(a) (taken from Fig. 1.15(b)) is seen to give rise to the spectrum of Fig. 1.16(b). Essentially the Fourier transform has converted the *time* domain plot of Fig. 1.16(a) into the *frequency* domain spectrum of Fig. 1.16(b). The process described above would have taken, perhaps, five seconds only. The detector collects *all* the spectral information virtually simultaneously and the computer 'decodes' that information into the conventional spectrum. It is in this way that the FT method speeds the collection of spectral data, typically, by factors of 10 to 1000.

We must consider one or two more points before we leave this discussion of basic FT spectroscopy. Firstly, recall that real samples do not emit radiation at precise frequencies; as we saw in the previous section, each emission is more or less broadened by various processes, and so each 'line' is really a package of slightly different frequencies. We show a typical peak in Fig. 1.17(a). In Fig. 1.17(b) we see that the peak can be considered as arising from a large number of sample molecules radiating at ν_{max}, the frequency maximum of the peak, with a smaller number radiating at frequencies away from that maximum, the number decreasing as the separation increases. If we wanted to discover the total signal emitted by such a peak we could, if we had time, plot out a cosine wave for each frequency, using an amplitude proportional to the number of molecules radiating at that frequency, and then add all the cosine waves together. To carry out this addition for the few 'packages' of radiating molecules shown in Fig. 1.17(b) would not be difficult; to do it for a real sample, where every single molecule might have a radiation frequency different from its neighbour, is clearly quite impracticable. Fortunately, however, the Fourier transform process is a reciprocal one—just as FT converts a time domain signal to a frequency domain spectrum (e.g. Fig. 1.16), so it will carry out the reverse conversion. Thus if we supply the frequency curve of Fig. 1.17(a) to a computer and carry out the FT, the resultant display will be exactly the same as adding the component cosine waves. The result is shown in Fig. 1.17(c).

We see that a detector receiving the total radiation from a single broad-line emission will show an oscillating signal whose overall amplitude decays smoothly to zero. The oscillation is the beat pattern set up by all the superimposed, but slightly different, cosine waves; the signal

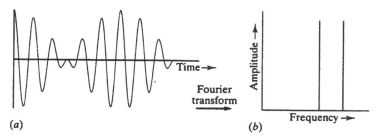

Figure 1.16 The use of the Fourier transform to convert the summed cosine waves of (a) (redrawn from Fig. 1.15(b)) into the frequency spectrum of (b).

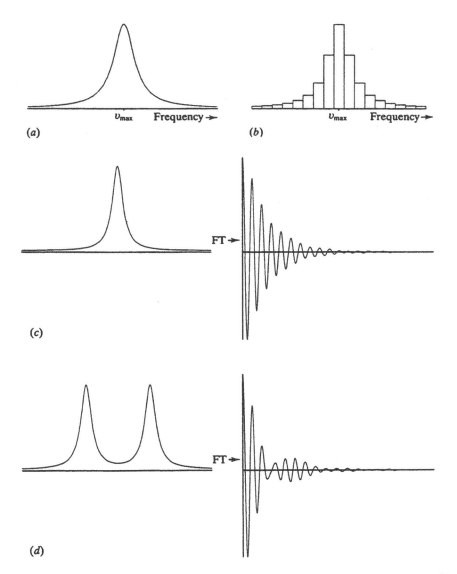

Figure 1.17 (*a*) The frequency distribution of a broad spectral line and (*b*) the histogram of its frequency distribution. The Fourier transform of (*c*) a single broad spectral line and (*d*) a pair of broad lines.

decays because, if we imagine all the waves in the peak to be 'in step' initially, after some time has elapsed the many different frequencies will be very much out of step, and on average half will have their amplitudes in the positive sense and half in the negative, giving a resultant of zero. Another way to think of this is to remember that two waves setting out in step with an infinitesimally small difference in frequency will take an infinite time to get back in step again, i.e. they will never do so. The frequency distribution of Fig. 1.17(*a*) has many infinitely close frequencies within it and so, after a few cycles, none of the individual waves ever get back into step again. If the band had been *infinitely* broad, i.e. containing an infinite number of infinitesimally close neighbours, none would ever have been in step after the first instant, and the FT of such a 'white' source is a single decaying signal with no beats. We shall return to this in a moment.

The corollary of these arguments is that the *rate of decay* of the overall signal depends on the *width* of the original frequency peak. A broader peak contains a larger number of different frequencies and so will decay more rapidly. Clearly both the position and the width of a frequency peak can be recovered from the time domain signal.

Next we should briefly consider the situation when a sample can undergo more than one spectroscopic transition, so that its frequency spectrum shows more than one peak. As before, the overall time domain signal is the sum of the various beat patterns set up by adding cosine waves from each broad peak. We show some typical patterns in Fig. 1.18, all taken from the type of frequency spectra typically found in nuclear magnetic resonance spectroscopy where the FT technique has found particularly wide application. Clearly the complexity of the time domain signal increases as more peaks are added; the point to remember is that all the information regarding the position, intensity, and width of each spectral line on the left of the figure is contained in the time domain pattern on the right; after experimentally observing the latter directly for a brief period, the frequency spectrum can be quickly recovered by Fourier transformation.

Although we stated initially that the FT process is most easily visualized in terms of the emission of radiation, the technique is just as readily applied to absorption. We have already seen that a 'white' source would show a single decay signal with no beats; an approximation to this is shown in Fig. 1.19(*a*), where a *very* broad emission line (which can be considered as a white source covering a limited region of the spectrum) and its Fourier transform are shown. Although the time domain signal decays very rapidly, it does show some beats; this is because the broad emission line is only an approximation to a proper white source. We can now imagine an absorbing sample making some 'holes' in this radiation, as shown on the left-hand side of Fig. 1.19(*b*), with its resulting FT signal on the right. Although we may find it difficult to imagine Fourier-transforming (or even just adding up) the *absence* of radiation at some

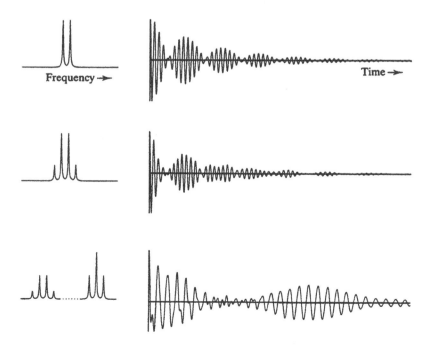

Figure 1.18 The Fourier transform of some selected spectra.

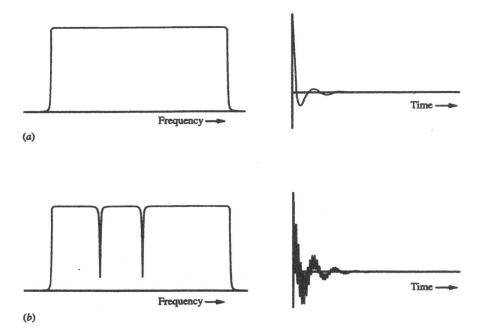

Figure 1.19 (*a*) An approximation to a 'white' source and its Fourier transform. (*b*) Some absorptions from a 'white' source and their Fourier transform.

frequencies, in practice a detector will collect a perfectly sensible signal which can be stored by a computer, transformed, and displayed as the normal absorption spectrum.

Finally we should point out that, although this introduction to the FT process has concentrated on *time* domain signals, monitoring the detector output as a function of time is not the only way in which the total spectrum from a sample can be collected. The student of physics will have recognized that the patterns of Fig. 1.15 can equally be described in terms of interference phenomena. In Chapter 3 we shall find that FT spectroscopy in the infra-red region operates by observing the interference pattern set up when the radiation beam is reflected between a pair of parallel mirrors which are steadily moved towards or away from each other; it is this pattern which is collected, recorded as a function of *distance* between mirrors, and later Fourier-transformed into the corresponding frequency spectrum. Whether to use interferometry or time domain signals in a particular application is based on purely practical grounds; it is difficult to find an infra-red detector with a response time sufficiently fast to be able to record a time domain signal, but it is very easy to reflect infra-red radiation between mirrors. Conversely, in the radiofrequency region, fast response detectors are no problem, but it is not at all easy to reflect the radiation between 'mirrors' in a sensible and coherent fashion.

We see then that the FT method allows us to record spectra much more rapidly than the conventional frequency sweep technique. This in itself is valuable; spectrometers are costly instruments, and the more work we can get from them in a given time the more justified the initial investment. However, rapid data collection brings other benefits, e.g. in being able to record the spectra of transient species such as unstable molecules or intermediates in a chemical reaction. Since the technique essentially reduces the time spent obtaining a spectrum from hours or minutes down to seconds or even fractions of a second, it vastly increases the range of materials which can be studied. There are other advantages of using FT instruments, but we shall leave discussion of them until the relevant chapters on magnetic resonance and infra-red spectroscopy.

1.9 ENHANCEMENT OF SPECTRA: COMPUTER AVERAGING

We have already mentioned, in Sec. 1.6, that the problem of background noise imposes a limitation on the sensitivity of any spectroscopic technique—unless a real signal peak stands out clearly from noise fluctuations it is impossible to be sure that it *is* a signal. A signal-to-noise (S/N) ratio of 3 or 4 is usually reckoned necessary for unambiguous recognition of a signal. There are several ways in which S/N can be improved for a given sample, but all require the expenditure of time. Thus it is possible electronically to damp out oscillations of the recorder pen so that it is less susceptible to high-frequency noise. The baseline of the spectrum will then be smoother, but, because the pen responds more slowly to any change (including changes in signal), one must sweep more slowly across the spectrum. Nor is FT spectroscopy immune from noise—detector and amplifier noises occur during the collection of data and are transformed into spurious frequencies in the spectrum.

The availability of cheaper and more powerful computers offers another method of signal enhancement, which involves recording the spectrum stepwise into a computer. Of course, this is already done if FT is intended, but it is just as easy to sample a frequency domain spectrum at, say, 2000 closely spaced points, and to store the intensity at each point in 2000 separate computer memory locations. This process may then be repeated as many times as we wish, but each time *adding* the new data into that already existing. Although in any one scan a weak signal may not be visible above the noise level, after n summed scans the signal will be n times larger in the store, whereas the noise, being random, will sometimes contribute to the store in a positive sense and sometimes negatively, so it will accumulate less rapidly. In fact it may be shown that n scans increase the noise level in the store by $n^{1/2}$, so the net gain in S/N is $n/n^{1/2} = n^{1/2}$.

If a single scan takes several minutes, as is usually the case in conventional frequency-sweep spectroscopy, the necessity to store 100 scans in order to give an improvement in S/N by a factor of 10 is rather costly in instrument time, so computer averaging is not often used in such measurements. However, the combination of computer averaging with Fourier transform is very powerful indeed. Here one time domain scan can be completed in a second or two, and 100 scans will only occupy a couple of minutes; thus a tenfold gain in S/N can be achieved in a total time often less than that required for a non-enhanced spectrum by ordinary sweep methods.

Other benefits follow from the addition of a computer to a spectrometer. The spectrum of a solvent or other background can be stored in the computer and subtracted from the observed spectrum in order to isolate the spectrum of the substance, or peak intensities can be automatically measured and converted to sample concentrations. Even the operation of the spectrometer itself can usefully be entrusted to the computer—samples can be changed automatically, and the optimum operating conditions can be determined and set for each new sample.

1.10 STIMULATED EMISSION: LASERS

We have already mentioned that, once radiation has been absorbed by a sample, the sample can lose its excess energy either by thermal collisions or by re-emission of radiation. In this section we shall consider the latter process in more detail, because it leads to the very important topic of laser radiation.

Radiation may be emitted by an excited molecule or atom either *spontaneously* or as the result of some stimulus acting on the molecule, called *stimulated emission*. Which of these two processes is most likely to occur in any given case depends on the energy jump involved, i.e. on the frequency of the radiation being emitted. For high-frequency transitions (infra-red, visible, and ultra-violet upwards) spontaneous emission is by far the most likely; conversely, for low-frequency changes (microwave and magnetic resonance) spontaneous emission is unlikely and, if the right conditions obtain, stimulated emission will occur.

Stimulated emission is a resonance phenomenon—an excited state drops to the ground state (emitting radiation of frequency $v = \Delta E/h$, where ΔE is the energy gap) only when a photon (i.e. radiation) of the same frequency v interacts with the system. We illustrate the situation in Fig. 1.20. On the left, in both (*a*) and (*b*), we show the excitation of a molecule by absorption of radiation of frequency v_{ex}. At the right in (*a*) we show spontaneous emission, when radiation $v_{emit.}$ is spontaneously given out, and in (*b*) stimulated emission where a photon of frequency $v_{stim.}$ interacts with the excited state and causes radiation of frequency $v_{emit.}$ to be released. Note particularly that, although we have given different subscripts to $v_{ex.}$, $v_{emit.}$, and $v_{stim.}$ in order to indicate their origins, they all represent *exactly the same frequency*, the frequency $\Delta E/h$.

Radiation emitted under stimulation of this sort has three very important qualities. Firstly, it is of a very *precisely defined* frequency: the excited state does not spontaneously decay, so it is inherently long-lived, which implies (see the discussion of Heisenberg uncertainty in Sec. 1.7.1) a narrow energy level. Secondly, the emitted radiation is *in phase* with the stimulating radiation: the excited state is stimulated to emit by interaction with the oscillating electromagnetic field of $v_{stim.}$, so it is not surprising that the maximum amplitude of the emitted wave coincides with that of $v_{stim.}$. Since the waves are of exactly the same frequency, they remain in phase as they leave the sample. Finally, the stimulating and emitted radiation are *coherent*, which means that they travel in precisely the same direction. In contrast *spontaneous* emission can occur at any time (so each emitted photon is not necessarily in phase with any other), in any direction, and within a more or less broad range of frequencies.

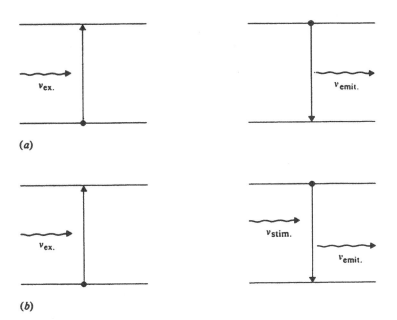

(*a*)

(*b*)

Figure 1.20 Showing (*a*) spontaneous and (*b*) stimulated emission from an excited energy state.

Of course the stimulating radiation of Fig. 1.20 is still present in the system after emission has occurred—it is in no way absorbed—so it can go on to interact with another excited molecule to induce more emission. Equally the emitted radiation has the right frequency to stimulate emission from yet another excited molecule. Clearly, all the time a supply of excited molecules exists, this process is likely to cascade and a great deal of radiation may be emitted coherently. This *amplification* of the original stimulating photon is reflected in the name of the process—*l*ight *a*mplification by *s*timulated *e*mission of *r*adiation, or *laser*.

In fact, as we have said, light (or, more properly, visible radiation) is far more likely to be emitted *spontaneously*, and so not to have the coherent properties of laser radiation. It was in the microwave region that the first successful amplification by stimulated emission was performed (and the process was therefore christened *maser*, standing for *m*icrowave *a*mplification by *s*timulated *e*mission of *r*adiation). For the process to be possible in higher-frequency regions it is necessary to find systems with long-lived excited states so that stimulated, rather than spontaneous, emission may predominate, and this may only be achieved if more than two energy levels are involved.

Consider the three energy levels of Fig. 1.21(*a*). Excitation from the ground state, level 1, to the normal excited state, level 2, can occur by absorption of radiation, as usual. Provided that, as well as emitting spontaneously, level 2 can transfer some molecules into a metastable state (level 3) which cannot easily revert spontaneously to the ground state, then the population of level 3 builds up, and laser action becomes feasible. The ruby laser is an example of this type of three-level system. Ruby is basically aluminium oxide containing a trace (about 0.05 per cent) of chromium ions, which gives it its characteristic colour. A discharge tube wound round a rod of ruby is flashed very briefly to raise the chromium ions into an excited electronic state; they drop rapidly by thermal, non-radiative processes, into a metastable state some $14\,000\,\text{cm}^{-1}$ above the ground state, and they then revert to the ground state, by laser action initiated by one of the infrequent spontaneously emitted photons emitting radiation at about 690 nm wavelength. The decay from excited to metastable state releases quite large amounts of heat, so the ruby must be allowed to cool before another excitation cycle is commenced; it is thus operated as a pulsed laser.

In some cases the laser emission arises by reversion of level 3 to a lower state other than the original level 1, as shown in the four-level system of Fig. 1.21(*b*). This situation occurs particularly when, as is quite possible, levels 3 and 4 belong to an entirely *different* molecular species from levels 1 and 2. For example, in the helium–neon laser it is the helium atoms which are

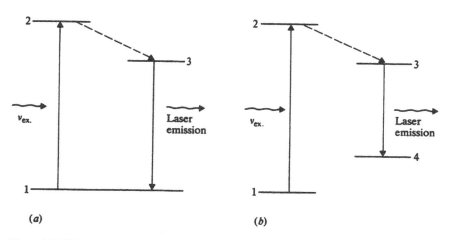

(*a*) (*b*)

Figure 1.21 The energy levels of (*a*) a three-level and (*b*) a four-level system operating as a laser.

initially excited (level 1 to 2), and which then transfer their excitation energy to neon atoms by collisions; this can happen only because neon has an excited state with almost exactly the same energy as the excited state of helium, so a resonance transfer of energy is possible. It also happens that the excited state of neon does not readily undergo an ordinary spectroscopic transition back to its ground state, so the conditions are ideal for laser action. Provided the exciting radiation for helium is maintained, so replenishing the population of excited helium atoms, this type of laser can operate continuously. We shall discuss the precise electronic energy levels involved in more detail in Chapter 5.

The extreme coherence of laser radiation makes it ideal in applications like communications or distance measurement. From the spectroscopic point of view it is becoming increasingly important as new types of spectroscopy are developed but, for the purposes of this book, its most far-reaching application is in the area of vibrational spectroscopy. Raman spectroscopy, discussed at length in Chapter 4, requires an intense monochromatic source for which a laser in the visible or near-infra-red region is ideal.

For virtually all other spectroscopic measurements, however, either a wide-band or a tunable source is desirable. Lasers cannot be wide-band (although the CO_2 laser, to be described in Sec. 3.8.4, comes close to this since it gives a series of closely spaced lasing bands), but recent developments have led to their becoming tunable. To change the emission frequency of a laser it is necessary to be able to modify the energy levels between which laser transitions take place. Solids emitting laser radiation can be subjected to varying temperatures or pressures in order smoothly to change the relevant energies, but the extent of such changes is relatively small. More usefully, lasers made from coloured organic substances in solution—the so-called dye lasers—are widely tunable. In these the active material is usually a rare earth ion held in the centre of an organic 'ligand' which complexes firmly to the ion. The system can be tuned, within a continuum corresponding to a broad region of fluorescence, by building a spectrum dispersing element into the laser cavity, and it is now possible to produce laser emission at any frequency from the near-infra-red to the ultra-violet. Such intense, sharply defined but variable-frequency sources will certainly find application in many areas of spectroscopy.

1.11 SYNCHROTRON RADIATION

As spectroscopic measurements become increasingly demanding, means to overcome sensitivity problems are developed and adapted. Sensitivity can often be improved by using a very powerful radiation source to increase the signal-to-noise ratio and so greatly reduce scanning time. Synchrotron radiation, first observed in the late 1940s at the GEC synchrotron at Schenectady, New York, is such a source, and it has the huge additional advantage of being readily tunable. Since 1940 a number of synchrotrons have been constructed in various locations around the world, and the advantages to spectroscopists are such that there is considerable competition for 'beamtime'.

An electron synchrotron consists of a large ring-shaped vacuum chamber into which pulses of high-velocity electrons are injected after acceleration by electric fields. Magnetic fields, provided by magnets arranged around the ring, force the electrons to follow the curved path. At velocities close to that of light, electrons emit radiation when they are accelerated so that a narrow beam of radiation is produced tangentially to their path.

This so-called synchrotron radiation can be tuned over a wide range of frequencies by changing the accelerating magnetic field. A number of different bending magnets (the two main types of which are called 'wigglers' and 'undulators') can be used on a single synchrotron ring to deliver radiation of a particular frequency range via a beam line to a spectrometer placed outside the ring. The tunability and brilliance of these sources enable many otherwise extremely difficult experiments to be carried out. Some newer experimental techniques, such as EXAFS, that are described in Chapter 8 have been developed using synchrotron sources.

BIBLIOGRAPHY

George, W. O., and H. O. Willis: *Computer Methods in Ultraviolet, Visible and Infra-red Spectroscopy*, Royal Society of Chemistry, 1990

PROBLEMS

(Useful constants: $N = 6.023 \times 10^{23} \, \text{mol}^{-1}$; $k = 1.381 \times 10^{-23} \, \text{J K}^{-1}$; $h = 6.626 \times 10^{-34} \, \text{J s}$; $c = 2.998 \times 10^8 \, \text{m s}^{-1}$.)

1.1 Convert the following spectroscopic quantities as indicated:

$2000 \, \text{cm}^{-1}$	to μm
$0.15 \, \text{nm}$	to Hz
$500 \, \text{nm}$	to cm^{-1}
$9 \, \text{GHz}$	to cm^{-1}

State in which region of the electromagnetic spectrum you would expect each to appear, and what sort of transition each corresponds to.

1.2 The wavelength of the radiation absorbed during a particular spectroscopic transition is observed to be $10 \, \mu\text{m}$. Express this in frequency (Hz) and in wavenumber (cm^{-1}), and calculate the energy change during the transition in both joules per molecule and joules per mole. If the energy change were twice as large, what would be the wavelength of the corresponding radiation?

1.3 Which of the following molecules would show (*a*) a microwave (rotational) spectrum, (*b*) an infra-red (vibrational) spectrum: Br_2, HBr, CS_2?

1.4 How would you expect widening of the slits in a spectrometer to affect its resolution and its sensitivity?

1.5 State the Beer–Lambert law. The transmittance of an aqueous solution of $KMnO_4$ at a certain wavelength is 1 per cent (that is 0.01) for a 10^{-3} molar solution in a 1 cm cell. What is (*a*) its absorbance and (*b*) the molar absorption coefficient of $KMnO_4$?

1.6 A particular molecule is known to undergo spectroscopic transitions between the ground state and two excited states, (*a*) and (*b*), its lifetime in (*a*) being about 10 s and in (*b*) about 0.1 s. Calculate the approximate uncertainty in the excited state energy levels and the widths of the associated spectral 'lines' in hertz.

1.7 A certain transition involves an energy change of $4.005 \times 10^{-22} \, \text{J molecule}^{-1}$. If there are 1000 molecules in the ground state, what is the approximate equilibrium population of the excited state at temperatures of (*a*) 29 K, (*b*) 290 K, and (*c*) 2900 K? What would your answer have been if the energy change were 10 times greater?

1.8 Sketch the curve you would expect from Fourier transformation of a sine wave.

MICROWAVE SPECTROSCOPY

2.1 THE ROTATION OF MOLECULES

We saw in the previous chapter that spectroscopy in the microwave region is concerned with the study of rotating molecules. The rotation of a three-dimensional body may be quite complex and it is convenient to resolve it into rotational components about three mutually perpendicular directions through the centre of gravity—the principal axes of rotation. Thus a body has three principal *moments of inertia*, one about each axis, usually designated I_A, I_B, and I_C.

Molecules may be classified into groups according to the relative values of their three principal moments of inertia—which, it will be seen, is tantamount to classifying them according to their shapes. We shall describe this classification here before discussing the details of the rotational spectra arising from each group.

1. *Linear molecules.* These, as the name implies, are molecules in which all the atoms are arranged in a straight line, such as hydrogen chloride HCl or carbon oxysulphide OCS, illustrated below

$$H—Cl$$

$$O—C—S$$

The three directions of rotation may be taken as (*a*) about the bond axis, (*b*) end-over-end rotation in the plane of the paper, and (*c*) end-over-end rotation at right angles to the plane. It is self-evident that the moments of (*b*) and (*c*) are the same (that is $I_B = I_C$) while that of (*a*) is very small. As an approximation we may say that $I_A = 0$, although it should be noted that this *is* only an approximation (see Sec. 2.3.1).

Thus for linear molecules we have:

$$I_B = I_C \qquad I_A = 0 \tag{2.1}$$

2. *Symmetric tops.* Consider a molecule such as methyl fluoride, where the three hydrogen atoms are bonded tetrahedrally to the carbon, as shown below:

As in the case of linear molecules, the end-over-end rotation in, and out of, the plane of the paper are still identical and we have $I_B = I_C$. The moment of inertia about the C—F bond axis (chosen as the main rotational axis since the centre of gravity lies along it) is now not negligible, however, because it involves the rotation of three comparatively massive hydrogen atoms off this axis. Such a molecule spinning about this axis can be imagined as a top, and hence the name of the class. We have then:

$$\text{Symmetric tops: } I_B = I_C \neq I_A \qquad I_A \neq 0 \tag{2.2}$$

There are two subdivisions of this class which we may mention: if, as in methyl fluoride above, $I_B = I_C > I_A$, then the molecule is called a *prolate* symmetric top; whereas if $I_B = I_C < I_A$, it is referred to as *oblate*. An example of the latter type is boron trichloride, which, as shown, is planar and symmetrical. In this case $I_A = 2I_B = 2I_C$.

3. *Spherical tops.* When a molecule has all three moments of inertia identical, it is called a spherical top. A simple example is the tetrahedral molecule methane CH_4. We have then:

$$\text{Spherical tops: } I_A = I_B = I_C \tag{2.3}$$

In fact these molecules are only of academic interest in this chapter. Since they can have no dipole moment owing to their symmetry, rotation alone can produce no dipole change and hence no rotational spectrum is observable.

4. *Asymmetric tops.* These molecules, to which the majority of substances belong, have all three moments of inertia different:

$$I_A \neq I_B \neq I_C \tag{2.4}$$

Simple examples are water H_2O and vinyl chloride $CH_2{=}CHCl$.

Perhaps it should be pointed out that one can (and often does) describe the classification of molecules into the four rotational classes in far more rigorous terms than have been used above (see, for example, Herzberg, *Molecular Spectra and Molecular Structure*, vol. II). However, for the purposes of this book the above description is adequate.

2.2 ROTATIONAL SPECTRA

We have seen that rotational energy, along with all other forms of molecular energy, is quantized: this means that a molecule cannot have any arbitrary amount of rotational energy (i.e. any arbitrary value of angular momentum) but its energy is limited to certain definite values depending on the shape and size of the molecule concerned. The permitted energy values—the so-called rotational energy *levels*—may in principle be calculated for any molecule by solving the Schrödinger equation for the system represented by that molecule. For simple molecules the mathematics involved is straightforward for tedious, while for complicated systems it is probably impossible without gross approximations. We shall not concern ourselves unduly with this, however, being content merely to accept the results of existing solutions and to point out where reasonable approximations may lead.

We shall consider each class of rotating molecule in turn, discussing the linear molecule in most detail, because much of its treatment can be directly extended to symmetrical and unsymmetrical molecules.

2.3 DIATOMIC MOLECULES

2.3.1 The Rigid Diatomic Molecule

We start with this, the simplest of all linear molecules, shown in Fig. 2.1. Masses m_1 and m_2 are joined by a rigid bar (the bond) whose length is

$$r_0 = r_1 + r_2 \tag{2.5}$$

The molecule rotates end-over-end about a point C, the centre of gravity: this is defined by the moment, or balancing, equation:

$$m_1 r_1 = m_2 r_2 \tag{2.6}$$

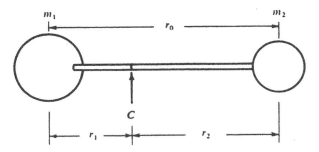

Figure 2.1 A rigid diatomic molecule treated as two masses, m_1 and m_2, joined by a rigid bar of length $r_0 = r_1 + r_2$.

The moment of inertia about C is defined by:

$$
\begin{aligned}
I &= m_1 r_1^2 + m_2 r_2^2 \\
&= m_2 r_2 r_1 + m_1 r_1 r_2 \qquad \text{(from Eq. (2.6))} \\
&= r_1 r_2 (m_1 + m_2)
\end{aligned}
\tag{2.7}
$$

However, from Eqs (2.5) and (2.6):

$$
m_1 r_1 = m_2 r_2 = m_2 (r_0 - r_1)
$$

Therefore,

$$
r_1 = \frac{m_2 r_0}{m_1 + m_2} \quad \text{and} \quad r_2 = \frac{m_1 r_0}{m_1 + m_2}
\tag{2.8}
$$

Replacing (2.8) in (2.7):

$$
I = \frac{m_1 m_2}{m_1 + m_2}\, r_0^2 = \mu r_0^2
\tag{2.9}
$$

where we have written $\mu = m_1 m_2 / (m_1 + m_2)$, and μ is called the *reduced mass* of the system. Equation (2.9) defines the moment of inertia conveniently in terms of the atomic masses and the bond length.

By the use of the Schrödinger equation it may be shown that the rotational energy levels allowed to the rigid diatomic molecule are given by the expression:

$$
E_J = \frac{h^2}{8\pi^2 I}\, J(J+1) \quad \text{joules} \quad \text{where } J = 0, 1, 2, \ldots
\tag{2.10}
$$

In this expression h is Planck's constant, and I is the moment of inertia, either I_B or I_C, since both are equal. The quantity J, which can take integral values from zero upwards, is called the *rotational quantum number*: its restriction to integral values arises directly out of the solution to the Schrödinger equation and is by no means arbitrary, and it is this restriction which effectively allows only certain discrete rotational energy levels to the molecule.

Equation (2.10) expressed the allowed energies in joules; we, however, are interested in differences between these energies, or, more particularly, in the corresponding frequency, $v = \Delta E / h$ Hz, or wavenumber, $\bar{v} = \Delta E / hc$ cm^{-1}, of the radiation emitted or absorbed as a consequence of changes between energy levels. In the rotational region spectra are usually discussed in terms of wavenumber, so it is useful to consider energies expressed in these units. We write:

$$
\varepsilon_J = \frac{E_J}{hc} = \frac{h}{8\pi^2 Ic}\, J(J+1) \quad \text{cm}^{-1} \quad (J = 0, 1, 2, \ldots)
\tag{2.11}
$$

where c, the velocity of light, is here expressed in cm s^{-1}, since the unit of wavenumber is reciprocal *centimetres*.

Equation (2.11) is usually abbreviated to:

$$
\varepsilon_J = BJ(J+1) \quad \text{cm}^{-1} \quad (J = 0, 1, 2, \ldots)
\tag{2.12}
$$

where B, the *rotational constant*, is given by

$$
B = \frac{h}{8\pi^2 I_B c} \quad \text{cm}^{-1}
\tag{2.13}
$$

in which we have used explicitly the moment of inertia I_B. We might equally well have used I_C and a rotational constant C, but the notation of (2.13) is conventional.

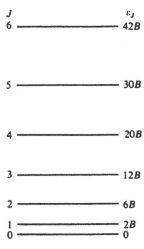

Figure 2.2 The allowed rotational energies of a rigid diatomic molecule.

From Eq. (2.12) we can show the allowed energy levels diagrammatically as in Fig. 2.2. Plainly for $J = 0$ we have $\varepsilon_J = 0$ and we would say that the molecule is not rotating at all. For $J = 1$, the rotational energy is $\varepsilon_1 = 2B$ and a rotating molecule then has its lowest angular momentum. We may continue to calculate ε_J with increasing J values and, in principle, there is no limit to the rotational energy the molecule may have. In practice, of course, there comes a point at which the centrifugal force of a rapidly rotating diatomic molecule is greater than the strength of the bond, and the molecule is disrupted, but this point is not reached at normal temperatures.

We now need to consider *differences* between the levels in order to discuss the spectrum. If we imagine the molecule to be in the $J = 0$ state (the *ground rotational state*, in which no rotation occurs), we can let incident radiation be absorbed to raise it to the $J = 1$ state. Plainly the energy absorbed will be:

$$\varepsilon_{J=1} - \varepsilon_{J-0} = 2B - 0 = 2B \quad \text{cm}^{-1}$$

and, therefore,

$$\bar{v}_{J=0 \to J=1} = 2B \quad \text{cm}^{-1} \tag{2.14}$$

In other words, an absorption line will appear at $2B\,\text{cm}^{-1}$. If now the molecule is raised from the $J = 1$ to the $J = 2$ level by the absorption of more energy, we see immediately:

$$\begin{aligned}
\bar{v}_{J=1 \to J=2} &= \varepsilon_{J=2} - \varepsilon_{J=1} \\
&= 6B - 2B = 4B \quad \text{cm}^{-1}
\end{aligned} \tag{2.15}$$

In general, to raise the molecule from the state J to state $J + 1$, we would have:

$$\begin{aligned}
\bar{v}_{J \to J+1} &= B(J+1)(J+2) - BJ(J+1) \\
&= B[J^2 + 3J + 2 - (J^2 + J)]
\end{aligned}$$

or

$$\bar{v}_{J \to J+1} = 2B(J+1) \quad \text{cm}^{-1} \tag{2.16}$$

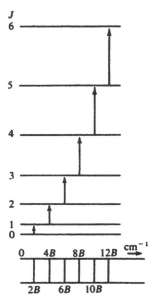

Figure 2.3 Allowed transitions between the energy levels of a rigid diatomic molecule and the spectrum which arises from them.

Thus a stepwise raising of the rotational energy results in an absorption spectrum consisting of lines at $2B$, $4B$, $6B$, ... cm^{-1}, while a similar lowering would result in an identical emission spectrum. This is shown at the foot of Fig. 2.3.

In deriving this pattern we have made the assumption that a transition can occur from a particular level only to its immediate neighbour, either above or below: we have not, for instance, considered the sequence of transitions $J = 0 \rightarrow J = 2 \rightarrow J = 4$ ... In fact, a rather sophisticated application of the Schrödinger wave equation shows that, for this molecule, we need only consider transitions in which J changes by one unit—all other transitions being spectroscopically *forbidden*. Such a result is called a *selection rule*, and we may formulate it for the rigid diatomic rotator as:

$$\text{Selection rule:} \quad \Delta J = \pm 1 \tag{2.17}$$

Thus Eq. (2.16) gives the *whole* spectrum to be expected from such a molecule.

Of course, only if the molecule is asymmetric (heteronuclear) will this spectrum be observed, since if it is homonuclear there will be no dipole component change during the rotation, and hence no interaction with radiation. Thus molecules such as HCl and CO will show a rotational spectrum, while N_2 and O_2 will not. Remember, also, that rotation about the bond axis was rejected in Sec. 2.1: we can now see that there are two reasons for this. Firstly, the moment of inertia is very small about the bond so, applying Eqs (2.10) or (2.11) we see that the energy levels would be extremely widely spaced: this means that a molecule requires a great deal of energy to be raised from the $J = 0$ to the $J = 1$ state, and such transitions do not occur under normal spectroscopic conditions. Thus diatomic (and all linear) molecules are in the $J = 0$ state for rotation about the bond axis, and they may be said to be not rotating. Secondly, even if such a transition should occur, there will be no dipole change and hence no spectrum.

To conclude this section we shall apply Eq. (2.16) to an observed spectrum in order to determine the moment of inertia and hence the bond length. Gilliam *et al.*† have measured the first line ($J = 0$) in the rotation spectrum of carbon monoxide as $3.842\,35\,\text{cm}^{-1}$. Hence, from Eq. (2.16):

$$\bar{v}_{0 \to 1} = 3.842\,35 = 2B \quad \text{cm}^{-1}$$

or
$$B = 1.921\,18 \quad \text{cm}^{-1}$$

Rewriting Eq. (2.13) as: $I = h/8\pi^2 Bc$, we have:

$$I_{CO} = \frac{6.626 \times 10^{-34}}{8\pi^2 \times 2.997\,93 \times 10^{10} \times B} = \frac{27.9907 \times 10^{-47}}{B} \quad \text{kg m}^2$$

$$= 14.5695_4 \times 10^{-47} \text{ kg m}^2$$

where we express the velocity of light in cm s^{-1}, since B is in cm^{-1}. However, the moment of inertia is μr^2 (cf. Eq. (2.9)) and, knowing the relative atomic weights (H = 1.0080) to be C = = 12.0000, O = 15.9994, and the absolute mass of the hydrogen atom to be $1.673\,43 \times 10^{-27}\,\text{kg}$, we can calculate the masses of carbon and oxygen, respectively, as 19.921 68 and $26.561\,36 \times 10^{-27}\,\text{kg}$. The reduced mass is then:

$$\mu = \frac{19.921\,68 \times 26.561\,36 \times 10^{-54}}{46.483\,03 \times 10^{-27}} = 11.383\,65 \times 10^{-27} \text{ kg}$$

Hence:

$$r^2 = \frac{I}{\mu} = 1.2799 \times 10^{-20} \text{ m}^2$$

and
$$r_{CO} = 0.1131 \text{ nm (or } 1.131 \text{ Å)}$$

2.3.2 The Intensities of Spectral Lines

We want now to consider briefly the relative intensities of the spectral lines of Eq. (2.16); for this a prime requirement is plainly a knowledge of the relative probabilities of transition between the various energy levels. Does, for instance, a molecule have more or less chance of making the transition $J = 0 \to J = 1$ than the transition $J = 1 \to J = 2$? We mentioned above calculations which show that a change of $\Delta J = \pm 2$, ± 3, etc., was forbidden—in other words, the transition probability for all these changes is zero. Precisely similar calculations show that the probability of all changes with $\Delta J = \pm 1$ is almost the same—all, to a good approximation, are equally likely to occur.

This does not mean, however, that all spectral lines will be equally intense. Although the intrinsic probability that a single molecule in the $J = 0$ state, say, will move to $J = 1$ is the same as that of a single molecule moving from $J = 1$ to $J = 2$, in an assemblage of molecules, such as in a normal gas sample, there will be different numbers of molecules in each level to begin with, and therefore different total numbers of molecules will carry out transitions between the various levels. In fact, since the intrinsic probabilities are identical, the line intensities will be directly proportional to the initial numbers of molecules in each level.

The first factor governing the population of the levels is the Boltzmann distribution (cf. Sec. 1.7.2). Here we know that the rotational energy in the lowest level is zero, since $J = 0$, so, if we

† Gilliam, Johnson, and Gordy, *Physical Review*, **78**, 140 (1950).

have N_0 molecules in this state, the number in any higher state is given by:

$$N_J/N_0 = \exp(-E_J/kT) = \exp -[-BhcJ(J+1)/kT] \tag{2.18}$$

where, we must remember, c is the velocity of light in cm s^{-1} when B is in cm^{-1}. A very simple calculation shows how N_J varies with J; for example, taking a typical value of $B = 2$ cm^{-1}, and room temperature (say $T = 300$ K), the relative population in the $J = 1$ state is:

$$\frac{N_1}{N_0} = \exp\left(-\frac{2 \times 6.63 \times 10^{-34} \times 3 \times 10^{10} \times 1 \times 2}{1.38 \times 10^{-23} \times 300}\right)$$

$$= \exp(-0.019) \approx 0.98$$

and we see that there are almost as many molecules in the $J = 1$ state, at equilibrium, as in the $J = 0$. In a similar way the two graphs of Fig. 2.4 have been calculated, showing the more rapid decrease of N_J/N_0 with increasing J and with larger B.

A second factor is also required—the possibility of *degeneracy* in the energy states. Degeneracy is the existence of two or more energy states which have exactly the *same* energy. In the case of the diatomic rotator we may approach the problem in terms of its angular momentum.

The defining equations for the energy and angular momentum of a rotator are:

$$E = \frac{1}{2}I\omega^2 \quad \mathbf{P} = I\omega$$

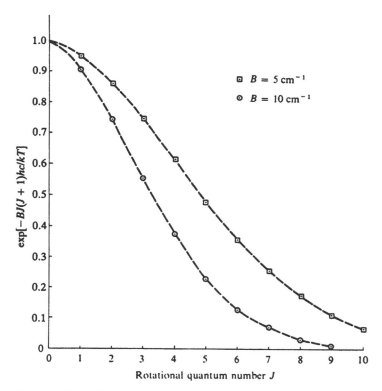

Figure 2.4 The Boltzmann populations of the rotational energy levels of Fig. 2.2. The diagram has been drawn taking values of $B = 5$ and 10 cm^{-1} and $T = 300$ K in Eq. (2.18).

where I is the moment of inertia, ω the rotational frequency (in radians per second), and $\mathbf{P}$ the angular momentum. Rearrangement of these gives:

$$\mathbf{P} = \sqrt{2EI}$$

The energy level expression of Eq. (2.10) can be rewritten:

$$2EI = J(J+1)\,\frac{h^2}{4\pi^2}$$

and hence

$$\mathbf{P} = \sqrt{J(J+1)}\,\frac{h}{2\pi} = \sqrt{J(J+1)}\ \text{units} \qquad (2.19)$$

where, following convention, we take $h/2\pi$ as the fundamental unit of angular momentum. Thus we see that $\mathbf{P}$, like E, is quantized.

Throughout the above derivation $\mathbf{P}$ has been printed in bold face type to show that it is a *vector*—i.e. it has *direction* as well as *magnitude*. The direction of the angular momentum vector is conventionally taken to be along the axis about which rotation occurs and it is usually drawn as an arrow of length proportional to the magnitude of the momentum. The number of different directions which an angular momentum vector may take up is limited by a quantum mechanical law which may be stated:

For integral values of the rotational quantum number (in this case J), the angular momentum vector may only take up directions such that its component along a given reference direction is zero or an integral multiple of angular momentum units.

We can see the implications of this most easily by means of a diagram. In Fig. 2.5 we show the case $J = 1$. Here $\mathbf{P} = \sqrt{1 \times 2}$ units $= \sqrt{2}$, and, as Fig. 2.5(a) shows, a vector of length $\sqrt{2}(= 1.41)$ can have only *three* integral or zero components along a reference direction (here assumed to be from top to bottom in the plane of the paper): $+1$, 0, and -1. Thus the angular

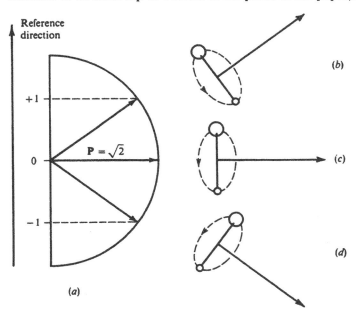

Figure 2.5 The three degenerate orientations of the rotational angular momentum vector for a molecule with $J = 1$.

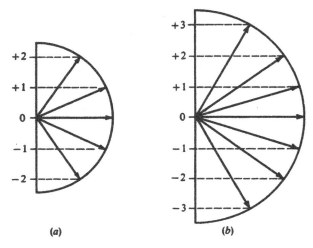

Figure 2.6 The five and seven degenerate rotational orientations for a molecule with $J = 2$ and $J = 3$, respectively.

momentum vector in this instance can be oriented in only three different directions (Fig. 2.5(b) to (d)) with respect to the reference direction. All three rotational directions are, of course, associated with the same angular momentum and hence the same rotational energy: the $J = 1$ level is thus threefold degenerate.

Figure 2.6(a) and (b) shows the situation for $J = 2 (P = \sqrt{6})$ and $J = 3 (P\ 2\sqrt{3})$ with fivefold and sevenfold degeneracy, respectively. In general it may readily be seen that *each energy level is $(2J + 1)$-fold degenerate.*

Thus we see that, although the molecular population in each level decreases exponentially (Eq. (2.18)), the number of degenerate levels available increases rapidly with J. The total relative population at an energy E_J will plainly be:

$$\text{Population} \propto (2J + 1)\exp(-E_J/kT) \qquad (2.20)$$

When this is plotted against J the points fall on a curve of the type shown in Fig. 2.7, indicating that the population rises to a maximum and then diminishes. Differentiation of Eq. (2.20) shows that the population is a maximum at the nearest integral J value to:

$$\text{Maximum population:} \quad J = \sqrt{\frac{kT}{2hcB}} - \frac{1}{2} \qquad (2.21)$$

We have seen that line intensities are directly proportional to the populations of the rotational levels; hence it is plain that transitions between levels with very low or very high J values will have small intensities while the intensity will be a maximum at or near the J value given by Eq. (2.21).

2.3.3 The Effect of Isotopic Substitution

When a particular atom in a molecule is replaced by its isotope—an element identical in every way except for its atomic mass—the resulting substance is identical chemically with the original. In particular there is no appreciable change in internuclear distance on isotopic substitution. There is, however, a change in total mass and hence in the moment of inertia and B value for the molecule.

Considering carbon monoxide as an example, we see that on going from $^{12}C^{16}O$ to $^{13}C^{16}O$ there is a mass increase and hence a decrease in the B value. If we designate the ^{13}C molecule

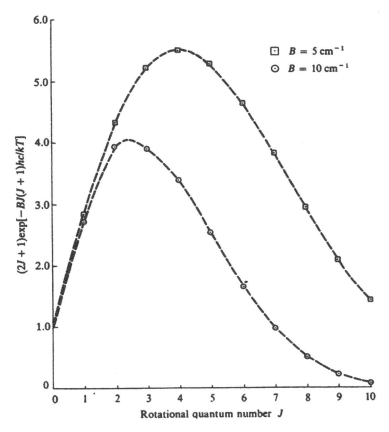

Figure 2.7 The relative populations, including degeneracy, of the rotational energy levels of a diatomic molecule. The diagram has been drawn for the same conditions as Fig. 2.4.

with a prime we have $B > B'$. This change will be reflected in the rotational energy levels of the molecule and Fig. 2.8 shows, much exaggerated, the relative lowering of the ^{13}C levels with respect to those of ^{12}C. Plainly, as shown by the diagram at the foot of Fig. 2.8, the spectrum of the heavier species will show a smaller separation between the lines ($2B'$) than that of the lighter one ($2B$). Again the effect has been much exaggerated for clarity, and the transitions due to the heavier molecule are shown dashed.

Observation of this decreased separation has led to the evaluation of precise atomic weights. Gilliam *et al.*, as already stated, found the first rotational absorption of ^{12}C^{16}O to be at $3.842\,35\,\text{cm}^{-1}$, while that of ^{13}C^{16}O was at $3.673\,37\,\text{cm}^{-1}$. The values of B determined from these figures are:

$$B = 1.921\,18\,\text{cm}^{-1} \quad \text{and} \quad B' = 1.836\,69\,\text{cm}^{-1}$$

where the prime refers to the heavier molecule. We have immediately:

$$\frac{B}{B'} = \frac{h}{8\pi^2 Ic} \cdot \frac{8\pi^2 I'c}{h} = \frac{I'}{I} = \frac{\mu'}{\mu} = 1.046$$

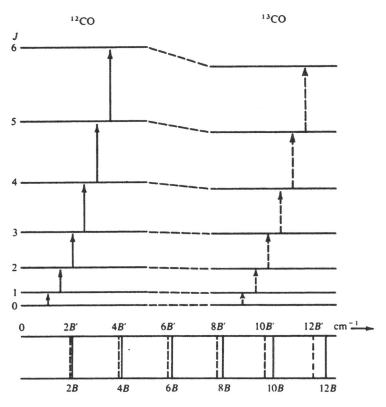

Figure 2.8 The effect of isotopic substitution on the energy levels and rotational spectrum of a diatomic molecule such as carbon monoxide.

where μ is the reduced mass, and the internuclear distance is considered unchanged by isotopic substitution. Taking the mass of oxygen to be 15.9994 and that of carbon-12 to be 12.00, we have:

$$\frac{\mu'}{\mu} = 1.046 = \frac{15.9994m'}{15.9994 + m'} \times \frac{12 + 15.9994}{12 \times 15.9994}$$

from which m', the atomic weight of carbon-13, is found to be 13.0007. This is within 0.02 per cent of the best value obtained in other ways.

It is noteworthy that the data quoted above were obtained by Gilliam *et al.* from $^{13}C^{16}O$ molecules in natural abundance (i.e. about 1 per cent of ordinary carbon monoxide). Thus, besides allowing an extremely precise determination of atomic weights, microwave studies can give directly an estimate of the abundance of isotopes by comparison of absorption intensities.

2.3.4 The Non-rigid Rotator

At the end of Sec. 2.3.1 we indicated how internuclear distances could be calculated from microwave spectra. It must be admitted that we selected our data carefully at this point—spectral lines for carbon monoxide, other than the first, would not have shown the constant $2B$ separation predicted by Eq. (2.16). This is shown by the spectrum of hydrogen fluoride given

Table 2.1 Rotation spectrum of hydrogen fluoride

J	$\bar{\nu}_{obs.}$† (cm^{-1})	$\bar{\nu}_{calc.}$‡ (cm^{-1})	$\Delta\bar{\nu}_{obs.}$ (cm^{-1})	B ($= \frac{1}{2}\Delta\bar{\nu}$)	r (nm)
0	41.08	41.11			
			41.11	20.56	0.0929
1	82.19	82.18			
			40.96	20.48	0.0931
2	123.15	123.14			
			40.85	20.43	0.0932
3	164.00	163.94			
			40.62	20.31	0.0935
4	204.62	204.55			
			40.31	20.16	0.0938
5	244.93	244.89			
			40.08	20.04	0.0941
6	285.01	284.93			
			39.64	19.82	0.0946
7	324.65	324.61			
			39.28	19.64	0.0951
8	363.93	363.89			
			38.89	19.45	0.0955
9	402.82	402.70			
			38.31	19.16	0.0963
10	441.13	441.00			
			37.81	18.91	0.0969
11	478.94	478.74			

† Lines numbered according to $\bar{\nu}_J = 2B(J+1)$ cm^{-1}. Observed data from 'An Examination of the Far Infra-red Spectrum of Hydrogen Fluoride' by A. A. Mason and A. H. Nielsen, published in Scientific Report No. 5, August 1963, Contract No. AF 19(604)-7981, by kind permission of the authors.

‡ See Sec. 2.3.5 for details of the calculation.

in Table 2.1; it is evident that the separation between successive lines (and hence the apparent B value) decreases steadily with increasing J.

The reason for this decrease may be seen if we calculate internuclear distances from the B values. The calculations are exactly similar to those of Sec. 2.3.1 and the results are shown in column 6 of Table 2.1. Plainly the bond length increases with J and we can see that our assumption of a *rigid* bond is only an approximation; in fact, of course, all bonds are elastic to some extent, and the increase in length with J merely reflects the fact that the more quickly a diatomic molecule rotates the greater is the centrifugal force tending to move the atoms apart.

Before showing how this elasticity may be quantitatively allowed for in rotational spectra, we shall consider briefly two of its consequences. Firstly, when the bond is elastic, a molecule may have vibrational energy—i.e. the bond will stretch and compress periodically with a certain fundamental frequency dependent upon the masses of the atoms and the elasticity (or force constant) of the bond. If the motion is simple harmonic (which, we shall see in Chapter 3, is usually a very good approximation to the truth) the force constant is given by:

$$k = 4\pi^2 \bar{\omega}^2 c^2 \mu \qquad (2.22)$$

where $\bar{\omega}$ is the vibration frequency (expressed in cm^{-1}), and c and μ have their previous definitions. Plainly the variation of B with J is determined by the force constant—the weaker the bond, the more readily will it distort under centrifugal forces.

The second consequence of elasticity is that the quantities r and B vary during a vibration. When these quantities are measured by microwave techniques many hundreds of vibrations occur during a rotation, and hence the measured value is an average. However, from the defining equation of B we have:

$$B = \frac{h}{8\pi^2 Ic} = \frac{h}{8\pi^2 c\mu r^2}$$

or
$$B \propto 1/r^2 \qquad (2.23)$$

since all other quantities are independent of vibration. Now, although in simple harmonic motion a molecular bond is compressed and extended an equal amount on each side of the equilibrium distance and the average value of the distance is therefore unchanged, the average value of $1/r^2$ is *not* equal to $1/r_e^2$, where r_e is the equilibrium distance. We can see this most easily by an example. Consider a bond of equilibrium length 0.1 nm vibrating between the limits 0.09 and 0.11 nm. We have:

$$\langle r \rangle_{av.} = \frac{0.09 + 0.11}{2} = 0.1 = r_e$$

but

$$\left\langle \frac{1}{r^2} \right\rangle_{av.} = \frac{(1/0.09)^2 + (1/0.11)^2}{2} = 103.05 \, \text{nm}^2$$

and therefore $\langle r \rangle_{av.} = \sqrt{1/103.5} = 0.0985$ nm. The difference, though small, is not negligible compared with the precision with which B can be measured spectroscopically. In fact the real situation is rather worse. We shall see in Chapter 3 that real vibrations are not simple harmonic, since a real bond may be stretched more easily than it may be compressed, and this usually results in $r_{av.}$ being greater than $r_{eq.}$.

It is usual, then, to define three different sets of values for B and r. At the equilibrium separation, r_e, between the nuclei, the rotational constant is B_e; in the vibrational ground state the average internuclear separation is r_0 associated with a rotational constant B_0; while if the molecule has excess vibrational energy the quantities are r_v and B_v, where v is the vibrational quantum number.

During the remainder of this chapter we shall ignore the small differences between B_0, B_e, and B_v—the discrepancy is most important in the consideration of vibrational spectra in Chapter 3.

We should note, in passing, that the rotational spectrum of hydrogen fluoride given in Table 2.1 extends from the microwave well into the infra-red region (cf. Fig. 1.4). This underlines the comment made in Chapter 1 that there is no fundamental distinction between spectral regions, only differences in technique. Since hydrogen fluoride, together with other diatomic hydrides, has a small moment of inertia and hence a large B value, the spacings between rotational energy levels become large and fall into the infra-red region after only a few transitions. Historically, indeed, the moments of inertia and bond lengths of these molecules were first determined from spectral studies using infra-red techniques.

2.3.5 The Spectrum of a Non-rigid Rotator

The Schrödinger wave equation may be set up for a non-rigid molecule, and the rotational energy levels are found to be:

$$E_J = \frac{h^2}{8\pi^2 I} J(J+1) - \frac{h^4}{32\pi^4 I^2 r^2 k} J^2(J+1)^2 \quad \text{J}$$

or
$$\varepsilon_J = E_J/hc = BJ(J+1) - DJ^2(J+1)^2 \quad \text{cm}^{-1} \tag{2.24}$$

where the rotational constant, B, is as defined previously, and the *centrifugal distortion constant*, D, is given by:

$$D = \frac{h^3}{32\pi^4 I^2 r^2 kc} \quad \text{cm}^{-1} \tag{2.25}$$

which is a positive quantity. Equation (2.24) applies for a simple harmonic force field only; if the force field is anharmonic, the expression becomes:

$$\varepsilon_J = BJ(J+1) - DJ^2(J+1)^2 + HJ^3(J+1)^3 + KJ^4(J+1)^4 \dots \quad \text{cm}^{-1} \tag{2.26}$$

where H, K, etc., are small constants dependent upon the geometry of the molecule. They are, however, negligible compared with D and most modern spectroscopic data are adequately fitted by Eq. (2.24).

From the defining equations of B and D it may be shown directly that:

$$D = \frac{16B^3\pi^2\mu c^2}{k} = \frac{4B^3}{\bar{\omega}^2} \tag{2.27}$$

where $\bar{\omega}$ is the vibrational frequency of the bond and k has been expressed according to Eq. (2.22). We shall see in Chapter 3 that vibrational frequencies are usually of the order of 10^3 cm^{-1}, while B we have found to be of the order of 10 cm^{-1}. Thus we see that D, being of the order 10^{-3} cm^{-1}, is very small compared with B. For small J, therefore, the correction term $DJ^2(J+1)^2$ is almost negligible, while for J values of 10 or more it may become appreciable.

Figure 2.9 shows, much exaggerated, the lowering of rotational levels when passing from the rigid to the non-rigid diatomic molecule. The spectra are also compared, the dashed lines connecting corresponding energy levels and transitions of the rigid and the non-rigid molecules. It should be noted that the selection rule for the latter is still $\Delta J = \pm 1$.

We may easily write an analytical expression for the transitions:

$$\begin{aligned}
\varepsilon_{J+1} - \varepsilon_J = \bar{\nu}_J &= B[(J+1)(J+2) - J(J+1)] \\
&\quad - D[(J+1)^2(J+2)^2 - J^2(J+1)^2] \\
&= 2B(J+1) - 4D(J+1)^3 \quad \text{cm}^{-1}
\end{aligned} \tag{2.28}$$

where $\bar{\nu}_J$ represents equally the upward transition from J to $J+1$, or the downward from $J+1$ to J. Thus we see analytically, and from Fig. 2.9, that the spectrum of the elastic rotor is similar to that of the rigid molecule except that each line is displaced slightly to low frequency, the displacement increasing with $(J+1)^3$.

A knowledge of D gives rise to two useful items of information. Firstly, it allows us to determine the J value of lines in an observed spectrum. If we have measured a few isolated transitions it is not always easy to determine from which J value they arise; however, fitting Eq. (2.28) to them—provided three consecutive lines have been measured—gives unique values for B, D, and J. The precision of such fitting is shown by Table 2.1 where the wavenumbers are calculated from the equation:

$$\bar{\nu}_J = 41.122(J+1) - 8.52 \times 10^{-3}(J+1)^3 \quad \text{cm}^{-1} \tag{2.29}$$

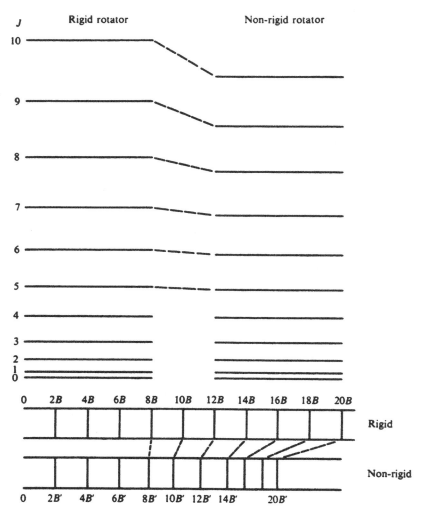

Figure 2.9 The change in rotational energy levels and spectrum when passing from a rigid to a non-rigid diatomic molecule. Energy levels on the right are calculated using $D = 10^{-3}B$.

Secondly, a knowledge of D enables us to determine—although rather inaccurately—the vibrational frequency of a diatomic molecule. From the above data for hydrogen fluoride and Eq. (2.27) we have:

$$\bar{\omega}^2 = \frac{4B^3}{D} = 16.33 \times 10^6 \ (\text{cm}^{-1})^2$$

i.e.

$$\bar{\omega} \approx 4050 \ \text{cm}^{-1}$$

In the next chapter we shall see that a more precise determination leads to the value $4138.3 \ \text{cm}^{-1}$; the 2 per cent inaccuracy in the present calculation is due partly to the assumption of simple harmonic motion and partly to the very small, and hence relatively inaccurate, value of D.

The force constant follows directly:

$$k = 4\pi^2 c^2 \bar{\omega}^2 \mu = 960 \, \mathrm{N \, m^{-1}}$$

which indicates, as expected, that H—F is a relatively strong bond.

2.4 POLYATOMIC MOLECULES

2.4.1 Linear Molecules

We consider first molecules such as carbon oxysulphide OCS or chloroacetylene HC≡CCl, where all the atoms lie on a straight line, since this type gives rise to particularly simple spectra in the microwave region. Since $I_B = I_C$; $I_A = 0$, as for diatomic molecules, the energy levels are given by a formula identical with Eq. (2.26), i.e.

$$\varepsilon_J = BJ(J+1) - DJ^2(J+1)^2 + \cdots \quad \mathrm{cm^{-1}} \tag{2.30}$$

and the spectrum will show the same $2B$ separation modified by the distortion constant. In fact, the whole of the discussion on diatomic molecules applies equally to all linear molecules; three points, however, should be underlined:

1. Since the moment of inertia for the end-over-end rotation of a polyatomic linear molecule is considerably greater than that of a diatomic molecule, the B value will be much smaller and the spectral lines more closely spaced. Thus B values for diatomic molecules are about $10 \, \mathrm{cm^{-1}}$, while for triatomic molecules they can be $1 \, \mathrm{cm^{-1}}$ or less, and for larger molecules smaller still.
2. The molecule must, as usual, possess a dipole moment if it is to exhibit a rotational spectrum. Thus OCS will be microwave active, while OCO (more usually written CO_2) will not. In particular, it should be noted that isotopic substitution does not lead to a dipole moment since the bond lengths and atomic charges are unaltered by the substitution. Thus $^{16}OC^{18}O$ is microwave inactive.
3. A non-cyclic polyatomic molecule containing N atoms has altogether $N - 1$ individual bond lengths to be determined. Thus in the triatomic molecule OCS there is the CO distance, r_{CO}, and the CS distance, r_{CS}. On the other hand, there is only *one* moment of inertia for the end-over-end rotation of OCS, and only this one value can be determined from the spectrum. Table 2.2 shows the data for this molecule. Over the four lines observed there is seen to be no

Table 2.2 Microwave spectrum of carbon oxysulphide

$J \to J+1$	$\bar{\nu}_{obs.}(\mathrm{cm^{-1}})$	$\Delta\bar{\nu}$	$B(\mathrm{cm^{-1}})$
$0 \to 1$	. . .		
		2×0.4055	0.2027
$1 \to 2$	0.8109		
		0.4054	0.2027
$2 \to 3$	1.2163		
		0.4054	0.2027
$3 \to 4$	1.6217		
		0.4054	0.2027
$4 \to 5$	0.0271		

appreciable centrifugal distortion, and, taking the value of B as $0.2027\,\text{cm}^{-1}$, we calculate:

$$I_B = \frac{h}{8\pi^2 Bc} = 137.95 \times 10^{-47}\ \text{kg m}^2$$

From this one observation it is plainly impossible to deduce the two unknowns, r_{CO} and r_{CS}. The difficulty can be overcome, however, if we study a molecule with different atomic masses but the *same* bond lengths—i.e. an isotopically substituted molecule—since this will have a different moment of inertia.

Let us consider the rotation of OCS in some detail. Figure 2.10 shows the molecule, where r_O, r_C, and r_S represent the distances of the atoms from the centre of gravity. Consideration of moments gives:

$$m_O r_O + m_C r_C = m_S r_S \tag{2.31}$$

where m_i is the mass of atom i. The moment of inertia is:

$$I = m_O r_O^2 + m_C r_C^2 + m_S r_S^2 \tag{2.32}$$

and we have the further equations:

$$r_O = r_{CO} + r_C \qquad r_S = r_{CS} - r_C \tag{2.33}$$

where r_{CO} and r_{CS} are the bond lengths of the molecule. It is these we wish to determine. Substituting (2.33) in (2.31) and collecting terms:

$$(m_C + m_O + m_S)r_C = m_S r_{CS} - m_O r_{CO}$$

or

$$M r_C = m_S r_{CS} - m_O r_{CO} \tag{2.34}$$

where we write M for the total mass of the molecule. Substituting (2.33) in (2.32):

$$I = m_O(r_{CO} + r_C)^2 + m_C r_C^2 + m_S(r_{CS} - r_C)^2$$
$$= M r_C^2 + 2r_C(m_O r_{CO} - m_S r_{CS}) + m_O r_{CO}^2 + m_S r_{CS}^2$$

and finally substituting for r_C from Eq. (2.34):

$$I = m_O r_{CO}^2 + m_S r_{CS}^2 - \frac{(m_O r_{CO} - m_S r_{CS})^2}{M} \tag{2.35}$$

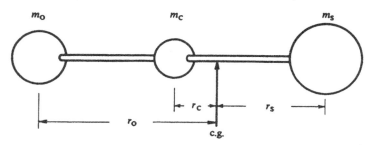

Figure 2.10 The molecule of carbon oxysulphide, OCS, showing the distances of each atom from the centre of gravity.

Considering now the isotopic molecule, ^{18}OCS, we may write m_O' for m_O throughout Eq. (2.35):

$$I' = m_O' r_{CO}^2 + m_S r_{CS}^2 - \frac{(m_O' r_{CO} - m_S r_{CS})^2}{M'} \qquad (2.36)$$

and we can now solve for r_{CO} and r_{CS}, provided we have extracted a value for I' from the microwave spectrum of the isotopic molecule. Note that we do *not* need to write r_{CO}', since we assume that the bond length is unaltered by isotopic substitution. This assumption may be checked by studying the molecules $^{16}OC^{34}S$ and $^{18}OC^{34}S$, since we would then have four moments of inertia. The bond distances found are quite consistent, and hence justify the assumption.

The extension of the above discussion to molecules with more than three atoms is straightforward; it suffices to say here that microwave studies have led to very precise determinations of many bond lengths in such molecules.

2.4.2 Symmetric Top Molecule

Although the rotational energy levels of this type of molecule are more complicated than those of linear molecules, we shall see that, because of their symmetry, their pure rotational spectra are still relatively simple. Choosing methyl fluoride again as our example we remember that

$$I_B = I_C \neq I_A \qquad I_A \neq 0$$

There are now two directions of rotation in which the molecule might absorb or emit energy—that about the main symmetry axis (the C—F bond in this case) and that perpendicular to this axis.

We thus need two quantum numbers to describe the degree of rotation, one for I_A and one for I_B or I_C. However, it turns out to be very convenient mathematically to have a quantum number to represent the *total* angular momentum of the molecule, which is the sum of the separate angular momenta about the two different axes. This is usually chosen to be the quantum number J. Reverting for a moment to *linear* molecules, remember that we there used J to represent the end-over-end rotation of a molecule; however, this was the *only* sort of rotation allowed, so it is quite consistent to use J, in general, to represent the *total angular momentum*. It is then conventional to use K to represent the angular momentum about the top axis—i.e. about the C—F bond in this case.

Let us briefly consider what values are allowed to K and J. Both must, by the conditions of quantum mechanics, be integral or zero. The total angular momentum can be as large as we like, that is J can be $0, 1, 2, \ldots, \infty$ (except, of course, for the theoretical possibility that a real molecule will be disrupted at very high rotational speeds). Once we have chosen J, however, K is rather more limited. Let us consider the case when $J = 3$. Plainly the rotational energy can be divided in several ways between motion about the main symmetry axis and motion perpendicular to this. If *all* the rotation is about the axis, $K = 3$; but note that K cannot be greater than J since J is the *total* angular momentum. Equally we could have $K = 2, 1$, or 0, in which case the motion perpendicular to the axis increases accordingly. Additionally, however, K can be negative—we can imagine positive and negative values of K to correspond with clockwise and anticlockwise rotation about the symmetry axis—and so can have values $-1, -2$, or -3.

In general, then, for a total angular momentum, J, we see that K can take values:

$$K = J, J-1, J-2, \ldots, 0, \ldots, -(J-1), -J \qquad (2.37)$$

which is a total of $2J + 1$ values altogether. This figure of $2J + 1$ is important and will recur.

If we take first the case of a *rigid* symmetric top—i.e. one in which the bonds are supposed not to stretch under centrifugal forces—the Schrödinger equation may be solved to give the allowed energy levels for rotation as:

$$\varepsilon_{J,K} = E_{J,K}/hc = BJ(J+1) + (A-B)K^2 \quad \text{cm}^{-1} \tag{2.38}$$

where, as before,

$$B = \frac{h}{8\pi^2 I_B c} \quad \text{and} \quad A = \frac{h}{8\pi^2 I_A c}.$$

Note that the energy depends on K^2, so that it is immaterial whether the top spins clockwise or anticlockwise: the energy is the same for a given angular momentum. For all $K > 0$, therefore, the rotational energy levels are *doubly degenerate*.

The selection rules for this molecule may be shown to be:

$$\Delta J = \pm 1 \text{ (as before)} \quad \text{and} \quad \Delta K = 0 \tag{2.39}$$

and, when these are applied to Eq. (2.38), the spectrum is given by:

$$\begin{aligned} \varepsilon_{J+1,K} - \varepsilon_{J,K} = \bar{v}_{J,K} &= B(J+1)(J+2) + (A-B)K^2 \\ &\quad - [BJ(J+1) + (A-B)K^2] \\ &= 2B(J+1) \quad \text{cm}^{-1} \end{aligned} \tag{2.40}$$

Thus the spectrum is independent of K, and hence rotational changes about the symmetry axis do not give rise to a rotational spectrum. The reason for this is quite evident—rotation about the symmetry axis does not change the dipole moment perpendicular to the axis (which always remains zero), and hence the rotation cannot interact with radiation. Equation (2.40) shows that the spectrum is just the same as for a linear molecule and that only one moment of inertia—that for end-over-end rotation—can be measured.

Equations (2.38) and (2.40) both apply to a rigid molecule. A real molecule, as we have seen, has elastic bonds and the picture inevitably becomes more complicated when the possible distortions of each individual bond in the symmetric top are allowed for. The details need not concern us here, but the microwave spectra of such molecules have, in the past, given very precise estimates of their bond lengths and angles. Table 2.3 collects some representative data for these, and for some linear and asymmetric top molecules (see the next section); this table illustrates the great precision which has been achieved.

Table 2.3 Some molecular data determined by microwave spectroscopy

Molecule	Type	Bond length (nm)	Bond angle (deg)
NaCl	Diatomic	0.236 06 ± 0.000 01	—
COS	Linear	0.116 4 ± 0.000 1 (CO) 0.155 9 ± 0.000 1 (CS)	—
HCN	Linear	0.106 317 ± 0.000 005 (CH) 0.115 535 ± 0.000 006 (CN)	—
NH_3	Sym. top	0.100 8 ± 0.000 4	107.3 ± 0.2
CH_3Cl	Sym. top	0.109 59 ± 0.000 05 (CH) 0.178 12 ± 0.000 05 (CCl)	108.0 ± 0.2 (HCH)
H_2O	Asym. top	0.095 84 ± 0.000 05	104.5 ± 0.3
O_3	Asym. top	0.127 8 ± 0.000 2	116.8 ± 0.5

2.4.3 Asymmetric Top Molecules

Since spherical tops show no microwave spectrum (cf. Sec. 2.1(3)), the only other class of molecule of interest here is the asymmetric top. These molecules, having three different moments of inertia, also have much more complicated rotational energy levels and spectra. No simple general expression corresponding to Eqs (2.24) or (2.28) can be derived for them, and they are usually treated by approximative methods, much computation being required before agreement between observed and calculated spectra is achieved. However, such methods have been very successful for small molecules and much accurate bond length and bond angle data have been derived. Some examples are included in Table 2.3 above.

2.5 TECHNIQUES AND INSTRUMENTATION

It is not proposed to give more than a brief outline here of the techniques of microwave spectroscopy since detailed accounts are available elsewhere, and the technique is rarely used as a routine analysis tool now as it has been largely superseded by other methods. Figure 2.11 shows, very diagrammatically, how the various components discussed below are linked together.

1. *The source and monochromator.* The usual source in this region is the klystron valve which, since it emits radiation over only a very narrow frequency range, is called 'monochromatic' and acts as its own monochromator. The actual emission frequency is variable electronically, and so a spectrum may be scanned over a limited range of frequencies using a single klystron.

   One slight disadvantage of this source is that the total energy radiated is very small—of the order of milliwatts only. However, since this is all concentrated into a very narrow range of frequencies, the electric field generated at these frequencies is more intense than might be expected from the radiative power. This means that a sharply tuned detector can be sufficiently sensitive to detect the signal.

2. *Beam direction.* This is achieved by the use of 'waveguides'—hollow tubes of copper or silver, often of rectangular cross-section—inside which the radiation is confined. The tubing may be gently tapered or bent to focus or direct the radiation. Atmospheric absorption is considerable, so the whole system must be efficiently evacuated.

3. *Sample and sample space.* Normally the sample is gaseous and is retained in a piece of evacuated waveguide by very thin mica windows. Very small pressures are sufficient to obtain a spectrum, and many substances which are normally thought of as liquid or solid have sufficient vapour pressure to allow them to be studied by the technique.

4. *Detector.* It is possible to use a radio receiver as detector, provided it can be tuned to the appropriate high frequency; however, a simple crystal detector is found to be more sensitive

Figure 2.11 Schematic diagram of a microwave spectrometer.

and easier to use. This detects the radiation focused on it by the waveguide, and the signal it gives is amplified for display on an oscilloscope or for permanent record on paper.

2.6 CHEMICAL ANALYSIS BY MICROWAVE SPECTROSCOPY

Improvements and simplifications in the techniques of microwave spectroscopy are now allowing it to move away from being purely a specialist research instrument towards becoming a technique for routine analysis. Even though effectively limited to gaseous samples, it has much to offer in this respect, since it is a highly sensitive (0.01 mmHg pressure is adequate) and specific analytical tool.

The microwave spectrum of a substance is very rich in lines since many rotational levels are populated at room temperatures, but since the lines are very sharp and their positions can be measured with great accuracy, observation of just a few of them is sufficient, after comparison with tabulated data, to establish the presence of a previously examined substance in a sample. The technique is also quantitative, since the intensity of a spectrum observed under given conditions is directly dependent on the amount of substance present. Thus mixtures can be readily analysed.

It is the *whole* molecule, by virtue of its moment(s) of inertia, which is examined by microwave spectroscopy. This means that the technique cannot detect the presence of particular molecular groupings in a sample, like —OH or —CH₃ (cf. the chapters on infra-red, Raman, and magnetic resonance spectroscopy later), but it can readily distinguish the presence of isotopes in a sample, and it can even detect different conformational isomers, provided they have different moments of inertia.

One fascinating area where microwave analysis is being used is in the chemical examination of interstellar space. Electronic spectroscopy has long been able to detect the presence of various atoms, ions, and a few radicals (for example —OH) in the light of stars but recently use of microwaves has extended the analysis to the detection of simple stable molecules in space. Some 80 or so molecules have already been characterized in this way, the earliest among them (water, ammonia, and formaldehyde) giving new impetus to speculations regarding the origins of biological molecules and of life itself. Such observations concern the emission of microwaves by these molecules and, by comparing the relative intensities of various rotational transitions, particularly in the spectrum of ammonia, accurate estimates can be made of the temperature of interstellar material.

2.7 THE MICROWAVE OVEN

One area where microwave radiation has become very familiar in recent years is the kitchen, in the shape of the microwave oven. While obviously not nearly as sophisticated as a spectrometer, its mode of operation depends entirely upon the absorption by the food of the microwave radiation in which it is bathed. In fact, it is the *water* molecules only which absorb the radiation and so become raised into high rotational states—the biological molecules in food are far too large to be able to rotate. As with many other excited states, the excess rotational energy of the water molecules is re-emitted as heat and the food becomes cooked.

The efficiency of the oven lies in the fact that this heating is *internal*. In a conventional hot oven a piece of meat or a cake is heated from the outside, and it must be left to cook until its centre has been raised to a sufficiently high temperature. In microwave heating, however, water molecules throughout the whole bulk of the food are simultaneously excited and 'heated', so cooking times are drastically reduced.

The effect of such concentrated microwave radiation on the human body, unfortunately, is similar—whatever is exposed to the radiation is rapidly heated and cooked from the inside! It is essential, therefore, to ensure that the door seal on a microwave is in good condition, so that no radiation is allowed to leak out.

BIBLIOGRAPHY

Chantry, G. W.: *Modern Aspects of Microwave Spectroscopy*, Academic Press, 1979.
Ebsworth, E. A. V., D. W. H. Rankin, and S. Cradock: *Structural Methods in Inorganic Chemistry*, 2nd ed., Blackwell Scientific Publications, 1991.
Gordy, W., and R. L. Cook: *Microwave Molecular Spectra (Techniques of Organic Chemistry*, vol. 9), John Wiley, 1970.
Gordy, W., W. V. Smith, and R. Trambarulo: *Microwave Spectroscopy*, John Wiley, 1966.
Ingram, D. J. E.: *Spectroscopy at Radio and Microwave Frequencies*, 2nd ed., Butterworth, 1967.
Kroto, H. W.: *Molecular Rotation Spectra*, Wiley–Interscience, 1975.
Sugden, T. M., and N. C. Kenney: *Microwave Spectroscopy of Gases*, Van Nostrand, 1965.
Townes, C. H., and A. L. Schawlow: *Microwave Spectroscopy*, McGraw-Hill, 1955.
Wollrab, J. E.: *Rotational Spectra and Molecular Structure*, Academic Press, 1967.

PROBLEMS

(Useful constants: $h = 6.626 \times 10^{-34}$ J s; $k = 1.381 \times 10^{-23}$ J K^{-1}; $c = 2.998 \times 10^8$ m s^{-1}; $8\pi^2 \times 78.956$; atomic masses (in kg): ^{1}H $= 1.673 \times 10^{-27}$, ^{2}D $= 3.344 \times 10^{-27}$, ^{19}F $= 31.55 \times 10^{-27}$, ^{35}Cl $= 58.06 \times 10^{-27}$, ^{37}Cl $= 61.38 \times 10^{-27}$, ^{79}Br $= 131.03 \times 10^{-27}$, ^{127}I $= 210.7 \times 10^{-27}$.)

2.1 Which of the following molecules will show a microwave rotational spectrum:

$$H_2, \ HCl, \ CH_4, \ CH_3Cl, \ CH_2Cl_2, \ H_2O, \ SF_6?$$

2.2 The rotational spectrum of ^{79}Br^{19}F shows a series of equidistant lines 0.71433 cm^{-1} apart. Calculate the rotational constant, B, and hence the moment of inertia and bond length of the molecule. Determine the wavenumber of the $J = 9 \rightarrow J = 10$ transition, and find which transition gives rise to the most intense spectral line at room temperature (say 300 K). Calculate the number of revolutions per second which the BrF molecule undergoes when in (*a*) the $J = 0$ state, (*b*) the $J = 1$ state, and (*c*) the $J = 10$ state.
Hint: Use $E = \frac{1}{2}I\omega^2$ in conjunction with Eqs (2.10) and (2.13), but remember that here ω is in *radians* per second.

2.3 The rotational constant for H^{35}Cl is observed to be 10.5909 cm^{-1}. What are the values of B for H^{37}Cl and for ^{2}D^{35}Cl?

2.4 A microwave spectrometer capable of operating only between 60 and 90 cm^{-1} was used to observe the rotational spectra of HI and DI. Absorptions were measured as follows:

HI (cm^{-1})	DI (cm^{-1})
64.275	65.070
77.130	71.577
89.985	78.094
	84.591

Find B, I and r for each molecule, and determine the J values between which transitions occur for the first line listed above for each. Do your results support the usual assumption that bond length is unchanged by isotopic substitution?

2.5 Sketch a diagram similar to that of Fig. 2.7, using $B = 5$ cm^{-1} and a temperature of 1600 K. (*Note*: Find the maximum and calculate two or three points on either side only—do not attempt to carry out the calculation for every value of J.)

2.6 A space probe was designed to seek CO in the atmosphere of Saturn by looking for lines in its rotational spectrum. If the bond length of CO is 112.8 pm, at what wavenumbers do the first three rotational transitions appear? What resolution would be needed to determine the isotopic ratio of ^{13}C to ^{12}C on Saturn by observing the first three ^{13}CO rotational lines as well? How could the experiment be extended to estimate the temperature of Saturn's atmosphere?

2.7 HCl has a B value of $10.593 \, \text{cm}^{-1}$ and a centrifugal distortion constant D of $5.3 \times 10^{-4} \, \text{cm}^{-1}$. Estimate the vibrational frequency and force constant of the molecule. The observed vibrational frequency is $2991 \, \text{cm}^{-1}$; explain, qualitatively, the discrepancy.

THREE

INFRA-RED SPECTROSCOPY

We saw in the previous chapter how the elasticity of chemical bonds led to anomalous results in the rotational spectra of rapidly rotating molecules—the bonds stretched under centrifugal forces. In this chapter we consider another consequence of this elasticity—the fact that atoms in a molecule do not remain in fixed relative positions but vibrate about some mean position. We consider first the case of a diatomic molecule and the spectrum which arises if its only motion is vibration; then we shall deal with the more practical case of a diatomic molecule undergoing vibration and rotation simultaneously; finally we shall extend the discussion to more complex molecules.

3.1 THE VIBRATING DIATOMIC MOLECULE

3.1.1 The Energy of a Diatomic Molecule

When two atoms combine to form a stable covalent molecule, for example HCl, they may be said to do so because of some internal electronic rearrangement. We shall not here discuss the detailed mechanisms of chemical bond formation, but we can simply look on the phenomenon as a balancing of forces. On the one hand there is a repulsion between the positively charged nuclei of both atoms, and between their negative electron 'clouds'; on the other there is an attraction between the nucleus of one atom and the electrons of the other, and vice versa. The two atoms settle at a mean internuclear distance such that these forces are just balanced and the total energy of the whole system is a minimum. Squeezing the atoms more closely together will cause the repulsive force to rise rapidly, while pulling them apart is resisted by the attractive force. Any attempt to distort the bond length requires an input of energy and we may plot energy against internuclear distance as in Fig. 3.1, where we have 'anchored' the chlorine atom on one axis and imagine pushing and pulling the hydrogen atom closer to or further from the chlorine—a bigger push or pull results in raising the energy more. At the energy minimum the internuclear distance is referred to as the equilibrium distance $r_{eq.}$, or more simply as the bond length.

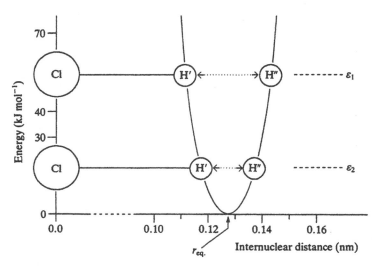

Figure 3.1 The energy of HCl as the bond is compressed or extended.

The compression and extension of a bond may be likened to the behaviour of a spring, and we may extend the analogy by assuming that the bond, like a spring, obeys Hooke's law. We may write:

$$f = -k(r - r_{eq.})$$ (3.1)

where f is the restoring force, k the force constant, and r the internuclear distance. In this case the energy curve is parabolic and has the form

$$E = \tfrac{1}{2} k(r - r_{eq.})^2$$ (3.2)

This model of a vibrating diatomic molecule—the so-called simple harmonic oscillator model—while only an approximation, forms an excellent starting point for the discussion of vibrational spectra.

3.1.2 The Simple Harmonic Oscillator

In Fig. 3.1 we have plotted the energy in the form of Eq. (3.2), i.e. as a parabola. The zero of curve and equation is at $r = r_{eq.}$, and any energy in excess of this, for example at ε_2, arises because of extension or compression of the bond. The figure shows that if one atom (Cl) is considered to be stationary on the $r = 0$ axis, the other (H) will oscillate between H′ and H″. In the case of HCl, it is a good approximation to say that, during vibrations, the heavy chlorine atom stays virtually still and it is the much lighter hydrogen which moves. However, only the *distance* between the two atoms is important and for any diatomic molecule we can always imagine ourselves to be sitting on one atom and watching the other move—from our point of view the atom we are on is stationary and can be assumed fixed on the $r = 0$ axis. Thus diagrams like Fig. 3.1 apply to *any* diatomic molecule.

If the energy of the HCl molecule of Fig. 3.1 is increased to ε_1 the oscillation will become more vigorous—that is to say the degree of compression and extension will be greater—but the vibrational frequency will not change. An elastic bond, like a spring, has a certain intrinsic vibrational frequency, dependent on the mass of the system and the force constant, but is independent of the amount of distortion. Classically it is simple to show that the oscillation frequency is:

$$\omega_{osc.} = \frac{1}{2\pi}\sqrt{\frac{k}{\mu}} \quad Hz \qquad (3.3)$$

where μ is the reduced mass of the system (cf. Eq. (2.9)). To convert this frequency to wave-numbers, the unit most usually employed in vibrational spectroscopy, we must divide by the velocity of light, c, expressed in $cm\,s^{-1}$ (cf. Sec. 1.1), obtaining:

$$\bar{\omega}_{osc.} = \frac{1}{2\pi c}\sqrt{\frac{k}{\mu}} \quad cm^{-1} \qquad (3.4)$$

Vibrational energies, like all other molecular energies, are quantized, and the allowed vibrational energies for any particular system may be calculated from the Schrödinger equation. For the simple harmonic oscillator these turn out to be:

$$E_v = (v + \tfrac{1}{2})h\omega_{osc.} \quad joules \quad (v = 0, 1, 2, \dots) \qquad (3.5)$$

where v is called the *vibrational quantum number*. Converting to the spectroscopic units, cm^{-1}, we have:

$$\varepsilon_v = \frac{E_v}{hc} = (v + \tfrac{1}{2})\bar{\omega}_{osc.} \quad cm^{-1} \qquad (3.6)$$

as the only energies allowed to a simple harmonic vibrator. Some of these are shown in Fig. 3.2.

In particular, we should notice that the *lowest* vibrational energy, obtained by putting $v = 0$ in Eq. (3.5) or (3.6), is

$$E_0 = \tfrac{1}{2}h\omega_{osc.} \quad joules \quad (\omega_{osc.} \text{ in Hz})$$

or

$$\varepsilon_0 = \tfrac{1}{2}\bar{\omega}_{osc.} \quad cm^{-1} \quad (\bar{\omega}_{osc.} \text{ in } cm^{-1}) \qquad (3.7)$$

The implication is that the diatomic molecule (and, indeed, *any* molecule) can never have zero vibrational energy; the atoms can never be completely at rest relative to each other. The quantity $\tfrac{1}{2}h\omega_{osc.}$ joules or $\tfrac{1}{2}\bar{\omega}_{osc.}$ cm^{-1} is known as the zero-point energy; it depends only on the classical vibration frequency and hence (Eq. (3.3) or (3.4)) on the strength of the chemical bond and the atomic masses.

The prediction of zero-point energy is the basic difference between the wave mechanical and classical approaches to molecular vibrations. Classical mechanics could find no objection to a molecule possessing no vibrational energy but wave mechanics insists that it must always vibrate to some extent; the latter conclusion has been amply borne out by experiment.

Further use of the Schrödinger equation leads to the simple *selection rule* for the harmonic oscillator undergoing vibrational changes:

$$\Delta v = \pm 1 \qquad (3.8)$$

To this we must, of course, add the condition that vibrational energy changes will only give rise to an observable spectrum if the vibration can interact with radiation i.e. (cf. Chapter 1) if the vibration involves a change in the dipole moment of the molecule. Thus vibrational spectra will be observable only in heteronuclear diatomic molecules since homonuclear molecules have no dipole moment.

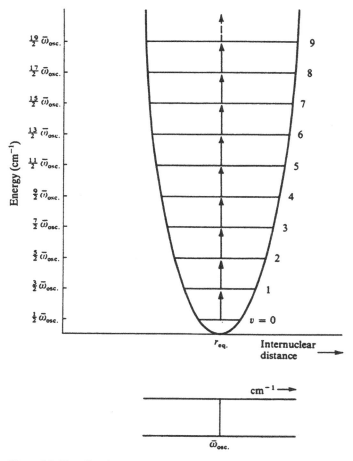

Figure 3.2 The vibrational energy levels and allowed transitions between them for a diatomic molecule undergoing simple harmonic motion.

Applying the selection rule we have immediately:

$$\varepsilon_{v+1\rightarrow v} = (v + 1 + \tfrac{1}{2})\bar{\omega}_{osc.} - (v + \tfrac{1}{2})\bar{\omega}_{osc.}$$
$$= \bar{\omega}_{osc.} \quad cm^{-1} \tag{3.9a}$$

for emission and

$$\varepsilon_{v\rightarrow v+1} = \bar{\omega}_{osc.} \quad cm^{-1} \tag{3.9b}$$

for absorption, whatever the initial value of v.

Such a simple result is also obvious from Fig. 3.2—since the vibrational levels are equally spaced, transitions between any two neighbouring states will give rise to the same energy change. Further, since the difference between energy levels expressed in cm^{-1} gives directly the wavenumber of the spectral line absorbed or emitted

$$\bar{\nu}_{spectroscopic} = \varepsilon = \bar{\omega}_{osc.} \quad cm^{-1} \tag{3.10}$$

This, again, is obvious if one considers the mechanism of absorption or emission in classical terms. In absorption, for instance, the vibrating molecule will absorb energy only from radiation with which it can coherently interact (cf. Fig. 1.8) and this must be radiation of its own oscillation frequency.

3.1.3 The Anharmonic Oscillator

Real molecules do not obey exactly the laws of simple harmonic motion; real bonds, although elastic, are not so homogeneous as to obey Hooke's law. If the bond between atoms is stretched, for instance, there comes a point at which it will break—the molecule dissociates into atoms. Thus although for small compressions and extensions the bond may be taken as perfectly elastic, for larger amplitudes—say greater than 10 per cent of the bond length—a much more complicated behaviour must be assumed. Figure 3.3 shows, diagrammatically, the shape of the energy curve for a typical diatomic molecule, together with (dashed) the ideal, simple harmonic parabola.

A purely empirical expression which fits this curve to a good approximation was derived by P. M. Morse, and is called the Morse function:

$$E = D_{eq.}[1 - \exp\{a(r_{eq.} - r)\}]^2 \tag{3.11}$$

where a is a constant for a particular molecule and $D_{eq.}$ is the dissociation energy.

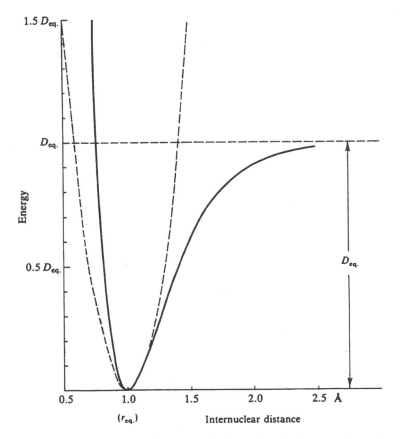

Figure 3.3 The Morse curve: the energy of a diatomic molecule undergoing anharmonic extensions and compressions.

When Eq. (3.11) is used instead of Eq. (3.2) in the Schrödinger equation, the pattern of the allowed vibrational energy levels is found to be:

$$\varepsilon_v = (v + \tfrac{1}{2})\bar{\omega}_e - (v + \tfrac{1}{2})^2 \bar{\omega}_e x_e \quad \text{cm}^{-1} \quad (v = 0, 1, 2, \dots) \tag{3.12}$$

where $\bar{\omega}_e$ is an oscillation frequency (expressed in wavenumbers) which we shall define more closely below, and x_e is the corresponding anharmonicity constant which, for bond stretching vibrations, is always small and positive ($\approx +0.01$), so that the vibrational levels crowd more closely together with increasing v. Some of these levels are sketched in Fig. 3.4.

It should be mentioned that Eq. (3.12), like (3.11), is an approximation only; more precise expressions for the energy levels require cubic, quartic, etc., terms in $(v + \tfrac{1}{2})$ with anharmonicity constants y_e, z_e, etc., rapidly diminishing in magnitude. These terms are important only at large values of v, and we shall ignore them.

If we rewrite Eq. (3.12), for the anharmonic oscillator, as:

$$\varepsilon_v = \bar{\omega}_e \{1 - x_e(v + \tfrac{1}{2})\}(v + \tfrac{1}{2}) \tag{3.13}$$

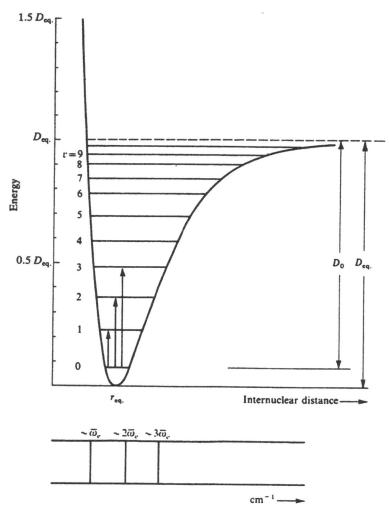

Figure 3.4 The vibrational energy levels and some transitions between them for a diatomic molecule undergoing anharmonic oscillations.

and compare with the energy levels of the *harmonic* oscillator (Eq. (3.6)), we see that we can write:

$$\bar{\omega}_{\text{osc.}} = \bar{\omega}_e\{1 - x_e(v + \tfrac{1}{2})\} \tag{3.14}$$

Thus the anharmonic oscillator behaves like the harmonic oscillator but with an oscillation frequency which decreases steadily with increasing v. If we now consider the hypothetical energy state obtained by putting $v = -\tfrac{1}{2}$ (at which, according to Eq. (3.13), $\varepsilon = 0$) the molecule would be at the equilibrium point with zero vibrational energy. Its oscillation frequency (in cm^{-1}) would be:

$$\bar{\omega}_{\text{osc.}} = \bar{\omega}_e$$

Thus we see that $\bar{\omega}_e$ may be defined as the (hypothetical) *equilibrium oscillation frequency* of the anharmonic system—the frequency for infinitely small vibrations about the equilibrium point. For any real state specified by a positive integral v the oscillation frequency will be given by Eq. (3.14). Thus in the ground state ($v = 0$) we would have:

$$\bar{\omega}_0 = \bar{\omega}_e(1 - \tfrac{1}{2}x_e) \quad \text{cm}^{-1}$$

and

$$\varepsilon_0 = \tfrac{1}{2}\bar{\omega}_e(1 - \tfrac{1}{2}x_e) \quad \text{cm}^{-1}$$

and we see that the zero-point energy differs slightly from that for the harmonic oscillator (Eq. (3.7)).

The selection rules for the anharmonic oscillator are found to be:

$$\Delta v = \pm 1, \pm 2, \pm 3, \ldots$$

Thus they are the same as for the harmonic oscillator, with the additional possibility of larger jumps. These, however, are predicted by theory and observed in practice to be of rapidly diminishing probability and normally only the lines of $\Delta v = \pm 1, \pm 2$, and ± 3, at the most, have observable intensity. Further, the spacing between the vibrational levels is, as we shall shortly see, of order 10^3 cm^{-1} and, at room temperature, we may use the Boltzmann distribution (Eq. (1.12)) to show

$$\frac{N_{v=1}}{N_{v=0}} = \exp\left\{-\frac{6.63 \times 10^{-34} \times 3 \times 10^{10} \times 10^3}{1.38 \times 10^{-23} \times 300}\right\}$$
$$\approx \exp(-4.8) \approx 0.008$$

In other words, the population of the $v = 1$ state is nearly 0.01 or some one per cent of the ground state population. Thus, to a very good approximation, we may ignore all transitions originating at $v = 1$ or more and restrict ourselves to the three transitions:

1. $v = 0 \rightarrow v = 1$, $\Delta v = +1$, with considerable intensity.

$$\Delta \varepsilon = \varepsilon_{v=1} - \varepsilon_{v=0}$$
$$= (1 + \tfrac{1}{2})\bar{\omega}_e - x_e(1 + \tfrac{1}{2})^2\bar{\omega}_e - \{\tfrac{1}{2}\bar{\omega}_e - (\tfrac{1}{2})^2 x_e\bar{\omega}_e\} \tag{3.15a}$$
$$= \bar{\omega}_e(1 - 2x_e) \quad \text{cm}^{-1}$$

2. $v = 0 \rightarrow v = 2$, $\Delta v = +2$, with small intensity.

$$\Delta \varepsilon = (2 + \tfrac{1}{2})\bar{\omega}_e - x_e(2 + \tfrac{1}{2})^2\bar{\omega}_e - \{\tfrac{1}{2}\bar{\omega}_e - (\tfrac{1}{2})^2 x_e\bar{\omega}_e\}$$
$$= 2\bar{\omega}_e(1 - 3x_e) \quad \text{cm}^{-1} \tag{3.15b}$$

3. $v = 0 \rightarrow v = 3$, $\Delta v = +3$, with normally negligible intensity.

$$\Delta \varepsilon = (3 + \tfrac{1}{2})\bar{\omega}_e - \{\tfrac{1}{2}\bar{\omega}_b - (\tfrac{1}{2})^2 x_e \bar{\omega}_e\}$$
$$= 3\bar{\omega}_e(1 - 4x_e) \quad \text{cm}^{-1} \tag{3.15c}$$

These three transitions are shown in Fig. 3.4. To a good approximation, since $x_e \approx 0.01$, the three spectral lines lie very close to $\bar{\omega}_e$, $2\bar{\omega}_e$, and $3\bar{\omega}_e$. The line near $\bar{\omega}_e$ is called the *fundamental absorption*, while those near $2\bar{\omega}_e$ *and* $3\bar{\omega}_e$ are called the *first* and *second overtones*, respectively. The spectrum of HCl, for instance, shows a very intense absorption at $2886\,\text{cm}^{-1}$, a weaker one at $5668\,\text{cm}^{-1}$, and a very weak one at $8347\,\text{cm}^{-1}$. If we wish to find the equilibrium frequency of the molecule from these data, we must solve any two of the three equations (cf. Eqs (3.15)):

$$\bar{\omega}_e(1 - 2x_e) = 2886 \quad \text{cm}^{-1}$$
$$2\bar{\omega}_e(1 - 3x_e) = 5668 \quad \text{cm}^{-1}$$
$$3\bar{\omega}_e(1 - 4x_e) = 8347 \quad \text{cm}^{-1}$$

and we find $\bar{\omega}_e = 2990\,\text{cm}^{-1}$, $x_e = 0.0174$. Thus we see that, whereas for the ideal harmonic oscillator the spectral absorption occurred *exactly* at the classical vibration frequency, for real, anharmonic molecules the observed fundamental absorption frequency and the equilibrium frequency may differ considerably.

The force constant of the bond in HCl may be calculated directly from Eq. (2.22) by inserting the value of $\bar{\omega}_e$:

$$k = 4\pi^2 \bar{\omega}_e^2 c^2 \mu$$
$$= 516\,\text{N}\,\text{m}^{-1}$$

when the fundamental constants and the reduced mass are inserted. These data, together with those for a few of the very many other diatomic molecules studied by infra-red techniques, are collected in Table 3.1.

Although we have ignored transitions from $v = 1$ to higher states, we should note that, if the temperature is raised or if the vibration has a particularly low frequency, the population of the $v = 1$ state may become appreciable. Thus at, say, $600\,\text{K}$ (i.e. about $300°\text{C}$) $N_{v=1}/N_{v=0}$ becomes $\exp(-2.4)$ or about 0.09, and transitions from $v = 1$ to $v = 2$ will be some 10 per cent the intensity of those from $v = 0$ to $v = 1$. A similar increase in the excited state population

Table 3.1 Some molecular data for diatomic molecules determined by infra-red spectroscopy

Molecule	Vibration (cm^{-1})	Anharmonicity constant x_e	Force constant (Nm^{-1})	Internuclear distance $r_{eq.}$ (nm)
HF	4138.5	0.0218	966	0.0927
HCl†	2990.6	0.0174	516	0.1274
HBr	2649.7	0.0171	412	0.1414
HI	2309.5	0.0172	314	0.1609
CO	2169.7	0.0061	1902	0.1131
NO	1904.0	0.0073	1595	0.1151
ICl†	384.2	0.0038	238	0.2321

† Data refers to the ^{35}Cl isotope

would arise if the vibrational frequency were 500 cm^{-1} instead of 1000 cm^{-1}. We may calculate the wavenumber of this transition as:

4. $v = 1 \rightarrow v = 2$, $\Delta v = +1$, normally very weak

$$\Delta\varepsilon = 2\tfrac{1}{2}\bar{\omega}_e - 6\tfrac{1}{4}x_e\bar{\omega}_e - (1\tfrac{1}{2}\bar{\omega}_e - 2\tfrac{1}{4}x_e\bar{\omega}_e)$$
$$= \bar{\omega}_e(1 - 4x_e) \quad \text{cm}^{-1} \tag{3.15d}$$

Thus, should this weak absorption arise, it will be found close to and at slightly *lower* wavenumbers than the fundamental (since x_e is small and positive). Such weak absorptions are usually called *hot bands* since a high temperature is one condition for their occurrence. Their nature may be confirmed by raising the temperature of the sample when a true hot band will increase in intensity.

We turn now to consider a diatomic molecule undergoing simultaneous vibration and rotation.

3.2 THE DIATOMIC VIBRATING ROTATOR

We saw in Chapter 2 that a typical diatomic molecule has rotational energy separations of 1–10 cm^{-1}, while in the preceding section we found that the vibrational energy separations of HCl were nearly 3000 cm^{-1}. Since the energies of the two motions are so different we may, as a first approximation, consider that a diatomic molecule can execute rotations and vibrations quite independently. This, which we shall call the Born–Oppenheimer approximation (although, cf. Eq. (6.1), this strictly includes electronic energies), is tantamount to assuming that the combined rotational–vibrational energy is simply the sum of the separate energies:

$$E_{\text{total}} = E_{\text{rot.}} + E_{\text{vib.}} \quad \text{(joules)}$$
$$\varepsilon_{\text{total}} = \varepsilon_{\text{rot.}} + \varepsilon_{\text{vib.}} \quad (\text{cm}^{-1}) \tag{3.16}$$

We shall see later in what circumstances this approximation does not apply.

Taking the separate expressions for $\varepsilon_{\text{rot.}}$ and $\varepsilon_{\text{vib.}}$ from Eqs (2.26) and (3.12), respectively, we have:

$$\varepsilon_{J,v} = \varepsilon_J + \varepsilon_v$$
$$= BJ(J+1) - DJ^2(J+1)^2 + HJ^3(J+1)^3 + \cdots \tag{3.17}$$
$$+ (v+\tfrac{1}{2})\bar{\omega}_e - x_e(v+\tfrac{1}{2})^2\bar{\omega}_e \quad \text{cm}^{-1}$$

Initially, we shall ignore the small centrifugal distortion constants D, H, etc., and hence write:

$$\varepsilon_{\text{total}} = \varepsilon_{J,v} = BJ(J+1) + (v+\tfrac{1}{2})\bar{\omega}_e - x_e(v+\tfrac{1}{2})^2\bar{\omega}_e \tag{3.18}$$

Note, however, that it is not logical to ignore D since this implies that we are treating the molecule as rigid, yet vibrating! The retention of D would have only a very minor effect on the spectrum.

The rotational levels are sketched in Fig. 3.5 for the two lowest vibrational levels, $v = 0$ and $v = 1$. There is, however, no attempt at scale in this diagram since the separation between neighbouring J values is, in fact, only some 1/1000 of that between the v values. Note that since the rotational constant B in Eq. (3.18) is taken to be the same for all J and v (a consequence of the Born–Oppenheimer assumption), the separation between two levels of given J is the same in the $v = 0$ and $v = 1$ states.

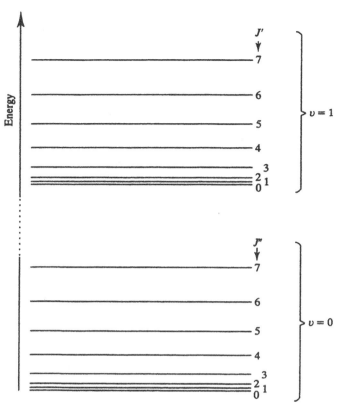

Figure 3.5 Some of the rotational energy levels for the first two vibrational states of a diatomic molecule.

It may be shown that the selection rules for the combined motions are the same as those for each separately; therefore we have:

$$\Delta v = \pm 1, \pm 2, \text{ etc.} \quad \Delta J = \pm 1 \tag{3.19}$$

Strictly speaking we may also have $\Delta v = 0$, but this corresponds to the purely rotational transitions already dealt with in Chapter 2. Note carefully, however, that a *diatomic* molecule, except under very special and rare circumstances, may *not* have $\Delta J = 0$; in other words a vibrational change *must* be accompanied by a simultaneous rotational change.

In Fig. 3.6 we have drawn some of the relevant energy levels and transitions, designating rotational quantum numbers in the $v = 0$ state as J'' and in the $v = 1$ state as J'. The use of a single prime for the upper state and a double for the lower state is conventional in all branches of spectroscopy.

Remember (and cf. Eq. (2.20)) that the rotational levels J'' are filled to varying degrees in any molecular population, so the transitions shown will occur with varying intensities. This is indicated schematically in the spectrum at the foot of Fig. 3.6.

An analytical expression for the spectrum may be obtained by applying the selection rules (Eq. (3.19)) to the energy levels (Eq. (3.18)). Considering only the $v = 0 \rightarrow v = 1$ transition we have in general:

$$\begin{aligned}
\Delta\varepsilon_{J,v} &= \varepsilon_{J',v=1} - \varepsilon_{J'',v=0} \\
&= BJ'(J'+1) + 1\tfrac{1}{2}\bar{\omega}_e - 2\tfrac{1}{4}x_e\bar{\omega}_e - \{BJ''(J''+1) + \tfrac{1}{2}\bar{\omega}_e - \tfrac{1}{4}x_e\bar{\omega}_e\} \\
&= \bar{\omega}_o + B(J'-J'')(J'+J''+1) \quad \text{cm}^{-1}
\end{aligned}$$

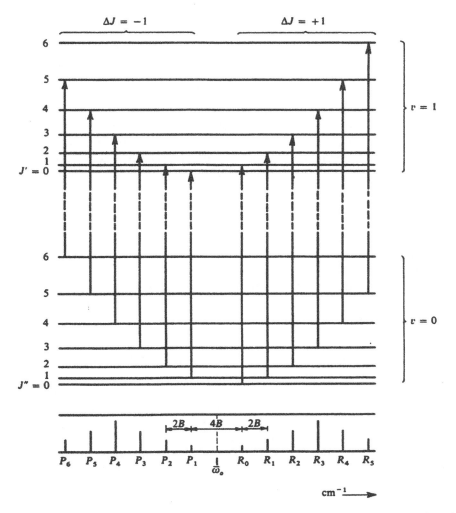

Figure 3.6 Some transitions between the rotational–vibrational energy levels of a diatomic molecule, together with the spectrum arising from them.

where, for brevity, we write $\bar{\omega}_o$ for $\bar{\omega}_e(1 - 2x_e)$.

We should note that taking B to be identical in the upper and lower vibrational states is a direct consequence of the Born–Oppenheimer approximation—rotation is unaffected by vibrational changes.

Now we can have:

1. $\Delta J = +1$, that is $J' = J'' + 1$ or $J' - J'' = +1$; hence

$$\Delta\varepsilon_{J,v} = \bar{\omega}_o + 2B(J'' + 1) \quad \text{cm}^{-1} \qquad J'' = 0, 1, 2, \ldots \qquad (3.20a)$$

2. $\Delta J = -1$, that is $J'' = J' + 1$ or $J' - J'' = -1$: hence

$$\Delta\varepsilon_{J,v} = \bar{\omega}_o - 2B(J' + 1) \quad \text{cm}^{-1} \qquad J' = 0, 1, 2, \ldots \qquad (3.20b)$$

These two expressions may conveniently be combined into:

$$\Delta\varepsilon_{J,v} = \bar{v}_{\text{spect.}} = \bar{\omega}_o + 2Bm \quad \text{cm}^{-1} \qquad m = \pm 1, \pm 2, \ldots \qquad (3.20c)$$

where m, replacing $J'' + 1$ in Eq. (3.20a) and $J' + 1$ in Eq. (3.20b) has positive values for $\Delta J = +1$ and is negative if $\Delta J = -1$. Note particularly that m *cannot be zero* since this would imply values of J' or J'' to be -1. The frequency $\bar{\omega}_o$ is usually called the *band origin* or *band centre*.

Equation (3.20c), then, represents the combined vibration–rotation spectrum. Evidently it will consist of equally spaced lines (spacing $= 2B$) on each side of the band origin $\bar{\omega}_o$, but, since $m \neq 0$, the *line at $\bar{\omega}_o$ itself will not appear*. Lines to the low-frequency side of $\bar{\omega}_o$, corresponding to negative m (that is $\Delta J = -1$) are referred to as the *P branch*, while those to the high-frequency side (m positive, $\Delta J = +1$) are called the *R branch*. This apparently arbitrary notation may become clearer if we state here that later, in other contexts, we shall be concerned with ΔJ values of 0 and ± 2, in addition to ± 1 considered here; the labelling of line series is then quite consistent:

$$\text{Lines arising from } \Delta J = -2 \quad -1 \quad 0 \quad +1 \quad +2$$
$$\text{called:} \qquad\qquad\qquad O \quad\; P \quad Q \quad R \quad\; S \quad \text{branch}$$

The P and R notation, with the *lower J (J'')* value as a suffix, is illustrated on the diagrammatic spectrum of Fig. 3.6. This is the conventional notation for such spectra.

It is readily shown that the inclusion of the centrifugal distortion constant D leads to the following expression for the spectrum:

$$\Delta\varepsilon = \bar{\nu}_{\text{spect.}} = \bar{\omega}_o + 2Bm - 4Dm^3 \;\; \text{cm}^{-1} \qquad (m = \pm 1, \pm 2, \pm 3, \dots) \qquad (3.21)$$

However, we have seen in Chapter 2 that B is some $10 \,\text{cm}^{-1}$ or less, while D is only some 0.01 per cent of B. Since a good infra-red spectrometer has a resolving power of about $0.5 \,\text{cm}^{-1}$ it is obvious that D is negligible to a very high degree of accuracy.

The anharmonicity factor, on the other hand, is not negligible. It affects not only the position of the band origin (since $\bar{\omega}_o = \bar{\omega}_e(1 - 2x_e)$), but, by extending the selection rules to include $\Delta v = \pm 2, \pm 3$, etc., also allows the appearance of overtone bands having identical rotational structure. This is illustrated in Fig. 3.7(a), where the fundamental absorption and first overtone of carbon monoxide are shown. From the band centres we can calculate, as shown in the next section, the equilibrium frequency $\bar{\omega}_e$ and the anharmonicity constant x_e.

3.3 THE VIBRATION–ROTATION SPECTRUM OF CARBON MONOXIDE

In Fig. 3.7(b) we show the fundamental vibration–rotation band of carbon monoxide under high resolution, with some lines in the P and R branches numbered according to their J'' values. Table 3.2 gives the observed wavenumbers of the first five lines in each branch. We shall discuss shortly the slight decrease in separation between the rotational lines as the wavenumber increases; this decrease is apparent from the table and from a close inspection of the 'wings' of the spectrum.

From the table we see that the band centre is at about $2143 \,\text{cm}^{-1}$ while the average line separation near the centre is $3.83 \,\text{cm}^{-1}$. This immediately gives:

$$2B = 3.83 \,\text{cm}^{-1} \qquad B = 1.915 \,\text{cm}^{-1}$$

This is in satisfactory agreement with the value $B = 1.921\,18 \,\text{cm}^{-1}$ derived by microwave studies (cf. Sec. 2.3.1) and we could, therefore, have obtained quite good values for the rotational constant and hence the moment of inertia and bond length from infra-red data alone. Historically, of course, the infra-red values came first, the more precise microwave values following much later.

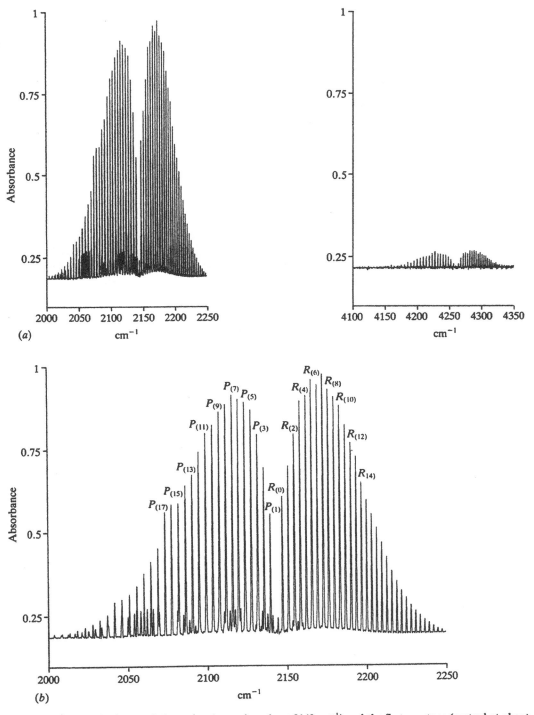

Figure 3.7 (*a*) The fundamental absorption (centred at about 2143 cm^{-1}) and the first overtone (centred at about 4260 cm^{-1}) of carbon monoxide. (*b*) The fundamental band under higher resolution. The lines are labelled according to their J'' values. The P branch is complicated by the presence of a band centred at about 2100 cm^{-1} due to the 1 per cent of ^{13}CO in the sample; some rotational lines from this band appear between P branch lines, others are overlapped by a P branch line and give it an enhanced intensity (e.g. lines $P_{(16)}$, $P_{(23)}$, and $P_{(24)}$. (*Thanks are due to Miss J. Cook of the University of York for providing this spectrum.*)

Table 3.2 Part of the infra-red spectrum of carbon monoxide

Line	$\bar{\nu}$	Separation $\Delta\bar{\nu}$	Line	$\bar{\nu}$	Separation $\Delta\bar{\nu}$
$P_{(1)}$	2139.43		$R_{(0)}$	2147.08	
		3.88			3.78
$P_{(2)}$	2135.55		$R_{(1)}$	2150.86	
		3.92			3.73
$P_{(3)}$	2131.63		$R_{(2)}$	2154.59	
		3.95			3.72
$P_{(4)}$	2127.68		$R_{(3)}$	2158.31	
		3.98			3.66
$P_{(5)}$	2123.70		$R_{(4)}$	2161.97	

It is worth noting at this point that approximate rotational data are obtainable from spectra even if the separate rotational lines are not resolved. Thus Fig. 3.8 shows the spectrum of carbon monoxide under much poorer resolution, when the rotational fine structure is blurred out to an envelope. Now we saw in Eq. (2.21) that the maximum population of levels, and hence the maximum intensity of transition, occurs at a J value of $\sqrt{kT/2Bhc} - \frac{1}{2}$. Remembering that $m = J + 1$ we substitute in Eq. (3.20c) $m = \pm\sqrt{kT/2Bhc} + \frac{1}{2}$ and obtain:

$$\bar{\nu}_{\text{max. intensity}} = \bar{\omega}_0 \pm 2B(\sqrt{kT/2Bhc} + \tfrac{1}{2})$$

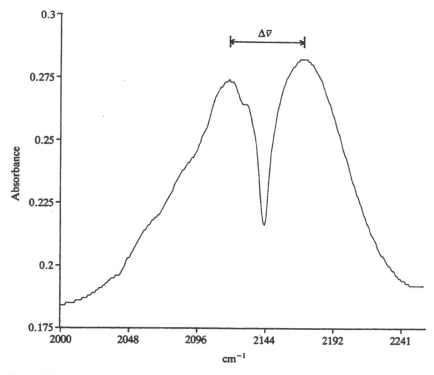

Figure 3.8 The fundamental band of Fig. 3.7(*b*) under very low resolution. All rotational fine structure has been lost and a typical *PR* contour is seen. (*Thanks are due to Miss J. Cook of the University of York for providing this spectrum.*)

where the $+$ and $-$ signs refer to the R and P branches, respectively. The *separation* between the two maxima, $\Delta\bar\nu$, is then:

$$\Delta\bar\nu = 4B(\sqrt{kT/2Bhc} + \tfrac{1}{2}) = \sqrt{8kTB/hc} + 2B$$

or, since B is small compared with $\Delta\bar\nu$, we can write

$$\Delta\bar\nu \approx \sqrt{8kTB/hc} \qquad B \approx hc(\Delta\bar\nu)^2/8kT \quad \text{cm}^{-1}$$

where c is expressed in cm s^{-1}. In the case of carbon monoxide the separation is about $55\,\text{cm}^{-1}$ (Fig. 3.8), while the temperature at which the spectrum was obtained was about $300\,\text{K}$. We are led, then, to a B value of about $1.8\,\text{cm}^{-1}$, which is in fair agreement with the earlier values, but much less precise.

From Table 3.2 we see that the band origin, at the midpoint of $P_{(1)}$ and $R_{(0)}$, is at $2143.26\,\text{cm}^{-1}$. This, then, is the fundamental vibration frequency of carbon monoxide, if anharmonicity is ignored. The latter can be taken into account, however, since the first overtone is found to have its origin at $4260.04\,\text{cm}^{-1}$. We have:

$$\bar\omega_e(1 - 2x_e) = \bar\omega_o = 2143.26$$
$$2\bar\omega_e(1 - 3x_e) = 4260.04$$

from which $\omega_e = 2169.74\,\text{cm}^{-1}$, $x_e = 0.0061$.

3.4 BREAKDOWN OF THE BORN–OPPENHEIMER APPROXIMATION: THE INTERACTION OF ROTATIONS AND VIBRATIONS

So far we have assumed that vibration and rotation can proceed quite independently of each other. A molecule vibrates some 10^3 times during the course of a single rotation, however, so it is evident that the bond length (and hence the moment of inertia and B constant) also changes continually during the rotation. If the vibration is simple harmonic the mean bond length will be the same as the equilibrium bond length and it will not vary with vibrational energy; this is seen in Fig. 3.1. However, the rotational constant B depends on $1/r^2$ and, as shown by an example in Sec. 2.3.4, the average value of this quantity is not the same as $1/r_{\text{eq.}}^2$, where $r_{\text{eq.}}$ is the equilibrium length. Further, an increase in the vibrational energy is accompanied by an increase in the vibrational amplitude and hence the value of B will depend on the v quantum number.

In the case of anharmonic vibrations the situation is rather more complex. Now an increase in vibrational energy will lead to an increase in the average bond length—this is perhaps most evident from Fig. 3.4. The rotational constant then varies even more with vibrational energy.

In general, it is plain that, since $r_{\text{av.}}$ *increases* with the vibrational energy, B is *smaller* in the upper vibrational state than in the lower. In fact an equation of the form:

$$B_v = B_e - \alpha(v + \tfrac{1}{2}) \tag{3.22}$$

gives, to a high degree of approximation, the value of B_v, the rotational constant in vibrational level v in terms of the equilibrium value B_e and α, a small positive constant for each molecule.

Here we restrict our discussion to the fundamental vibrational change, i.e. the change $v = 0 \to v = 1$, and we may take the respective B values as B_0 and B_1 with $B_0 > B_1$. For this transition:

$$\Delta\varepsilon = \varepsilon_{J',v=1} - \varepsilon_{J'',v=0}$$
$$= \bar\omega_o + B_1 J'(J' + 1) - B_0 J''(J'' + 1) \quad \text{cm}^{-1}$$

where, as before, $\bar\omega_o = \bar\omega_e(1 - 2x_e)$.

We then have the two cases:

1.
$$\Delta J = +1 \quad J' = J'' + 1$$

$$\Delta\varepsilon = \bar{\nu}_R = \bar{\omega}_o + (B_1 + B_0)(J'' + 1) + (B_1 - B_0)(J'' + 1)^2 \quad \mathrm{cm}^{-1}$$
$$(J'' = 0, 1, 2, \ldots)$$
(3.23a)

and

2.
$$\Delta J = -1 \quad J'' = J' + 1$$

$$\Delta\varepsilon = \bar{\nu}_P = \bar{\omega}_o - (B_1 + B_0)(J' + 1) + (B_1 - B_0)(J'+1)^2 \quad \mathrm{cm}^{-1}$$
$$(J' = 0, 1, 2, \ldots)$$
(3.23b)

where we have written $\bar{\nu}_P$ and $\bar{\nu}_R$ to represent the wavenumbers of the P and R branch lines, respectively. These two equations can be combined into the expression:

$$\bar{\nu}_{P,R} = \bar{\omega}_o + (B_1 + B_0)m + (B_1 - B_0)m^2 \quad \mathrm{cm}^{-1} \qquad (m = \pm1, \pm2, \ldots)$$
(3.23c)

where positive m values refer to the R branch and negative to P.

We see that ignoring vibration–rotation interaction involves setting $B_1 = B_0$, when Eq. (3.23c) immediately simplifies to (3.20c). Since $B_1 < B_0$ the last term of (3.23c) is always negative, irrespective of the sign of m, and the effect on the spectrum of a diatomic molecule is to crowd the rotational lines more closely together with increasing m on the R branch side, while the P branch lines become more widely spaced as (negative) m increases. Normally B_1 and B_0 differ only slightly and the effect is marked only for high m values. This is exactly the situation shown in the spectrum of carbon monoxide (Fig. 3.7(b)).

In Table 3.3 some of the data for carbon monoxide are tabulated, together with the positions of lines calculated from the equation:

$$\bar{\nu}_{\text{spect.}} = 2143.28 + 3.813m - 0.0175m^2 \quad \mathrm{cm}^{-1}$$

From this we see that, for this molecule:

$$B_1 = 1.898 \, \mathrm{cm}^{-1} \qquad B_0 = 1.915 \, \mathrm{cm}^{-1}$$

Table 3.3 Observed and calculated wavenumbers of some lines in the spectrum of carbon monoxide

m	J''	$\bar{\nu}_{\text{obs.}}$	$\bar{\nu}_{\text{calc.}}$†
30	29	2241.64	2241.91
25	24	2227.63	2227.65
20	19	2212.62	2212.54
15	14	2196.66	2196.53
10	9	2179.77	2179.66
5	4	2161.97	2161.90
0	–	(Band centre)	2143.28
−5	5	2123.70	2123.78
−10	10	2103.27	2103.40
−15	15	2082.01	2082.15
−20	20	2059.91	2060.02
−25	25	2037.03	2037.02
−30	30	2013.35	2013.14

† Values calculated from : $\bar{\nu} = 2143.28 + 3.813m - 0.0175m^2$.

INFRA-RED SPECTROSCOPY **71**

and hence, using Eq. (3.22), we have:

$$\alpha = 0.018 \qquad B_e = 1.924\,\text{cm}^{-1}$$

Further, we can calculate the equilibrium bond length and the bond lengths in the $v = 0$ and $v = 1$ states (cf. Sec. 2.3.1) to be:

$$r_{\text{eq.}} = 0.1130\,\text{nm} \qquad r_0 = 0.1133\,\text{nm} \qquad r_1 = 0.1136\,\text{nm}$$

3.5 THE VIBRATIONS OF POLYATOMIC MOLECULES

In this section and the next, just as in the corresponding one dealing with the pure rotational spectra of polyatomic molecules, we shall find that although there is an increase in the complexity, only slight and quite logical extensions to the simple theory are adequate to give us an understanding of the spectra. We shall need to discuss:

1. The number of fundamental vibrations and their symmetry.
2. The possibility of overtone and combination bands.
3. The influence of rotation on the spectra.

3.5.1 Fundamental Vibrations and Their Symmetry

Consider a molecule containing N atoms: we can refer to the position of each atom by specifying three coordinates (e.g. the x, y, and z cartesian coordinates). Thus the total number of coordinate values is $3N$ and we say the molecule has $3N$ *degrees of freedom* since each coordinate value may be specified quite independently of the others. However, once all $3N$ coordinates have been fixed, the bond distances and bond angles of the molecule are also fixed and no further arbitrary specifications can be made.

Now the molecule is free to move in three-dimensional space, as a whole, without change of shape. We can refer to such movement by noting the position of its centre of gravity at any instant—to do this requires a statement of three coordinate values. This translational movement uses three of the $3N$ degrees of freedom leaving $3N - 3$. In general, also, the rotation of a non-linear molecule can be resolved into components about three perpendicular axes (cf. Sec. 1.1). Specification of these axes also requires three degrees of freedom, and the molecule is left with $3N - 6$ degrees of freedom. The only other motion allowed to it is internal vibration, so we know immediately that a non-linear N-atomic molecule can have $3N - 6$ different internal vibrations:

$$\text{Non-linear:} \quad 3N - 6 \text{ fundamental vibrations} \qquad (3.24a)$$

If, on the other hand, the molecule is linear, we saw in Chapter 2 that there is no rotation about the bond axis; hence only two degrees of rotational freedom are required, leaving $3N - 5$ degrees of vibrational freedom—one more than in the case of a non-linear molecule:

$$\text{Linear:} \quad 3N - 5 \text{ fundamental vibrations} \qquad (3.24b)$$

In both cases, since an N-atomic molecule has $N - 1$ bonds (for acyclic molecules) between its atoms, $N - 1$ of the vibrations are bond-stretching motions; the other $2N - 5$ (non-linear) or $2N - 4$ (linear) are bending motions.

Let us look briefly at examples of these rules. First, we see that for a diatomic molecule (perforce linear) such as we have already considered in this chapter, $N = 2$, $3N - 5 = 1$ and thus there can be only one fundamental vibration. Note, however, that the $3N - 5$ rule says nothing

(a) Symmetric stretch
3651.7 cm^{-1}
ν_1, parallel ($\parallel$)

(b) Symmetric bend
1595.0 cm^{-1}
ν_2, parallel ($\parallel$)

(c) Antisymmetric stretch
3755.8 cm^{-1}
ν_3, perpendicular ($\perp$)

Figure 3.9 The symmetry of the water molecule and its three fundamental vibrations.

about the presence, absence, or intensity of overtone vibrations—these are governed solely by anharmonicity.

Next, consider water, H_2O. This (Fig. 3.9) is non-linear and triatomic. Also in the figure are the $3N - 6 = 3$ allowed vibrational modes, the arrows attached to each atom showing the direction of its motion during half of the vibration. Each motion is described as stretching or bending depending on the nature of the change in molecular shape.

These three vibrational motions are also referred to as the *normal modes of vibration* (or *normal vibrations*) of the molecule; in general a normal vibration is defined as a molecular motion in which all the atoms oscillate with the same frequency and pass through their equilibrium positions simultaneously.

Further, each motion of Fig. 3.9 is labelled either symmetric or antisymmetric. It is not necessary here to go far into the matter of general molecular symmetry since other excellent texts already exist for the interested student, but we can see quite readily that the water molecule contains some elements of symmetry. In particular consider the dashed line at the top of Fig. 3.9 which bisects the HOH angle; if we rotate the molecule about this axis by 180° its final appearance is identical with the initial one. This axis is thus referred to as a C_2 axis since twice in every complete revolution the molecule presents an identical aspect to an observer. This particular molecule has only the one rotational symmetry axis, and it is conventional to refer the molecular vibrations to this axis. Thus consider the first vibration (Fig. 3.9(a)). If we rotate the *vibrating* molecule by 180° the vibration is quite unchanged in character—we call this a symmetric vibration. The bending vibration, ν_2, is also symmetric. Rotation of the stretching motion of Fig. 3.9(c) about the C_2 axis, however, produces a vibration which is in antiphase with the original and so this motion is described as the antisymmetric stretching mode.

In order to be infra-red active, as we have seen, there must be a dipole change during the vibration and this change may take place either along the line of the symmetry axis (parallel to it, or $\parallel$) or at right angles to the line (perpendicular, $\perp$). Figure 3.10 shows the nature of the dipole changes for the three vibrations of water, and justifies the labels parallel or perpendicular attached to them in Fig. 3.9. We shall see later that the distinction is important when considering the influence of *rotation* on the spectrum.

Finally the vibrations are labelled in Fig. 3.9 as ν_1, ν_2, and ν_3. By convention it is usual to label vibrations in decreasing frequency within their symmetry type. Thus the symmetric vibrations of

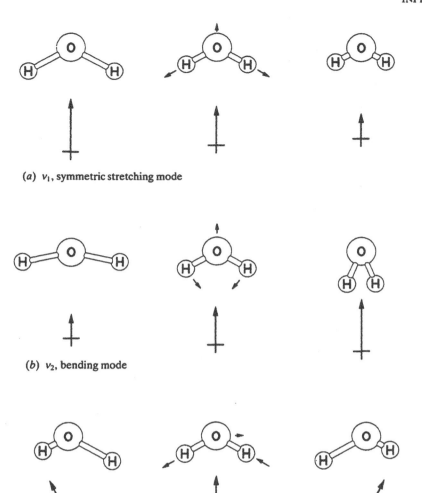

(a) ν_1, symmetric stretching mode

(b) ν_2, bending mode

(c) ν_3, asymmetric stretching mode

Figure 3.10 The change in the electric dipole moment produced by each vibration of the water molecule. This is seen to occur either along ($\parallel$) or across ($\perp$) the symmetry axis.

H_2O are labelled ν_1 for the highest fully symmetric frequency ($3651.7\,cm^{-1}$) and ν_2 for the next highest ($1595.0\,cm^{-1}$); the antisymmetric vibration at $3755.8\,cm^{-1}$ is then labelled ν_3.

Our final example is of the linear triatomic molecule CO_2, for which the normal vibrations are shown in Fig. 3.11. For this molecule there are two different sets of symmetry axes. There is an infinite number of twofold axes (C_2) passing through the carbon atom at right angles to the bond direction, and there is an ∞-fold axis (C_∞) passing through the bond axis itself (this is referred to as ∞-fold since rotation of the molecule about the bond axis through *any* angle gives an identical aspect). The names symmetric stretch and antisymmetric stretch are self-evident, but it should be noted that the symmetric stretch produces no change in the dipole moment (which remains zero) so that this vibration is not infra-red active; the vibration frequency may be obtained in other ways, however, which we shall discuss in the next chapter.

For linear triatomic molecules, $3N - 5 = 4$, and we would expect four vibrational modes instead of the three shown in Fig. 3.11. However, consideration shows that ν_2 in fact consists of

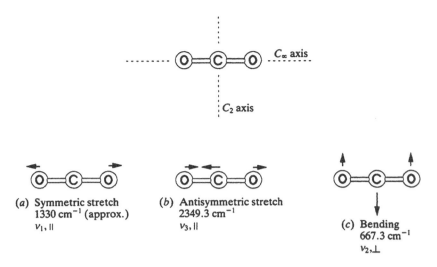

Figure 3.11 The symmetry and fundamental vibrations of the carbon dioxide molecule.

two vibrations—one in the plane of the paper as drawn and the other in which the oxygen atoms move simultaneously into and out of the plane. The two sorts of motion are, of course, identical in all respects except direction and are termed *degenerate*; they must, nevertheless, be considered as separate motions, and it is always in the degeneracy of a bending mode that the extra vibration of a linear molecule over a non-linear one is to be found.

It might be thought that v_2 of H_2O (Fig. 3.9(*b*)) could occur by the hydrogens moving simultaneously in and out of the plane of the paper. Such a motion is not a vibration, however, but a *rotation*. As the molecule approaches linearity this rotation degenerates into a vibration, and the molecule loses one degree of rotational freedom in exchange for one of vibration.

3.5.2 Overtone and Combination Frequencies

If one were able to observe the molecules of H_2O or CO_2 directly their overall vibrations would appear extremely complex; in particular, each atom would not follow tidily any one of the separate paths depicted in Figs 3.9 or 3.11, but its motion would essentially be a superposition of all such paths, since every possible vibration is always excited, at least to the extent of its zero-point energy. However, such superposition could be resolved into its components if, for instance, we could examine the molecules under stroboscopic light flashing at each fundamental frequency in turn. This is, so to speak, the essence of infra-red spectroscopy—instead of flashing we have the radiation frequency, and the 'examination' is a sensing of dipole alteration. Thus, as we would expect, the infra-red spectrum of a complex molecule consists essentially of an absorption band at each of the $3N - 6$ (non-linear) or $3N - 5$ (linear) fundamental frequencies.

This is, of course, an over-simplification, in which two approximations are implicit: (1) that each vibration is simple harmonic, (2) that each vibration is quite independent and unaffected by the others. We shall consider (2) in more detail later; for the moment we can accept it as a good working approximation.

When the restriction to simple harmonic motion is lifted we have again, as in the case of the diatomic molecule (Sec. 3.1.3), the possibility of first, second, etc., overtones occurring at frequencies near $2v_1, 3v_1, \ldots, 2v_2, 3v_2, \ldots, 2v_3, \ldots$, etc., where each v_i is a fundamental mode. The intensities fall off rapidly. However, in addition, the selection rules now permit *combination bands* and *difference bands*. The former arise simply from the addition of two or more fundamental frequencies or overtones. Such combinations as $v_1 + v_2$, $2v_1 + v_2$, $v_1 + v_2 + v_3$, etc., become allowed, although their intensities are normally very small. Similarly the difference bands, for example, $v_1 - v_2$, $2v_1 - v_2$, $v_1 + v_2 - v_3$, have small intensities but are often to be found in a complex spectrum.

The intensities of overtone or combination bands may sometimes be considerably enhanced by a resonance phenomenon. It may happen that two vibrational modes in a particular molecule have frequencies very close to each other—they are described as *accidentally* degenerate. Note that we are not here referring to identical vibrations, such as the two identical v_2's of CO_2 (Fig. 3.11), but rather to the possibility of two quite different modes having similar energies. Normally the fundamental modes are quite different from each other and accidental degeneracy is found most often between a fundamental and some overtone or combination. A simple example is to be found in CO_2 where v_1, described as at about 1330 cm^{-1}, is very close to that of $2v_2 = 1334$ cm^{-1}. (As mentioned earlier, these bands are not observable in the infra-red, but both may be seen in the Raman spectrum discussed in the next chapter; the principles of resonance apply equally to both techniques.) Quantum mechanics shows that two such bands may interfere with each other in such a way that the higher is raised in frequency, the lower depressed—and in fact the Raman spectrum shows two bands, one at 1285 cm^{-1}, the other at 1385 cm^{-1}. Their mean is plainly at about 1330 cm^{-1}.

Note, however, that one of these bands arises from a fundamental mode (v_1), the other from the overtone $2v_2$, and we would normally expect the former to be much more intense than the latter. In fact, they are found to be of about the same intensity—the overtone has gained intensity at the expense of the fundamental. This is an extreme case—normally the overtone takes only a small part of the intensity from the fundamental. The situation is often likened to that of two pendulums connected to a common bar—when the pendulums have quite different frequencies they oscillate independently; when their frequencies are similar they can readily exchange energy, one with the other, and an oscillation given to one is transferred to and fro between them. They are said to resonate. Similarly two close molecular vibrational frequencies resonate and exchange energy—the phenomenon being known as *Fermi resonance* when a fundamental resonates with an overtone. In the spectrum of a complex molecule exhibiting many fundamentals and overtones, there is a good chance of accidental degeneracy, and Fermi resonance, occurring. However, it should be mentioned that not all such degeneracies lead to resonance. It is necessary, also, to consider the molecular symmetry and the type of degenerate vibrations; we shall not, however, pursue the topic further here.

3.6 THE INFLUENCE OF ROTATION ON THE SPECTRA OF POLYATOMIC MOLECULES

In Sec. 3.2 we found that the selection rule for the simultaneous rotation and vibration of a diatomic molecule was

$$\Delta v = \pm 1, \pm 2, \pm 3, \ldots \quad \Delta J = \pm 1 \quad \Delta J \neq 0$$

and that this gave rise to a spectrum consisting of approximately equally spaced line series on each side of a central minimum designated as the band centre.

Earlier in the present section we showed that the vibrations of complex molecules could be subdivided into those causing a dipole change either (1) parallel or (2) perpendicular to the major axis of rotational symmetry. The purpose of this distinction, and the reason for repeating it here, is that the selection rules for the *rotational* transitions of complex molecules depend, rather surprisingly, on the type of *vibration*, ∥ or ⊥, which the molecule is undergoing. Less surprisingly, the selection rules and the energies depend on the *shape* of the molecule also. We shall deal first with the linear molecule as the simplest, and then say a few words about the other types of molecule.

3.6.1 Linear Molecules

Parallel vibrations   The selection rule for these is identical with that for diatomic molecules, i.e.

$$\Delta J = \pm 1 \qquad \Delta v = \pm 1 \qquad\qquad \text{for simple harmonic motion} \qquad (3.25a)$$

$$\Delta J = \pm 1 \qquad \Delta v = \pm 1, \pm 2, \pm 3, \ldots \quad \text{for anharmonic motion} \qquad (3.25b)$$

(This is, in fact, as expected, since a diatomic molecule is linear and can undergo only parallel vibrations.) The spectra will thus be similar in appearance, consisting of P and R branches with lines about equally spaced on each side, no line occurring at the band centre. Now, however, the moment of inertia may be considerably larger, the B value correspondingly smaller, and the P or R line spacing will be less. Figure 3.12 shows part of the spectrum of HCN, a linear molecule whose structure is $H—C{\equiv}N$. The band concerned is the symmetric stretching frequency at about $3310\,\text{cm}^{-1}$ (corresponding to the v_1 mode of CO_2 in Fig. 3.11), and the spacing is observed to be about $2.8–3.0\,\text{cm}^{-1}$ near the band centre. This is to be compared, for example, with the spacing of about $4.0\,\text{cm}^{-1}$ in the case of CO.

For still larger molecules the value of B may be so small that separate lines can no longer be resolved in the P and R branches. In this case the situation is exactly analogous to that shown previously in Fig. 3.8 and the same remarks apply as to the possibility of deriving a rough value of B from the separation between the maxima of the P and R envelopes. We shall shortly see that a non-linear molecule cannot give rise to this type of band shape, so its observation somewhere within a spectrum is sufficient proof that a linear, or nearly linear, molecule is being studied.

Perpendicular vibrations   For these the selection rule is found to be:

$$\Delta v = \pm 1 \qquad \Delta J = 0, \pm 1 \qquad \text{for simple harmonic motion} \qquad (3.26)$$

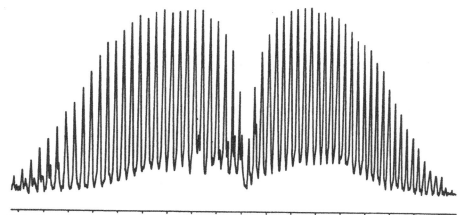

3220 3230 3240 3250 3260 3270 3280 3290 3300 3310 3320 3330 3340 3350 3360 3370 3380 3390

$$\text{cm}^{-1}$$

Figure 3.12 Spectrum of the symmetric stretching vibration of HCN, showing the P and R branch lines.

which implies that now, for the first time, a vibrational change can take place with *no simultaneous rotational transition*. The result is illustrated in Fig. 3.13, which shows the same energy levels and transitions as Fig. 3.6 with the addition of $\Delta J = 0$ transitions. If the oscillation is taken as simple harmonic the energy levels are identical with those of Eq. (3.18) and the P and R branch lines are given, as before, by Eqs (3.20) or (3.21). Transitions with $\Delta J = 0$, however, correspond to a Q *branch* whose lines may be derived from the equations:

$$
\begin{aligned}
\Delta\varepsilon &= \varepsilon_{J,v+1} - \varepsilon_{J,v} \\
&= 1\tfrac{1}{2}\bar{\omega}_e - 2\tfrac{1}{4}x_e\bar{\omega}_e + BJ(J+1) - \{\tfrac{1}{2}\bar{\omega}_e - \tfrac{1}{4}x_e\bar{\omega}_e + BJ(J+1)\} \\
&= \bar{\omega}_o \quad \text{cm}^{-1} \quad \text{for all } J
\end{aligned}
\tag{3.27}
$$

Thus the Q branch consists of lines superimposed upon each other at the band centre $\bar{\omega}_o$, one contribution arising for each of the populated J values. The resultant line is usually very intense.

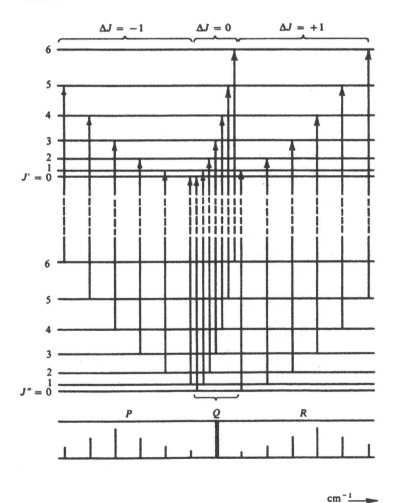

Figure 3.13 The rotational energy levels for two vibrational states, showing the effect on the spectrum of transitions for which $\Delta J = 0$.

If we take into account the fact that the B values differ slightly in the upper and lower vibrational states (cf. Sec. 3.4), we would write instead:

$$\Delta\varepsilon = \varepsilon_{J,v+1} - \varepsilon_{J,v}$$
$$= 1\tfrac{1}{2}\bar{\omega}_e - 2\tfrac{1}{4}x_e\bar{\omega}_e + B'J(J+1) - \{\tfrac{1}{2}\bar{\omega}_e - \tfrac{1}{4}x_e\omega_e + B(J+1)\} \qquad (3.28)$$
$$= \bar{\omega}_o + J(J+1)(B' - B'').$$

Further, if $B' < B''$, we see that the Q branch line would become split into a series of lines on the *low*-frequency side of $\bar{\omega}_o$ (since $B' - B''$ is negative). Normally, however, $B' - B''$ is so small that the lines cannot be resolved, and the Q branch appears as a somewhat broad absorption centred around $\bar{\omega}_o$. This is illustrated in Fig. 3.14, which is a spectrum of the bending mode of HCN (corresponding to v_2 of CO_2 in Fig. 3.11). Finally, if the rotational fine structure is unresolved, this type of band has the distinctive contour shown in Fig. 3.15.

It should be remembered (see Chapter 2) that polyatomic molecules with zero dipole moment do not give rise to pure rotation spectra in the microwave region (for example CO_2, $HC\equiv CH$, CH_4). Such molecules do, however, show vibrational spectra in the infra-red region (or Raman, cf. Chapter 4) and, if these spectra exhibit resolved fine structure, the moment of inertia of the molecule can be obtained.

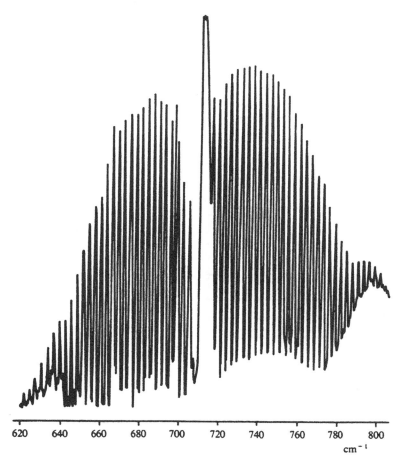

620	640	660	680	700	720	740	760	780	800	

$$cm^{-1}$$

Figure 3.14 Spectrum of the bending mode of HCN, showing the *PQR* structure. The broad absorption centred at $800\,cm^{-1}$ is due to an impurity.

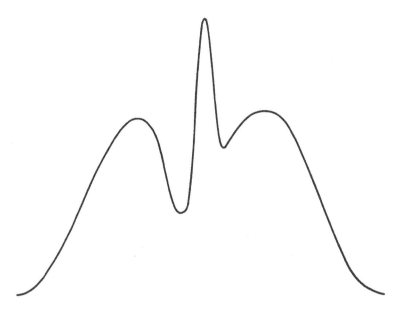

Figure 3.15 The contour of a *PQR* band under low resolution.

3.6.2 The Influence of Nuclear Spin

It is necessary here to say a brief word about the spectrum of carbon dioxide and other linear molecules possessing a centre of symmetry. A centre of symmetry means that identical atoms are symmetrically disposed with respect to the centre of gravity of the molecule. Thus, plainly both CO_2 [O=C=O] and ethyne [H—C≡C—H] possess a centre of symmetry, while HCN or N_2O [N≡N=O] do not.

The reader may have noticed that, although we used CO_2 as an example of a vibrating molecule in Fig. 3.11, we did not use it to illustrate real spectra in the subsequent discussion. This is because the centre of symmetry has an effect on the intensity of alternate lines in the P and R branches. The effect is due to the existence of nuclear spin (cf. Chapter 7) and is an additional factor determining the populations of rotational levels. In the case of CO_2 every alternate rotational level is completely unoccupied and so alternate lines in the P and R branches have zero intensity. This leads to a line spacing of $4B$ instead of the usual $2B$ discussed above. That the spacing is indeed $4B$ (and not $2B$ with an unexpectedly large value of B) can be shown in several ways, perhaps the most convincing of which is to examine the spectrum of the isotopic molecule ^{18}O—C—^{16}O. Here there is no longer a centre of symmetry, nuclear spin does not now affect the spectrum and the line spacing is found to be just half that for 'normal' CO_2.

In the case of acetylene, alternate levels have populations which differ by a factor of 3:1 (this, due to nuclear spin alone, is superimposed on the normal thermal distribution and degeneracy) so that the P and R branch lines show a strong, weak, strong, weak, . . . alternation in intensity, as shown in Fig. 3.16.

3.6.3 Symmetric Top Molecules

Following the Born–Oppenheimer approximation we can take the vibrational–rotational energy levels for this type of molecule to be the sum of the vibrational levels:

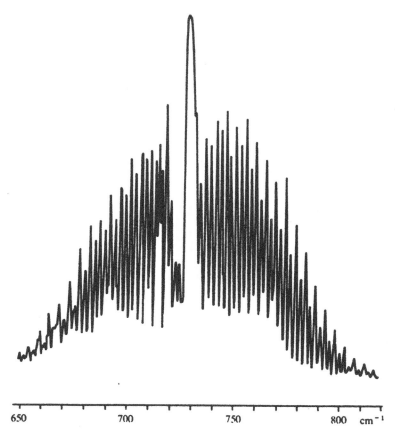

Figure 3.16 The spectrum of the bending mode of ethyne, HC≡CH, showing the strong, weak, strong, weak, . . . intensity alternation in the rotational fine structure due to the nuclear spin of the hydrogen atoms.

$$\varepsilon_{\text{vib.}} = (v + \tfrac{1}{2})\bar{\omega}_e - (v + \tfrac{1}{2})^2 x_e\bar{\omega}_e \quad \text{cm}^{-1} \quad (v = 0, 1, 2, 3, \dots)$$

and the rotational levels (cf. Eq. (2.38)):

$$\varepsilon_{\text{rot.}} = BJ(J+1) + (A - B)K^2 \quad \text{cm}^{-1}$$
$$[J = 0, 1, 2, \dots; \quad K = J, (J-1), (J-2), \dots, -J]$$

Thus

$$\varepsilon_{J,v} = \varepsilon_{\text{vib.}} + \varepsilon_{\text{rot.}} = (v + \tfrac{1}{2})\bar{\omega}_e - (v + \tfrac{1}{2})^2 x_e\bar{\omega}_e$$
$$+ BJ(J+1) + (A - B)K^2 \quad \text{cm}^{-1} \tag{3.29}$$

This equation assumes, of course, that centrifugal distortion is negligible.

Again it is necessary to divide the vibrations into those which change the dipole (1) parallel and (2) perpendicular to the main symmetry axis—which is nearly always the axis about which the 'top' rotates. The rotational selection rules differ for the two types.

Parallel vibrations Here the selection rule is:

$$\Delta v = \pm 1 \quad \Delta J = 0, \pm 1 \quad \Delta K = 0 \tag{3.30}$$

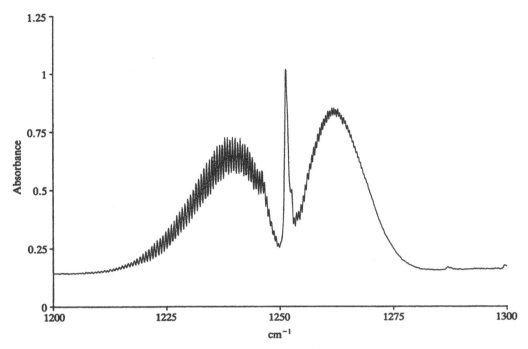

Figure 3.17 The parallel stretching vibration, centred at $1251\,\mathrm{cm}^{-1}$, of the symmetric top molecule methyl iodide, CH_3I, showing the typical PQR contour. (*Thanks are due to Mr J. Camplin of the University of York for providing this spectrum.*)

Since here $\Delta K = 0$, terms in K will be identical in the upper and lower state and so the spectral frequencies will be independent of K. Thus the situation will be identical to that discussed for the *perpendicular* vibrations of a linear molecule. The spectrum will contain P, Q, and R branches with a P, R line spacing of $2B$ (which is unlikely to be resolved) and a strong central Q branch. Such a spectrum, a ‖ band of methyl iodide, CH_3I, is shown in Fig. 3.17. The intensity of the Q branch (relative to lines in the P and R branches) varies with the ratio I_A/I_B; in the limit, when $I_A \to 0$, the symmetric top becomes a linear molecule and the Q branch has zero intensity, as discussed earlier.

Perpendicular vibrations For these the selection rule is:

$$\Delta v = \pm 1 \quad \Delta J = 0, \pm 1 \quad \Delta K = \pm 1 \tag{3.31}$$

Each of the following expressions is readily derivable for the spectral lines, taking the energy levels of Eq. (3.29).

1. $\Delta J = \pm 1$, $\Delta K = \pm 1$ (R branch lines):

$$\Delta\varepsilon = \bar{\nu}_{\text{spect.}} = \bar{\omega}_o + 2B(J+1) + (A-B)(1 \pm 2K) \quad \mathrm{cm}^{-1} \tag{3.32a}$$

2. $\Delta J = -1$, $\Delta K = \pm 1$ (P branch lines):

$$\bar{\nu}_{\text{spect.}} = \bar{\omega}_o - 2B(J+1) + (A-B)(1 \pm 2K) \quad \mathrm{cm}^{-1} \tag{3.32b}$$

3. $\Delta J = 0$, $\Delta K = \pm 1$ (Q branch lines):

$$\bar{\nu}_{\text{spect.}} = \bar{\omega}_o + (A-B)(1 \pm 2K) \quad \mathrm{cm}^{-1} \tag{3.32c}$$

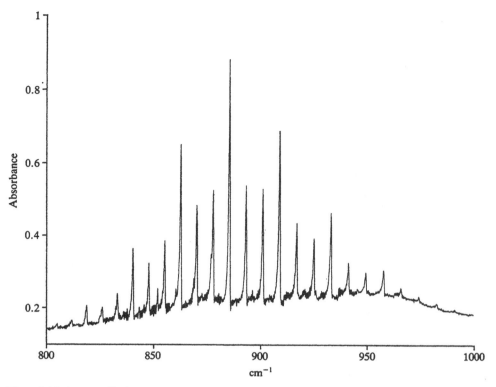

Figure 3.18 A perpendicular stretching vibration of methyl iodide showing the typical Q branch sequence. (*Thanks are due to Mr J. Camplin of the University of York for providing this spectrum.*)

We see, then, that this type of vibration gives rise to many sets of P and R branch lines since for each J value there are many allowed values of K ($K = J, J - 1, \ldots, -J$). The wings of the spectrum will thus be quite complicated and will not normally be resolvable into separate lines. The Q branch is also complex, since it too will consist of a series of lines on both sides of $\bar{\omega}_o$ separated by $2(A - B)$. This latter term may not be small (and is equal to zero only for *spherical top* molecules which have all their moments of inertia equal). For $A \gg B$ (for example CH_3I) the Q branch lines will be well separated and will appear as a series of maxima above the P, R envelope. This spectrum is shown in Fig. 3.18.

It will be noted in this figure that the lines have a distinct periodical variation in intensity—strong, weak, weak, strong, weak, weak, ... This behaviour reminds us of CO_2 and C_2H_2, discussed earlier, in which the presence or absence of nuclear spin altered the relative populations of the rotational levels. In that case, where the molecule had a twofold axis of symmetry, the periodicity also was two—strong, weak, strong, weak, ... It is not surprising, therefore, that the threefold periodicity, strong, weak, weak, strong, . . . , seen in CH_3I, arises because of its threefold axis of symmetry to rotations about the C—I axis. The appearance of such a spectrum confirms immediately that we are dealing with a molecule containing an XY_3 grouping.

3.6.4 Other Polyatomic Molecules

We shall not go further with the discussion of their detailed spectra here—it suffices to state that the complexity increases, naturally, with the molecular complexity. An excellent treatment is to be found in Herzberg's book, but the subject is not for the beginner in spectroscopy.

3.6.5 Summary

We have seen that the infra-red spectrum of even a simple diatomic molecule may contain a great many lines, while that of a polyatom may be extraordinarily complex, even though some of the details of fine structure are blurred by insufficient resolving power. Although in favourable cases much information may be obtained about bond lengths and angles or at least the general shape of a molecule, in others even the assignment of observed bands to particular molecular vibrations is not trivial. Assignments are based mainly on experience with related molecules, on the band contour (from which the type of vibration, $\parallel$ or $\perp$, can usually be deduced), and on the use of Raman spectra (see Chapter 4). Consideration of the symmetry of the molecule is also important because this determines which vibrations are likely to be infra-red active.

Fortunately the usefulness of infra-red spectroscopy extends far beyond the measurement of precise vibrational frequencies and molecular structural features. In the next section we discuss briefly the application of infra-red techniques to chemical analysis—a branch of the subject where it is by no means essential always to be able to assign observed bands precisely.

3.7 ANALYSIS BY INFRA-RED TECHNIQUES

Because of the $3N - 6$ and $3N - 5$ rules it is evident that a complex molecule is likely to have an infra-red spectrum exhibiting a large number of normal vibrations. Each normal mode involves some displacement of all, or nearly all, the atoms in the molecule, but while in some of the modes all atoms may undergo approximately the same displacement, in others the displacements of a small group of atoms may be much more vigorous than those of the remainder. Thus we may divide the normal modes into two classes: the *skeletal vibrations*, which involve many of the atoms to much the same extent, and the *characteristic group vibrations*, which involve only a small portion of the molecule, the remainder being more or less stationary. We deal with these classes separately.

3.7.1 Skeletal Vibrations

For organic molecules these usually fall in the range 1400–700 cm^{-1} and arise from linear or branched chain structures in the molecule. Thus such groups as

$$-\text{C}-\text{O}-\text{C}- \qquad -\text{C}-\overset{\overset{\textstyle \text{C}}{|}}{\underset{\underset{\textstyle \text{C}}{|}}{\text{C}}}-\text{C}- \qquad \bigcirc$$

etc., each give rise to several skeletal modes of vibration and hence several absorption bands in the infra-red. It is seldom possible to assign particular bands to specific vibrational modes, but the whole complex of bands observed is highly typical of the molecular structure under examination. Further, changing a substituent (on the chain, or in the ring) usually results in a marked change in the pattern of the absorption bands. Thus these bands are often referred to as the 'fingerprint' bands, because a molecule or structural moiety may often be recognized merely from the appearance of this part of the spectrum.

One example of infra-red (IR) spectroscopy as a fingerprint technique is shown in Fig. 3.19(*a*). The late 1970s saw the launch of *Voyager* space probes to observe the giant planets Jupiter, Saturn, Uranus, and Neptune. The probes were equipped with a number of spectrometers

Figure 3.19(a) The infra-red emission spectrum of the radiation from Jupiter and Saturn recorded with a spectrometer on *Voyager*. (*Reproduced, with permission, from Hanel, R. A., B. J. Conrath, D. E. Jennings, and R. E. Samuelson, in* Exploration of the Solar System by Infra-red Remote Sensing, *Cambridge University Press, 1992.*)

in order to gather as much information as possible. The atmospheres of these planets are mostly composed of hydrogen and helium, but IR spectroscopy proved to be extremely useful for the determination of some of the other constituents. Figure 3.19(a) shows the IR emission spectra of Saturn and Jupiter, together with the assignment of the major peaks.

Another excellent example of IR as a fingerprint technique is shown in Fig. 3.19(b), which compares the spectra of natural and synthetic thymidine. The remarkably exact correlation between the spectra proves that the synthetic product does not differ in the slightest degree from the natural substance.

3.7.2 Group Frequencies

Group frequencies, on the other hand, are usually almost independent of the structure of the molecule as a whole and, with a few exceptions, fall in the regions well above and well below that of the skeletal modes. Table 3.4 collects some of the data, of which a much more complete selection is to be found in the book by Williams and Fleming mentioned in the bibliography at the end of this chapter. We see that the vibrations of light atoms in terminal groups (for example —CH_3, —OH, —C≡N, >C=O, etc.) are of high frequency, while those of heavy atoms (—C—Cl, —C—Br, metal–metal, etc.) are low in frequency. Their frequencies, and consequently their spectra, are highly characteristic of the group, and can be used for analysis. For example, the —CH_3 group gives rise to a symmetric C—H stretching absorption

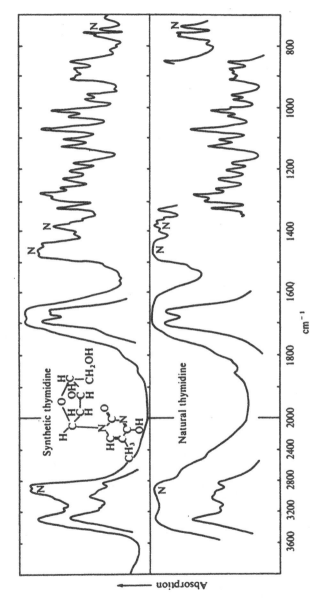

Figure 3.19(b) Comparison of the infra-red spectra of natural and synthetic thymidine. (N = absorption from liquid paraffin (nujol) in which the solid thymidine is suspended.) *(Reproduced by permission of Professor N. Sheppard FRS of the University of East Anglia, Norwich.)*

Table 3.4 Characteristic stretching frequencies of some molecular groups

Group	Approximate frequency (cm⁻¹)	Group	Approximate frequency (cm⁻¹)
—OH	3600	>C=O	1750–1600
—NH₂	3400	>C=C<	1650
≡CH	3300	>C=N<	1600
(benzene ring with H)	3060	≥C—C≤ / ≥C—N< / ≥C—O<	1200–1000
=CH₂	3030		
		>C=S	1100
—CH₃	2970 (asym. stretch)	≥C—F	1050
	2870 (sym. stretch)		
	1460 (asym. deform.)	≥C—Cl	725
	1375 (sym. deform.)		
—CH₂—	2930 (asym. stretch)	≥C—Br	650
	2860 (sym. stretch)		
	1470 (deformation)	≥C—I	550
—SH	2580		
—C≡N	2250		
—C≡C—	2220		

invariably falling between 2850 and 2890 cm⁻¹, an asymmetric stretching frequency at 2940–2980 cm⁻¹, a symmetric deformation (i.e. the opening and closing of the

$$\underset{\text{H} \quad \text{H} \quad \text{H}}{\overset{|}{\text{C}}}$$

'umbrella') at about 1375 cm⁻¹, and an asymmetric deformation at about 1470 cm⁻¹. Again, the $>C=O$ group shows a very sharp and intense absorption between 1600 and 1750 cm⁻¹, depending largely on the other substituents of the group. An example of the application of group frequency data is shown in Fig. 3.19(c); this is the spectrum of thioacetic acide—acetic acid in which one oxygen atom has been replaced by sulphur. The question might be asked: is the molecule $CH_3CO.SH$ or $CH_3CS.OH$? The infra-red spectrum gives a very clear answer. It shows a very sharp absorption at about 1730 cm⁻¹, and one at about 2600 cm⁻¹, and these are consistent with the presence of $>C=O$ and —SH groups, respectively (cf. Table 3.4). Also there is no strong absorption at 1100 cm⁻¹, thus indicating the absence of $>C=S$.

The idea of group vibrations also covers the motions of isolated features of a molecule which have frequencies not too near those of the skeletal vibrations. Thus isolated multiple bonds (for example $>C=C<$ or —C≡C—) have frequencies which are highly characteristic. When, however, two such groups which, in isolation, have comparable frequencies occur together in a molecule, resonance occurs and the group frequencies may be shifted considerably from the expected value. Thus the isolated carbonyl in a ketone

$$\left(\overset{R}{\underset{R}{>}} C=O \right)$$

Figure 3.19(*c*) Part of the infra-red spectrum of thioacetic acid, CH₃COSH, to illustrate the use of infra-red group frequencies.

and the $>$C$=$C$<$ double bond have group frequencies of 1715 and 1650 cm^{-1}, respectively; however, when the grouping

$$>\text{C}=\text{C}-\overset{|}{\underset{|}{\text{C}}}=\text{O}$$

occurs, their separate frequencies are shifted to 1675 and about 1600 cm^{-1}, respectively, and the intensity of the $>$C$=$C$<$ absorption increases to become comparable with that of the inherently strong $>$C$=$O band (cf. Fermi resonance, Sec. 3.5.2). Closer coupling of the two groups, as in the ketene radical, $>$C$=$C$=$O, gives rise to absorptions at about 2100 and 1100 cm^{-1}, which are very far removed from the 'characteristic' frequencies of the separate groups.

Shifts in group frequencies can arise in other ways too, particularly as the result of interactions between different molecules. Thus the $-$OH stretching frequency of alcohols is very dependent on the degree of hydrogen bonding, which lengthens and weakens the $-$OH bond, and hence lowers its vibrational frequency. If the hydrogen bond is formed between the $-$OH and, say, a carbonyl group, the the latter frequency is also lowered, although to a less extent than the $-$OH, since hydrogen bonding weakens the $>$C$=$O linkage also. However, shifts in group frequency position caused by resonance or intermolecular effects are in themselves highly characteristic and very useful for diagnostic purposes.

In a similar way a change of physical state may cause a shift in the frequency of a vibration, particularly if the molecule is rather polar. In general the more condensed phase gives a lower frequency: $\nu_{\text{gas}} > \nu_{\text{liquid}} \approx \nu_{\text{solution}} > \nu_{\text{solid}}$. Thus in the relatively polar molecule HCl there is a shift of some 100 cm^{-1} in passing from vapour to liquid and a further decrease of 20 cm^{-1} on solidification. Non-polar CO_2, on the other hand, shows negligible shifts in its symmetric vibrations (Fig. 3.11(*a*) and (*b*)) but a lowering of some 60 cm^{-1} in ν_3 on solidification.

Examination of Table 3.4 shows that there are logical trends in group frequencies, since Eq. (3.4):

$$\bar{\omega} = \frac{1}{2\pi c} \sqrt{\frac{k}{\mu}} \quad \text{cm}^{-1}$$

is approximately obeyed. Thus we see that increasing the mass of the atom undergoing oscillation within the group (i.e. increasing μ) tends to decrease the frequency —cf. the series CH, CF, CCl, CBr, or the values for $>C=O$ and $>C=S$. Also, increasing the strength of the bond, and hence increasing the force constant k, tends to increase the frequency, e.g. the series $-C-X$, $-C=X$, $-C\equiv X$, where X is C, N, or (in the first two fragments) O.

We should at this point consider very briefly the intensities of infra-red bands. We have seen that an infra-red spectrum only appears if the vibration produces a change in the permanent electric dipole of the molecule. It is reasonable to suppose, then, that the more polar a bond, the more intense will be the infra-red spectrum arising from vibrations of that bond. This is generally borne out in practice. Thus the intensities of the $>C=O$, $>C=N-$, and $>C=C<$ bands decrease in that order, as do those of the $-OH$, $>NH$, and $\geqq CH$ bands. For this reason, too, the vibrations of ionic crystal lattices often give rise to very strong absorptions. We shall see in the next chapter that the reverse is true in Raman spectroscopy—there the less polar (and hence usually more *polarizable*) bonds give the most intense spectral lines.

In summary, then, experience coupled with comparison spectra of known compounds enables one to deduce a considerable amount of structural information from an infra-red spectrum. It should perhaps be mentioned that the *complete* interpretation of the spectrum of a complex molecule can be a very difficult or impossible task. One is usually content to assign the strongest bands and to be able to explain some of the weaker ones as overtones or combinations.

3.8 TECHNIQUES AND INSTRUMENTATION

3.8.1 Outline

We first deal briefly with each component of the spectrometer as it is usually assembled for infra-red work.

1. *Source*. The source is always some form of filament which is maintained at red- or white-heat by an electric current. Two common sources are the Nernst filament, consisting of a spindle of rare earth oxides about 1 inch long and 0.1 inch in diameter, and the 'globar' filament, a rod of carborundum, somewhat thicker and longer than the Nernst. The Nernst requires to be pre-heated before it will conduct electricity, but once red-heat is reached the temperature is maintained by the current.
2. *Optical path and monochromator*. The beam is guided and focused by mirrors aluminized or silvered on their surfaces. Normally a focus is produced at the point where the sample is to be placed. Ordinary lenses and mirrors are not suitable as glass absorbs strongly over most of the frequencies used. Any windows which are essential (e.g. to contain a sample, or to protect the detector) must be made of mineral salts transparent to infra-red radiation which have been highly polished in order to reduce scattering to a minimum. NaCl, which is transparent above $650\,\mathrm{cm}^{-1}$, and KBr, transparent above $400\,\mathrm{cm}^{-1}$, are much used. For aqueous samples, AgCl and CaF_2 sample cells can be used, and are infra-red transparent above 400 and $1200\,\mathrm{cm}^{-1}$, respectively.

   A rotatable grating is usually used to disperse the radiation, having largely superseded rotatable prisms, which have poorer resolving power. Where the latter are still in use, the prism is usually made of NaCl or KBr.
3. *Detector*. Two main types are in common use, one sensing the heating effect of the radiation, the other depending on photoconductivity. In both the greater the effect (temperature or

conductivity rise) at a given frequency, the greater the transmittance (and the less the absorbance) of the sample at that frequency.

An example of the temperature method is to be found in the Golay cell which is pneumatic in operation. The radiation falls on to a very small cell containing air, and temperature changes are measured in terms of pressure changes within the cell which can be recorded directly as 'transmittance'. Alternative examples of thermal detectors are small, sensitive thermocouples or bolometers.

Pyroelectric detectors such as deuterated triglycine sulphate (DTGS) are in common use in FT spectrometers. As they are also thermal detectors they are sensitive across the whole infra-red range, but have the rapid signal response needed in interferometry.

The phenomenon of conductivity in substances is thought to arise as a consequence of the movement of loosely held electrons through the lattice; insulators, on the other hand, have no such loosely bound electrons. Semiconductors are essentially midway between these materials, having no loosely bound electrons in the normal state, but having 'conduction bands' or raised electron energy levels into which electrons may be readily excited by the absorption of energy from an outside source. Photoconductors are a particular class of semiconductor in which the energy required comes from incident radiation, and some materials, such as lead sulphide, have been found sufficiently sensitive to infra-red radiation (although only above some $3000 \, cm^{-1}$) that they make excellent detectors. The conductivity of the material can be measured continuously by a type of Wheatstone bridge network and, when plotted against frequency, this gives directly the transmittance of the sample.

Other photoconductive detectors in common use are indium antimonide (InSb) which can be used above $1400 \, cm^{-1}$ and mercury cadmium telluride (MCT) used above $\sim 700 \, cm^{-1}$. These detectors operate at liquid nitrogen temperatures, which reduces noise, and tend to have a faster response time and sensitivity, compared even with, for example, the DTGS. Where extremely high sensitivity is required doped photoconductors such as copper- or gold-doped germanium detectors can be used—but these generally need to operate at liquid helium temperatures.

4. *Sample.* For reasons just stated, the sample is held between plates of polished mineral salt rather than glass. Pure liquids are studied in thicknesses of about 0.01 mm, while solutions are usually 0.1–10 mm thick, depending on the dilution. Gas samples at pressures of up to 1 atmosphere or greater are usually contained in glass cells either 5 or 10 cm long, closed at their ends with rock salt windows. Special long-path cells, in which the radiation is repeatedly reflected up and down the cell, may be used for gases at low pressure, perhaps less than 100 mmHg.

Solid samples are more difficult to examine because the particles reflect and scatter the incident radiation and transmittance is always low. If the solid cannot be dissolved in a suitable solvent, it is best examined by grinding it very finely in paraffin oil (nujol) and thus forming a suspension, or 'mull'. This can then be held between salt plates in the same way as a pure liquid or solvent. Provided the refractive indices of the liquid and solid phase are not very different, scattering will be slight.

Another technique for handling solids is to grind them very finely with potassium bromide. Under very high pressure this material will flow slightly, and the mixture can usually be pressed into a transparent disk. This may then be placed directly in the infra-red beam in a suitable holder. Although superficially attractive the method is not generally recommended because of the difficulty in obtaining really reproducible results.

A further technique which is often used to study otherwise intractable samples is known as attenuated total reflectance (ATR) spectroscopy. Consider a trapezoidal block of a transparent material, as in Fig. 3.20(*a*). If the chamfer angle is properly chosen, radiation shone into one end

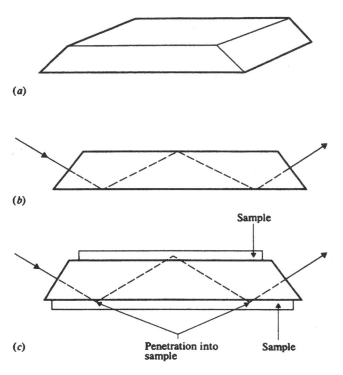

(a)

(b)

Sample

Penetration into
sample   Sample

(c)

Figure 3.20 Attenuated total reflection: (a) the transparent block, (b) internal reflection in the block, and (c) penetration into the sample pressed against the block.

will strike the flat surfaces at less than the critical angle and so will undergo total internal reflection to emerge, only slightly diminished in intensity, at the far end, as in Fig. 3.20(b). Now, although the internal reflection is conventionally called 'total', in fact the radiation beam penetrates slightly beyond the surface of the block during each reflection. If sample material is pressed closely to the outside of the block (Fig. 3.20(c)) the beam will travel a small distance through the sample at each reflection and so, on emerging at the far end of the block, it will 'carry' the absorption spectrum of the sample—the internal reflection is *attenuated*, or diminished by sample absorption, hence the name of this type of spectroscopy. The amount of penetration into the sample depends on the wavelength of the radiation and the angle of incidence, but it is of the order of 10^{-4}–10^{-3} cm for infra-red waves. During its passage through the block it may undergo some 10–20 reflections, so the total path length through the sample is 10^{-3}–10^{-2} cm, which is a short, but often perfectly adequate, path length for the production of a reasonable spectrum.

The block must be of material which is infra-red transparent, and must have a refractive index higher than that of the sample, otherwise internal reflection will not occur. Suitable materials are silver chloride, thallium halides, or germanium, and the block is typically some 5 cm long, 2 cm wide, and 0.5 cm thick. The sample material can be in any form (except gaseous, for which a path length of 10^{-2} cm is far too short) provided it can be kept in very close contact with the block, but the technique is usually reserved for samples difficult to study by ordinary means. Thus, it is virtually impossible to study fibrous material by transmission—the rough surface scatters all the radiation falling on it; but if the fibres are clamped firmly to the outside of an ATR block, quite acceptable spectra result. It is an excellent method of studying surface coatings, since it is only the surface of the sample which is penetrated by the radiation. Further, since the depth of penetration can, to some extent, be changed by varying the angle of incidence of the beam, the change in composition of a surface with depth can be studied. Thus one can measure the degree of oxidation of a polymer surface, or the diffusion of materials into a surface.

3.8.2 Double- and Single-Beam Operation

Figure 3.21 shows the spectrum of the atmosphere between 4000 and 400 cm^{-1} taken with a path length of some 2 m—this is not abnormally long for the beam paths in a spectrometer. It is evident that, although H_2O and CO_2 occur in air only in small percentages, their absorbance over much of the spectrum is considerable. Not only would this absorbance have to be subtracted from the spectrum of any sample run under comparable conditions but, since the percentage of water vapour in the atmosphere is variable, such a 'background' spectrum as Fig. 3.21 would have to be run afresh for each sample.

If the regions of these absorbances are not to be denied us in spectroscopic studies, some action must be taken either to remove the H_2O and CO_2 from the air, or to remove the effects of their spectra. It is possible to remove these gases either by complete evacuation of the spectrometer or by sweeping them out with a current of dry nitrogen or dry CO_2-free air. The first is not easy since a modern spectrometer may have a volume of some 0.3 m^3 and there will be a great many places in its container where leaks may occur. Nor is it ever completely effective, since water vapour proves to be remarkably tenacious and weeks of hard evacuation may be necessary before all the water vapour is desorbed from the surfaces inside the spectrometer. For this

Figure 3.21 The infra-red spectrum of atmospheric water vapour and carbon dioxide. (*Thanks are due to Miss J. Cook of the University of York for providing this spectrum.*)

reason, also, sweeping with a dry inert gas is not very effective. However, these methods do, quite rapidly, reduce the interference considerably.

The *effects* of this interference can be removed much more simply by using an instrument designed for *double-beam* operation. In this, the source radiation is divided into two by means of the mirrors M_1 and M_2 (Fig. 3.22). One beam is brought to a focus at the sample space, while the other follows an exactly equivalent path and is referred to as the reference beam. The two beams meet at the sector mirror M_3, which is sketched in plan view in Fig. 3.22(*b*). As this mirror rotates it alternately reflects the reference beam, or allows the sample beam through the spaces, into the monochromator. Thus the detector 'sees' the sample beam and reference beam alternately. Both beams have travelled the same distance through the atmosphere and thus both are reduced in energy to the same extent by absorption by CO_2 and H_2O.

If a sample, capable of absorbing energy from the beam at the particular frequency passed by the monochromator, is now placed in the sample beam, the detector will receive a signal alternating in intensity, since the sample beam carries less energy than the reference beam. It is a simple matter, electronically, to amplify this alternating signal and to arrange that a calibrated attenuator is driven into the reference beam until the signal is reduced to zero, i.e. until sample and reference beams are again balanced. The distance moved by the attenuator is a direct measure of the amount of energy absorbed by the sample.

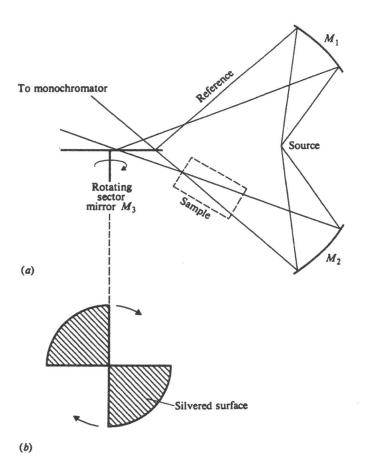

(a)

(b)

Figure 3.22 (*a*) Schematic diagram of a double-beam spectrometer; (*b*) a plan view of the rotating sector mirror, M_3.

By balancing sample and reference beams in this way, the absorption of atmospheric CO_2 and H_2O does not appear in the infra-red spectra since both beams are reduced in energy to the same extent. The double-beam spectrometer has other advantages, however.

Firstly, it is much simpler to amplify the alternating signal produced than the d.c. signal resulting from a single-beam detector.

Secondly, the sector mirror acts as a modulator since it interrupts the beam periodically and, by amplifying only that component of the signal having the sector mirror frequency (usually 10–100 rotations per second), a great improvement in the signal-to-noise ratio results.

Thirdly, when examining the spectra of solutions, one can put a cell containing the appropriate quantity of pure solvent into the reference beam, thus eliminating the solvent spectrum from the final trace. On a single-beam instrument the solvent spectrum must be taken separately and 'subtracted' from the solution spectrum in order to arrive at the spectrum of the substance of interest.

It should be pointed out, however, that a double-beam instrument is never *completely* effective in removing traces of water vapour or CO_2 from the spectra. No matter how carefully the instrument is assembled small differences occur in the beam paths and a small residual spectrum results. This can usually be removed, however, by sweeping with dry, inert gas as well as using the double-beam principle.

A further, more serious disadvantage which is not always appreciated by users of spectrometers is that the double-beam instrument only removes the spectral trace of CO_2 and H_2O; the very strong absorption of energy by these gases still remains in both beams. This means that at some parts of the spectrum the actual amount of energy reaching the detector may be extremely small. Under these conditions, unless the spectrometer is very carefully operated, the spectral trace of a substance may be quite false. Fortunately, regions of very high atmospheric absorption are few and narrow but they should be borne in mind when examining infra-red spectra. This disadvantage can only be removed by sweeping out or evacuating the spectrometer. Similar, but more pronounced, effects occur in regions of strong solvent absorbance when a compensating cell is put in the reference beam.

3.8.3 Fourier Transform Spectroscopy

Infra-red spectroscopy extends outside the limits we have discussed so far in this chapter, and in particular a good deal of useful molecular information is contained in spectra below $400 \, \text{cm}^{-1}$, i.e. the far infra-red region, from about $400 \, \text{cm}^{-1}$ to $20 \, \text{cm}^{-1}$ or $10 \, \text{cm}^{-1}$. Because sources are weak and detectors insensitive, this region is known as 'energy-limited' and difficulty is experienced in obtaining good signal-to-noise ratios by conventional means. The advent of Fourier transform spectroscopy, already introduced in Sec. 1.8, has made the far infra-red much more accessible, and has considerably speeded and improved spectroscopy in the infra-red region in general.

In this region Fourier transform (FT) methods are used in absorption. The apparatus derives from the classical attempt by Michelson to measure the 'ether wind' by determining the velocity of light in two perpendicular directions. A parallel beam of radiation is directed from the source to the interferometer, consisting of a beam splitter B and two mirrors M_1 and M_2 (Fig. 3.23). The beam splitter is a plate of suitably transparent material (e.g. potassium bromide) coated so as to reflect just 50 per cent of the radiation falling on it. Thus half the radiation goes to M_1, and half to M_2, returns from both these mirrors along the same path, and is then recombined to a single beam at the beam splitter (clearly half the total radiation is sent back to the source, but this is immaterial).

It is well known (and the essence of the Michelson experiment) that if *monochromatic* radiation is emitted by the source, the recombined beam leaving B shows constructive or

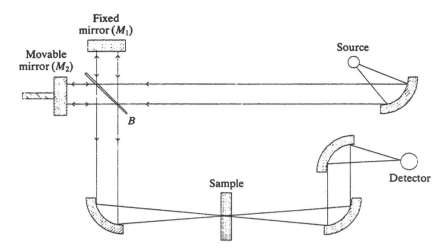

Figure 3.23 Schematic diagram of a Fourier transform infra-red spectrometer.

destructive interference, depending on the relative path lengths B to M_1 and B to M_2. Thus if the path lengths are identical or differ by an integral number of wavelengths, constructive interference gives a bright beam leaving B, whereas if the difference is a half-integral number of wavelengths, the beams cancel at B. As the mirror M_2 is moved smoothly towards or away from B, therefore, a detector sees radiation alternating in intensity. It is fairly easy to imagine that if the source emits *two* separate monochromatic frequencies, v_1 and v_2, then the interference pattern (beat pattern) of v_1 and v_2 would overlay the interference caused by M_1 and M_2; the detector would see a more complicated intensity fluctuation as M_2 is moved, but computing the Fourier transform of the resultant signal is a very rapid way of obtaining the original frequencies and intensities emitted by the source. Taking the process further, even 'white' radiation emitted by the source produces an interference pattern which can be transformed back to the original frequency distribution.

Figure 1.19(*a*) shows a typical interference pattern or interferogram for a 'white' source, where the wide range of frequencies causes a rapid diminishing of signal away from the position at which both mirrors are an equal distance from the beam splitter (the so-called zero retardation peak). No real source is truly white, of course, and Fig. 3.24 shows, schematically, the variation in total intensity caused by varying source output and beam splitter efficiency across the IR range for a typical FT spectrometer. Since FT infra-red spectroscopy is carried out as a single-beam technique, this background variation must be taken into account for each spectrum. If the beam from such a source is directed through a sample before reaching the detector, sample absorptions cause gaps in the frequency distribution which, after transformation, will appear as down-going peaks.

The production of a spectrum, then, is a two-stage process, which may be thought of as follows. Firstly, without a sample in the beam, mirror M_2 is moved smoothly over a period of time (e.g. one second) through a distance of about 1 cm, while the detector signal—the interferogram—is collected into a multi-channel computer (it may be, for instance, that the detector signal is monitored every thousandth of a second during the mirror traverse, and each piece of information put serially into one of a thousand different storage locations in the computer); the computer carries out the Fourier transformation of the stored data to produce the background spectrum. Secondly, a sample interferogram is recorded in exactly the same way, Fourier transformed, and then ratioed against the background spectrum for plotting as a transmittance spectrum. Alternatively, the sample and background spectra may each be calculated in

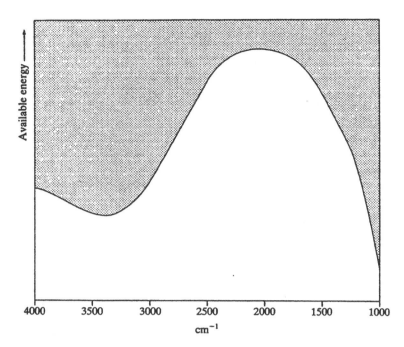

Figure 3.24 The single-beam profile of an infra-red interferometer.

absorbance forms and the latter simply subtracted from the former to give an absorbance spectrum of the sample alone.

There are a number of advantages to be gained by using Fourier transform rather than dispersive measurements in the infra-red.

1. The whole spectrum is obtained across the entire frequency range at once. In dispersive IR, it is common to have to change the dispersing grating during the scan, as one grating is not usually able to function sufficiently well over the whole range.
2. The multiplex (or Fellgett) advantage. This arises because the total scanning time for Fourier transform infra-red (FTIR) is considerably less than that required to produce a dispersive spectrum of the same sensitivity and resolution. For a dispersive spectrometer, each resolution element is scanned consecutively, so for a total scan time T, the time t taken to record one resolution element is T/n where n is the number of resolution elements. In FTIR, the entire frequency range is sampled for the whole total scan time T. In Sec. 1.9, we briefly discussed the signal-to-noise ratio, S/N. If t is the recording time, it is found that the signal (S) is proportional to t and the noise (N) is proportional to $\sqrt{t}$, or:

$$S \propto t \qquad N \propto \sqrt{t} \qquad \text{and so} \qquad S/N \propto \sqrt{t}$$

Therefore we may write:

$$(S/N)_{FTIR} = \sqrt{T} \qquad (S/N)_{DISP} = \sqrt{T/n}$$

or

$$(S/N)_{FTIR}/(S/N)_{DISP} = \sqrt{n}$$

Thus the multiplex advantage gives a gain of $\sqrt{n}$ in signal-to-noise for an FTIR spectrum recorded over the same total time as a dispersive spectrum. For example, for a spectrum recorded over the frequency range from 4000 to $400 \, cm^{-1}$, at $1 \, cm^{-1}$ resolution, $n = 3600$, giving an improvement of $\sqrt{n} = 60$. Thus, if it took 10 minutes to record a dispersive spectrum, an FTIR spectrum of the same S/N could be obtained in only 10 seconds.

3. The throughput (or Jacquinot) advantage. In a dispersive instrument the radiation is invariably brought to a focus on a slit, and it is essentially the image of the slit which is seen by the detector; a very fine slit gives good resolving power since only a narrow spread of frequencies falls on the detector at any one moment, but the total amount of energy passing through the instrument is severely limited, requiring high-gain and hence 'noisy' amplifiers. In FT work parallel beams are used throughout, and there is no need to bring the radiation to a focus except for convenience at the sample and at the detector—no slit is required and *all* the source energy passes through the instrument; consequently amplifiers are less critical and the resolving power is governed solely by the mirror traverse and computer capacity. It is for this reason that FT instruments were first developed for use in the energy-limited far-infra-red region. This is illustrated in Fig. 3.25, where we show schematically the relative areas of the infra-red beam for a dispersive and a Fourier transform instrument.

4. The Connes advantage. The position of the moving mirror is measured by counting interference fringes from a helium–neon laser. Because the frequency of the laser is known with great precision, it is possible to measure the mirror position, and hence the frequency, very accurately.

5. Resolving power. The resolving power of an FT instrument is constant over the entire spectrum; in a grating or prism instrument the resolving power depends on the angle which that component makes with the radiation beam, and hence varies with frequency—in particular it is usually especially poor at the ends of the spectrum.

The resolution of a Fourier transform spectrometer is related to the mirror travel by the equation:

$$\Delta\bar{\nu} = 1/\delta$$

where $\Delta\bar{\nu}$ is the resolution and δ the maximum travel of the mirror. Thus for a resolution of $1\,\text{cm}^{-1}$, the maximum mirror travel must be 1 cm, whereas for a resolution of $0.01\,\text{cm}^{-1}$, 100 cm travel is required. Obviously the ultimate limitation on resolution in FTIR is the ability to make a controlled bearing for the moving mirror to travel smoothly over long enough distances. Measurements at resolutions of 0.1–$2\,\text{cm}^{-1}$ are quite routine, however, and it is only in specialized, very high resolution work that difficulties arise.

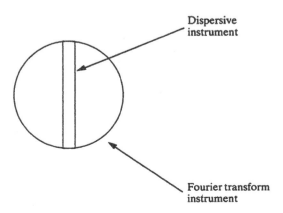

Dispersive
instrument

Fourier transform
instrument

Figure 3.25 Schematic comparison of the throughput of dispersive and Fourier transform instruments.

3.8.4 The Carbon Dioxide Laser

The carbon dioxide laser, actually made from a mixture of carbon dioxide and nitrogen, has potential for use in routine infra-red spectroscopy. The primary excitation step in this is the electron-impact excitation of nitrogen molecules into their first excited vibrational level, which, as Fig. 3.26 shows, is at about $2360 \, cm^{-1}$ above the ground state. This happens to be very close in energy to ν_3, the asymmetric stretching vibration of CO_2 at $2350 \, cm^{-1}$, and so resonance excitation of CO_2 to this level occurs by collision with excited N_2. Now both N_2 and CO_2 have rotational states associated with these vibrational energies, which are also indicated (schematically and not to scale) in Fig. 3.26. This means that many different rotational levels of N_2 will be populated in the initial excitation, and consequently many rotational levels of the ν_3 vibration of CO_2 will be collisionally activated. In fact, levels up to some $200 \, cm^{-1}$ above the lowest state of ν_3 are effectively populated in this way.

The excited CO_2 can, and of course does, decay spontaneously and directly to the ground state, liberating energy as heat. For this reason some helium is incorporated into the gas mixture, to help in transferring the heat to the walls of the containing tube, and the laser is normally operated in the pulsed mode, although pulses can be as high as several per second. In addition, however, stimulated emission is possible to ν_1, the symmetric stretching mode of CO_2 at about $1390 \, cm^{-1}$, and this mode also has associated rotational levels. The system, then, is

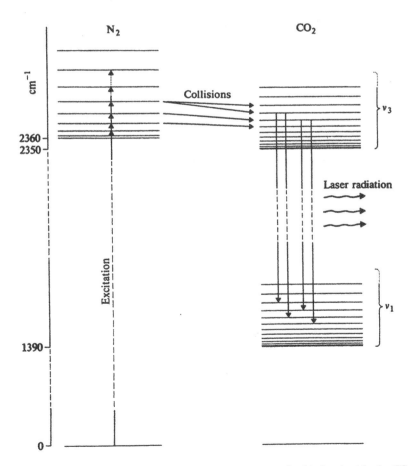

Figure 3.26 The energy levels of nitrogen and of carbon dioxide involved in the CO_2 laser.

basically a four-level laser (cf. Sec. 1.10), although the many additional rotational levels also play their part.

When acting as a laser, any activated molecules in v_3 can drop to any *allowed* levels of v_1, i.e. those levels to which transitions are allowed under the $\Delta J = \pm 1$ selection rule. This means that a large number (perhaps 80–100) of discrete laser frequencies are emitted over the range $900–1100 \, cm^{-1}$—the separation of the v_1 and v_3 vibrations of CO_2. The spacing between the emission lines is about $2 \, cm^{-1}$, the spacing of the rotational transitions. Now although this cannot be regarded as a 'continuous' source of radiation, and although its frequency range is very limited compared with the useful infra-red spectrum (covering $3000–400 \, cm^{-1}$ or lower), nonetheless it has some potential as an infra-red source, as we illustrate in the following paragraphs.

Most organic materials, especially when in solution, have fairly broad infra-red absorptions, with a bandwidth of some $5–20 \, cm^{-1}$; in this respect a source with discrete lines spaced at only $2 \, cm^{-1}$ is essentially continuous, in the sense that there is little loss of spectral information between the lines. Equally, many such molecules do have useful diagnostic absorptions in the $900–1100 \, cm^{-1}$ region. So spectra obtained with a CO_2 laser, although perhaps looking slightly different from those obtained with a normal source, may well have useful 'fingerprinting' possibilities.

Further, the laser radiation is very intense—in fact, since its main industrial use is in cutting and welding materials, an unmodified CO_2 laser tends to melt the spectrometer. Once its intensity has been reduced so that it is 'only' some 10^3–10^5 times stronger than a normal infra-red source, however, it proves ideally suited to applications such as monitoring air pollution. Here one is looking for parts-per-million quantities of pollutants, and the best method is to shine the infra-red beam through a very long path of the material. Clearly an intense beam that does not diverge over a large distance is exactly what is required.

Finally, quantitative measurements of materials are most accurately obtained by using monochromatic radiation, and in this respect one of the emission lines from the CO_2 laser, selected by the use of a normal dispersion grating, is ideal, and potentially very useful for continuous monitoring of production materials. A similar CO gas laser, also activated by electron-impact excitation of N_2, gives multiple discrete lines in the $2000 \, cm^{-1}$ region.

BIBLIOGRAPHY

Bellamy, L. J.: *The Infrared Spectra of Complex Molecules*, Chapman and Hall, vol. 1, 1975; vol. 2, 1980.

Ebsworth, E. A. V., D. W. H. Rankin, and S. Cradock: *Structural Methods in Inorganic Chemistry*, 2nd ed., Blackwell Scientific Publications, 1991.

Ferraro, J. R., and L. J. Basilo: *Fourier Transform Infrared Spectroscopy: Applications to Chemical Systems*, Academic Press, vol. 1, 1978; vol. 2, 1979.

Griffiths, P. R., and J. A. de Hareth: *Fourier Transform Infrared Spectroscopy*, Wiley, 1986.

Harris, D. C., and M. D. Bertolucci: *Symmetry and Spectroscopy*, Dover Publications, 1989.

Herzberg, G.: *Molecular Structure and Molecular Spectra:* vol. 1, *Spectra of Diatomic Molecules*, 2nd ed., Van Nostrand, 1950.

Herzberg, G.: *Molecular Structure and Molecular Spectra:* vol. 2, *Infrared and Raman Spectra of Polyatomic Molecules*, Van Nostrand, 1945.

Kemp, W.: *Organic Spectroscopy*, 3rd ed., Macmillan, 1991.

Nakamoto, K.: *Infrared and Raman Spectra of Inorganic and Coordination Compounds*, 4th ed., Wiley–Interscience, 1986.

Williams, D. H., and I. Fleming: *Spectroscopic Methods in Organic Chemistry*, 4th ed., McGraw-Hill, 1989.

PROBLEMS

(Useful constants: $h = 6.626 \times 10^{-34}\,\text{J s}$; $c = 2.998 \times 10^8\,\text{m s}^{-1}$; $N = 6.023 \times 10^{23}\,\text{mol}^{-1}$; $k = 1.381 \times 10^{-23}\,\text{J K}^{-1}$; $4\pi^2 = 39.478$; $1\,\text{cm}^{-1} \equiv 11.958\,\text{J mol}^{-1}$; atomic masses: $^{14}\text{N} = 23.25 \times 10^{-27}\,\text{kg}$, $^{16}\text{O} = 26.56 \times 10^{-27}\,\text{kg}$.)

3.1 The fundamental and first overtone transitions of $^{14}\text{N}^{16}\text{O}$ are centred at $1876.06\,\text{cm}^{-1}$ and $3724.20\,\text{cm}^{-1}$, respectively. Evaluate the equilibrium vibration frequency, the anharmonicity, the exact zero-point energy, and the force constant of the molecule. Assuming that in Eq. (3.12) v is a continuous variable, use calculus to determine the maximum value of ε_v, and hence calculate a value for the dissociation energy of NO. Criticize this method.

3.2 The vibrational wavenumbers of the following molecules in their $v = 0$ states are: HCl: $2885\,\text{cm}^{-1}$; DCl: $1990\,\text{cm}^{-1}$; D_2: $2990\,\text{cm}^{-1}$; and HD: $3627\,\text{cm}^{-1}$. Calculate the energy change, in kJ mol^{-1}, of the reaction

$$\text{HCl} + \text{D}_2 \rightarrow \text{DCl} + \text{HD}$$

and determine whether energy is liberated or absorbed.

Hint: Consider the zero-point energies of the four molecules concerned.

3.3 The equilibrium vibration frequency of the iodine molecule I_2 is $215\,\text{cm}^{-1}$, and the anharmonicity constant x is 0.003; what, at $300\,\text{K}$, is the intensity of the 'hot band' ($v = 1 \rightarrow v = 2$ transition) relative to that of the fundamental ($v = 0 \rightarrow v = 1$)?

3.4 An infra-red spectrum of OCS is obtained in which the rotational fine structure is not resolved. Using data from Table 2.2, calculate the separation between the P and R branch maxima at $T = 300\,\text{K}$.

3.5 How many normal modes of vibration are possible for the following molecules: HBr, O_2, OCS (linear), SO_2 (bent), BCl_3, $\text{HC} \equiv \text{CH}$, CH_4 CH_3I, C_6H_6?

3.6 Estimate, using data from Table 3.4, the vibrational wavenumber of (a) $\equiv$CD, (b) —OD, (c) $\gtrless$C—S—. (Relative atomic masses are: H = 1, D = 2, C = 12, O = 16, S = 32.)

3.7 The vibration frequency of $^1\text{H}^{35}\text{Cl}$ is $2990.6\,\text{cm}^{-1}$; *without* calculating the bond force constant, estimate the frequency for $^1\text{H}^{37}\text{Cl}$, $^2\text{D}^{35}\text{Cl}$, and $^2\text{D}^{37}\text{Cl}$.

3.8 In the vibration–rotation spectrum ($v = 0 \rightarrow 1$) of HF, the rotational constants are slightly different in the $v = 0$ and $v = 1$ states; experimentally, their values are found to be $B_{v=0} = 20.6\,\text{cm}^{-1}$ and $B_{v=1} = 19.8\,\text{cm}^{-1}$. Calculate the percentage increase in bond length on going from $v = 0$ to $v = 1$. What effect does this lengthening of the bond have on the spacing of the rotational lines?

3.9 Assume the following data for the molecule $^1\text{H}^{35}\text{Cl}$:
> Bond length: $127.5\,\text{pm}$
> Bond force constant: $516.3\,\text{Nm}^{-1}$
> Atomic masses: $^1\text{H} = 1.673 \times 10^{-27}\,\text{kg}$, $^{35}\text{Cl} = 58.066 \times 10^{-27}\,\text{kg}$

Do the following, giving answers in cm^{-1}:
(a) Calculate the zero-point energy and the energy of the fundamental vibration ν_0.
(b) Calculate the rotational constant B.
(c) Calculate the wavenumbers of the lines $P_{(1)}$, $P_{(2)}$, $P_{(3)}$, $R_{(0)}$, $R_{(1)}$, and $R_{(2)}$.
(d) Sketch the expected vibration–rotation spectrum of HCl, including the approximate intensity distribution.
(e) Suggest two differences which you would expect to find between the spectrum you have sketched in (d) and that which is actually observed for HCl, giving your reasons.

3.10 Explain why the $\text{C}{=}\text{O}$ stretching vibration of an aldehyde gives rise to a strong absorption in the infra-red, yet the absorption due to the $\text{C}{=}\text{C}$ vibration in an alkene is normally very weak.

RAMAN SPECTROSCOPY

4.1 INTRODUCTION

When a beam of light is passed through a transparent substance, a small amount of the radiation energy is scattered, the scattering persisting even if all dust particles or other extraneous matter are rigorously excluded from the substance. If monochromatic radiation, or radiation of a very narrow frequency band, is used, the scattered energy will consist almost entirely of radiation of the incident frequency (the so-called *Rayleigh scattering*) but, in addition, certain discrete frequencies above and below that of the incident beam will be scattered; it is this which is referred to as *Raman scattering*.

4.1.1 Quantum Theory of Raman Effect

The occurrence of Raman scattering may be most easily understood in terms of the quantum theory of radiation. This treats radiation of frequency v as consisting of a stream of particles (called photons) having energy hv where h is Planck's constant. Photons can be imagined to undergo collisions with molecules and, if the collision is perfectly elastic, they will be deflected unchanged. A detector placed to collect energy at right angles to an incident beam will thus receive photons of energy hv, i.e. radiation of frequency v. Elastic scattering can be likened to a ball bearing striking a rigid table—the ball bearing bounces off the table without any loss of energy.

However, it may happen that energy is exchanged between photon and molecule during the collision: such collisions are called 'inelastic'. The molecule can gain or lose amounts of energy only in accordance with the quantal laws; i.e. its energy change, ΔE joules, must be the difference in energy between two of its allowed states. That is to say, ΔE must represent a change in the vibrational and/or rotational energy of the molecule. If the molecule *gains* energy ΔE, the photon will be scattered with energy $hv - \Delta E$ and the equivalent radiation will have a frequency $v - \Delta E/h$. Conversely, if the molecule *loses* energy ΔE, the scattered frequency will be $v + \Delta E/h$. The inelastic process can be pictured in terms of a ball bearing striking a drum. If the surface of the drum is stationary when the ball bearing hits, it will start oscillating at its own normal frequency and the ball bearing will be reflected with less energy, having lost an amount of energy equal to that taken up by the oscillation of the drum. On the other hand, if the drum is already oscillating when the ball bearing strikes, and if the ball bearing hits at the right phase of the drum's oscillation, the drum will give energy to the ball bearing—rather like a catapult—and the ball bearing will be flung off with increased energy.

Radiation scattered with a frequency lower than that of the incident beam is referred to as Stokes' radiation, while that at higher frequency is called anti-Stokes' radiation. Since the former is accompanied by an *increase* in molecular energy (which can always occur, subject to certain selection rules) while the latter involves a *decrease* (which can only occur if the molecule is originally in an excited vibrational or rotational state), Stokes' radiation is generally more intense than anti-Stokes' radiation. Overall, however, the total radiation scattered at any but the incident frequency is extremely small, and sensitive apparatus is needed for its study.

4.1.2 Classical Theory of the Raman Effect: Molecular Polarizability

The classical theory of the Raman effect, while not wholly adequate, is worth consideration since it leads to an understanding of a concept basic to this form of spectroscopy—the polarizability of a molecule. When a molecule is put into a static electric field it suffers some distortion, the positively charged nuclei being attracted towards the negative pole of the field, the electrons to the positive pole. This separation of charge centres causes an *induced electric dipole moment* to be set up in the molecule and the molecule is said to be *polarized*. The size of the induced dipole μ, depends both on the magnitude of the applied field, E, and on the ease with which the molecule can be distorted. We may write

$$\mu = \alpha E \tag{4.1}$$

where α is the *polarizability* of the molecule.

Consider first the diatomic molecule H_2, which we show placed in an electric field in Fig. 4.1(a) and (b) in end-on and sideways orientation, respectively. The electrons forming the bond are more easily displaced by the field *along* the bond axis (Fig. 4.1(b)) than that *across* the bond (Fig. 4.1(a)), and the polarizability is thus said to be *anisotropic*. This fact may be confirmed

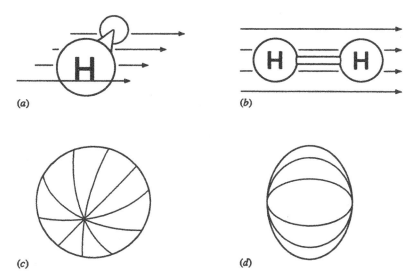

Figure 4.1 The hydrogen molecule in an electric field and its polarizability ellipsoid, seen along and across the bond axis.

experimentally (e.g. by a study of the intensity of lines in the Raman spectrum of H_2), when it is found that the induced dipole moment for a given field applied along the axis is approximately twice as large as that induced by the same field applied across the axis; fields in other directions induce intermediate dipole moments.

The polarizability of a molecule in various directions is conventionally represented by drawing a *polarizability ellipsoid*. We shall define this formally in the next paragraph, but for hydrogen its general shape is that of a squashed sphere, like a tangerine, and we have drawn this in two orientations in Fig. 4.1(*c*) and (*d*) (together with totally imaginary tangerine segment lines to make the picture clearer). In Fig. 4.1(*c*) we are looking down on to the top of the tangerine just off its axis and in Fig. 4.1(*d*) we look at it sideways; these viewpoints correspond to those for the hydrogen molecule in Fig. 4.1(*a*) and (*b*), respectively.

In general a polarizability ellipsoid is defined as a three-dimensional surface whose distance from the electrical centre of the molecule (in H_2 this is also the centre of gravity) is proportional to $1/\sqrt{a_i}$, where a_i is the polarizability along the line joining a point i on the ellipsoid with the electrical centre. Thus where the polarizability is *greatest*, the axis of the ellipsoid is *least*, and vice versa. (Historically this representation arose by analogy with the momentum of a body—the momental ellipsoid is defined similarly using $1/\sqrt{I_i}$, where I_i is the moment of inertia about an axis i.)

If we imagine applying an electric field across the bond axis of H_2, as in Fig. 4.1(*a*), a certain amount of polarization of the molecule will occur. If we also imagine the molecule rotating about its bond axis, it is obvious that it will present exactly the same aspect to the electric field at all orientations—i.e. its polarizability will be exactly the same in any direction across the axis. This means that a section through the polarizability ellipsoid will be circular, which is what we have drawn in Fig. 4.1(*c*).

If the field is applied *along* the bond axis, as in Fig. 4.1(*b*), the polarizability is greater, as we mentioned earlier. Thus the cross-section of the ellipsoid is *less*, as shown in Fig. 4.1(*d*).

The student must not make the mistake of confusing a polarizability ellipsoid with electron orbitals or electron clouds. In a sense the polarizability ellipsoid is the *inverse* of an electron cloud—where the electron cloud is largest the electrons are further from the nucleus and so are most easily polarized. This, as we have seen, is represented by a *small* axis for the polarizability ellipsoid.

All diatomic molecules have ellipsoids of the same general tangerine shape as H_2, as do linear polyatomic molecules, such as CO_2, $HC\equiv CH$, etc. They differ only in the relative sizes of their major and minor axes.

When a sample of such molecules is subjected to a beam of radiation of frequency v the electric field experienced by each molecule varies according to the equation (cf. Eq. (1.1)):

$$E = E_0 \sin 2\pi v t \qquad (4.2)$$

and thus the induced dipole also undergoes oscillations of frequency v:

$$\mu = \alpha E = \alpha E_0 \sin 2\pi v t \qquad (4.3)$$

Such an oscillating dipole emits radiation of its own oscillation frequency and we have immediately in Eq. (4.3) the classical explanation of Rayleigh scattering.

If, in addition, the molecule undergoes some internal motion, such as vibration or rotation, which *changes the polarizability* periodically, then the oscillating dipole will have superimposed upon it the vibrational or rotational oscillation. Consider, for example, a vibration of frequency $v_{vib.}$ which changes the polarizability: we can write

$$\alpha = \alpha_0 + \beta \sin 2\pi v_{vib.} t \qquad (4.4)$$

where α_0 is the equilibrium polarizability and β represents the rate of change of polarizability with the vibration. Then we have:

$$\mu = \alpha E = (\alpha_0 + \beta \sin 2\pi\nu_{\text{vib.}}t)E_0 \sin 2\pi\nu t$$

or, expanding and using the trigonometric relation:

$$\sin A \sin B = \tfrac{1}{2}\{\cos(A - B) - \cos(A + B)\}$$

we have

$$\mu = \alpha_0 E_0 \sin 2\pi\nu t + \tfrac{1}{2}\beta E_0\{\cos 2\pi(\nu - \nu_{\text{vib.}})t - \cos 2\pi(\nu + \nu_{\text{vib.}})t\} \qquad (4.5)$$

and thus the oscillating dipole has frequency components $\nu \pm \nu_{\text{vib.}}$ as well as the exciting frequency ν.

It should be carefully noted, however, that if the vibration does not alter the polarizability of the molecule (and we shall later give examples of such vibrations) then $\beta = 0$ and the dipole oscillates only at the frequency of the incident radiation; the same is true of a rotation. Thus we have the general rule:

> In order to be Raman active a molecular rotation or vibration must cause some change in a component of the molecular polarizability. A change in polarizability is, of course, reflected by a change in either the *magnitude* or the *direction* of the polarizability ellipsoid.

(This rule should be contrasted with that for infra-red and microwave activity, which is that the molecular motion must produce a change in the electric dipole of the molecule.)

Let us now consider briefly the shapes of the polarizability ellipsoids of more complicated molecules, taking first the bent triatomic molecule H_2O shown in Fig. 4.2(a). By analogy with the discussion for H_2 given above, we might expect the polarizability surface to be composed of *two* similar ellipsoids, one for each bond. While this may be correct in minute detail, we must remember that the oscillating electric field which we wish to apply for Raman spectroscopy is usually that of radiation in the visible or ultra-violet region, i.e. having a wavelength of some 1 μm–10 nm (cf. Fig. 1.4); molecular bonds, on the other hand, have dimensions of only some 0.1 nm, so we cannot expect our radiation to probe the finer details of bond polarizability—even the hardest of X-rays can scarcely do that. Instead the radiation can only sense the average polarizability in various directions through the molecule, and the polarizability ellipsoid, it may be shown, is always a true ellipsoid—i.e. a surface having *all* sections elliptical (or possibly circular). In the particular case of H_2O the polarizability is found to be different along all three of the major axes of the molecule (which lie along the line in the molecular plane bisecting the HOH angle, at right angles to this in the plane, and perpendicularly to the plane), and so all three of the ellipsoidal axes are also different; the ellipsoid is sketched in various orientations in Fig. 4.2(b). Other such molecules, for example H_2S or SO_2, have similarly shaped ellipsoids but with different dimensions.

Symmetric top molecules, because of their axial symmetry, have polarizability ellipsoids rather similar to those of linear molecules, i.e. with a circular cross-section at right angles to their axis of symmetry. It should be stressed, however, that sections in other planes are truly *elliptical*. For a molecule such as chloroform, $CHCl_3$ (Fig. 4.3(a)), where the chlorine atoms are bulky, the usual tendency is to draw the polarizability surface as egg-shaped, fatter at the chlorine-containing end. This is not correct; the polarizability ellipsoid for chloroform is shown at Fig. 4.3(b) where it will be seen that, since the polarizability is greater across the symmetry axis, the *minor* axis of the ellipsoid lies in this direction. Similar molecules are, for example, CH_3Cl and NH_3, etc. (although the latter fortuitously has a virtually spherical 'ellipsoid').

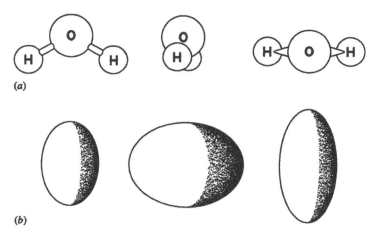

(a)

(b)

Figure 4.2 The water molecule and its polarizability ellipsoid, seen along the three main axes.

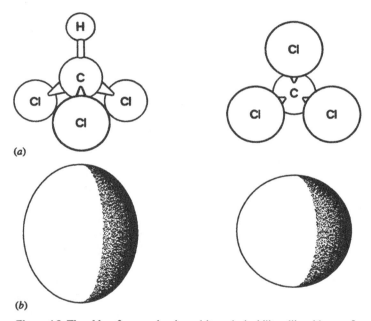

(a)

(b)

Figure 4.3 The chloroform molecule and its polarizability ellipsoid, seen from across and along the symmetry axis.

Finally, spherical top molecules, such as CH_4, CCl_4, SiH_4, etc., have spherical polarizability surfaces, since they are completely isotropic as far as incident radiation is concerned.

We are now in a position to discuss in detail the Raman spectra of various types of molecule. Since we shall be dealing with rotational and vibrational changes it is evident that expressions for the energy levels and for many of the allowed transitions will be identical with those already discussed in the previous two chapters. For clarity we shall repeat any such expressions but not rederive them, being content to give a cross-reference to where their derivation may be found.

4.2 PURE ROTATIONAL RAMAN SPECTRA

4.2.1 Linear Molecules

The rotational energy levels of linear molecules have already been stated (cf. Eq. (2.24)):

$$\varepsilon_J = BJ(J+1) - DJ^2(J+1)^2 \quad \text{cm}^{-1} \quad (J = 0, 1, 2, \ldots)$$

but, in Raman spectroscopy, the precision of the measurements does not normally warrant the retention of the term involving D, the centrifugal distortion constant. Thus we take the simpler expression:

$$\varepsilon_J = BJ(J+1) \quad \text{cm}^{-1} \quad (J = 0, 1, 2, \ldots) \tag{4.6}$$

to represent the energy levels.

Transitions between these levels follow the formal selection rule:

$$\Delta J = 0, \text{ or } \pm 2 \text{ only} \tag{4.7}$$

which is to be contrasted with the corresponding selection rule for microwave spectroscopy, $\Delta J = \pm 1$, given in Eq. (2.17). The fact that in Raman work the rotational quantum number changes by two units rather than one is connected with the symmetry of the polarizability ellipsoid. For a linear molecule, such as is depicted in Fig. 4.1, it is evident that during end-over-end rotation the ellipsoid presents the same appearance to an observer *twice* in every complete rotation. It is equally clear that rotation about the bond axis produces no change in polarizability and hence, as in infra-red and microwave spectroscopy, we need concern ourselves only with end-over-end rotations.

If, following the usual practice, we define ΔJ as $(J_{\text{upper state}} - J_{\text{lower state}})$ then we can ignore the selection rule $\Delta J = -2$ since, for a pure rotational change, the upper state quantum number must necessarily be greater than that in the lower state. Further, the 'transition' $\Delta J = 0$ is trivial since this represents no change in the molecular energy and hence Rayleigh scattering only.

Combining, then, $\Delta J = +2$ with the energy levels of Eq. (4.6) we have:

$$\begin{aligned} \Delta\varepsilon &= \varepsilon_{J'=J+2} - \varepsilon_{J''=J} \\ &= B(4J + 6) \quad \text{cm}^{-1} \end{aligned} \tag{4.8}$$

Since $\Delta J = +2$, we may label these lines S branch lines (cf. Sec. 3.2) and write

$$\Delta\varepsilon_S = B(4J + 6) \quad \text{cm}^{-1} \quad (J = 0, 1, 2, \ldots) \tag{4.9}$$

where J is the rotational quantum number in the *lower* state.

Thus if the molecule gains rotational energy from the photon during collision we have a series of S branch lines to the low wavenumber side of the exciting line (Stokes' lines), while if the molecule loses energy to the photon the S branch lines appear on the high wavenumber side (anti-Stokes' lines). The wavenumbers of the corresponding spectral lines are given by:

$$\bar{\nu}_S = \bar{\nu}_{\text{ex.}} \pm \Delta\varepsilon_S = \bar{\nu}_{\text{ex.}} \pm B(4J + 6) \quad \text{cm}^{-1} \tag{4.10}$$

where the plus sign refers to anti-Stokes' lines, the minus to Stokes' lines, and $\bar{\nu}_{\text{ex.}}$ is the wavenumber of the exciting radiation.

The allowed transitions and the Raman spectrum arising are shown schematically in Fig. 4.4. Each transition is labelled according to its lower J value and the relative intensities of the lines are indicated assuming that the population of the various energy levels varies according to Eq. (2.21) and Fig. 2.7. In particular it should be noted here that Stokes' and anti-Stokes' lines have comparable intensity because many rotational levels are populated and hence downward transitions are approximately as likely as upward ones.

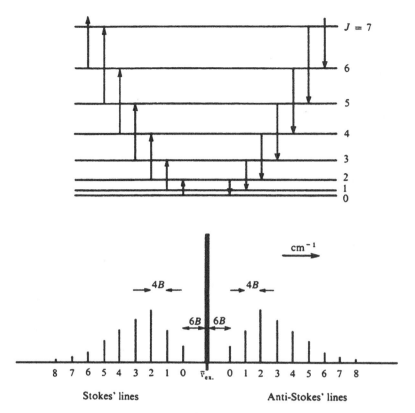

Figure 4.4 The rotational energy levels of a diatomic molecule and the rotational Raman spectrum arising from transitions between them. Spectral lines are numbered according to their lower J values.

When the value $J = 0$ is inserted into Eq. (4.10) it is seen immediately that the separation of the first line from the exciting line is $6B \, \text{cm}^{-1}$, while the separation between successive lines is $4B \, \text{cm}^{-1}$. For diatomic and light triatomic molecules the rotational Raman spectrum will normally be resolved and we can immediately obtain a value of B, and hence the moment of inertia and bond lengths for such molecules. If we recall that homonuclear diatomic molecules (for example O_2, H_2) give no infra-red or microwave spectra since they possess no dipole moment, whereas they *do* give a rotational Raman spectrum, we see that the Raman technique yields structural data unobtainable from the techniques previously discussed. It is thus complementary to microwave and infra-red studies, not merely confirmatory.

It should be mentioned that, if the molecule has a centre of symmetry (as, for example, do H_2, O_2, CO_2), then the effects of nuclear spin will be observed in the Raman as in the infra-red. Thus for O_2 and CO_2 (since the spin of oxygen is zero) every alternate rotational level is absent; for example, in the case of O_2, every level with *even* J values is missing, and thus every transition labelled $J = 0, 2, 4, \ldots$ in Fig. 4.4 is also completely missing from the spectrum. In the case of H_2, and other molecules composed of nuclei with non-zero spin, the spectral lines show an alternation of intensity.

Linear molecules with more than three heavy atoms have large moments of inertia and their rotational fine structure is often unresolved in the Raman spectrum. Direct structural information is not, therefore, obtainable, but we shall see shortly that, taken in conjunction with the infra-red spectrum, the Raman can still yield much very useful information.

4.2.2 Symmetric Top Molecules

The polarizability ellipsoid for a typical symmetric top molecule, for example $CHCl_3$, was shown in Fig. 4.3(b). Plainly rotation about the top axis produces no change in the polarizability, but end-over-end rotations will produce such a change.

From Eq. (2.38) we have the energy levels:

$$\varepsilon_{J,K} = BJ(J+1) + (A-B)K^2 \quad cm^{-1}$$
$$(J = 0, 1, 2, \ldots; K = \pm J, \pm(J-1), \ldots) \tag{4.11}$$

The selection rules for Raman spectra are:

$$\left.\begin{array}{l} \Delta K = 0 \\ \Delta J = 0, \pm 1, \pm 2 \quad \text{(except for } K = 0 \text{ states} \\ \qquad\qquad\qquad\qquad \text{when } \Delta J = \pm 2 \text{ only)} \end{array}\right\} \tag{4.12}$$

K, it will be remembered, is the rotational quantum number for axial rotation, so the selection rule $\Delta K = 0$ implies that changes in the angular momentum about the top axis will not give rise to a Raman spectrum—such rotations are, as mentioned previously, Raman inactive. The restriction of ΔJ to ± 2 for $K = 0$ states means effectively that ΔJ cannot be ± 1 for transitions involving the ground state ($J = 0$) since $K = \pm J, \pm(J-1), \ldots, 0$. Thus for all J values other than zero, K also may be different from zero and $\Delta J = \pm 1$ transitions are allowed.

Restricting ourselves, as before, to positive ΔJ we have the two cases:

1. $\Delta J = +1$ (R branch lines)

$$\Delta\varepsilon_R = \varepsilon_{J'=J+1} - \varepsilon_{J''=J}$$
$$= 2B(J+1) \quad cm^{-1} \qquad (J = 1, 2, 3, \ldots \text{ (but } J \neq 0)) \tag{4.13a}$$

2. $\Delta J = +2$ (S branch lines)

$$\Delta\varepsilon_S = \varepsilon_{J'=J+2} - \varepsilon_{J''=J}$$
$$= B(4J+6) \quad cm^{-1} \qquad (J = 0, 1, 2, \ldots) \tag{4.13b}$$

Thus we shall have two series of lines in the Raman spectrum:

$$\left.\begin{array}{l} \bar{\nu}_R = \bar{\nu}_{\text{ex.}} \pm \Delta\varepsilon_R = \bar{\nu}_{\text{ex.}} \pm 2B(J+1) \quad cm^{-1} \qquad (J = 1, 2, \ldots) \\ \bar{\nu}_S = \bar{\nu}_{\text{ex.}} \pm \Delta\varepsilon_S = \bar{\nu}_{\text{ex.}} \pm B(4J+6) \quad cm^{-1} \qquad (J = 0, 1, 2, \ldots) \end{array}\right\} \tag{4.14}$$

These series are sketched separately in Fig. 4.5(a) and (b), where each line is labelled with its corresponding *lower* J value. In the R branch, lines appear at $4B, 6B, 8B, 10B, \ldots$ cm^{-1} from the exciting line, while the S branch series occurs at $6B, 10B, 14B, \ldots$ cm^{-1}. The complete spectrum, shown in Fig. 4.5(c) illustrates how every alternate R line is overlapped by an S line. Thus a marked intensity alternation is to be expected which, it should be noted, is not connected with nuclear spin statistics.

4.2.3 Spherical Top Molecules: Asymmetric Top Molecules

Examples of spherical top molecules are those with tetrahedral symmetry such as methane, CH_4, or silane, SiH_4. The polarizability ellipsoid for such molecules is a spherical surface and it is evident that rotation of this ellipsoid will produce no change in polarizability. Therefore the pure rotations of spherical top molecules are completely inactive in the Raman.

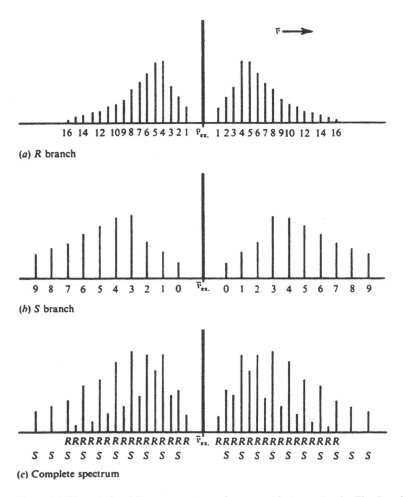

Figure 4.5 The rotational Raman spectrum of a symmetric top molecule. The *R* and *S* branch lines are shown separately in (*a*) and (*b*), respectively, with the total spectrum in (*c*).

Normally *all* rotations of asymmetric top molecules, on the other hand, are Raman active. Their Raman spectra are thus quite complicated and will not be dealt with in detail here; it suffices to say that, as in the microwave region, the spectra may often be interpreted by considering the molecule as intermediate between the oblate and prolate types of symmetric top.

4.3 VIBRATIONAL RAMAN SPECTRA

4.3.1 Raman Activity of Vibrations

If a molecule has little or no symmetry it is a very straightforward matter to decide whether its vibrational modes will be Raman active or inactive: in fact, it is usually correct to assume that *all* its modes are Raman active. However, when the molecule has considerable symmetry it is not always easy to make the decision, since it is sometimes not clear, without detailed consideration, whether or not the polarizability changes during the vibration.

We consider first the simple asymmetric top molecule H_2O whose polarizability ellipsoid was shown in Fig. 4.2. In Fig. 4.6 we illustrate in (a), (b), and (c), respectively, the three fundamental modes v_1, v_2, and v_3, sketching for each mode the equilibrium configuration in the centre with the extreme positions to right and left. The approximate shapes of the corresponding polarizability ellipsoids are also shown.

During the symmetric stretch, in Fig. 4.6(a), the molecule as a whole increases and decreases in size; when a bond is stretched, the electrons forming it are less firmly held by the nuclei and so the bond becomes more polarizable. Thus the polarizability ellipsoid of H_2O may be expected to decrease in size while the bonds stretch, and to increase while they compress, but to maintain an

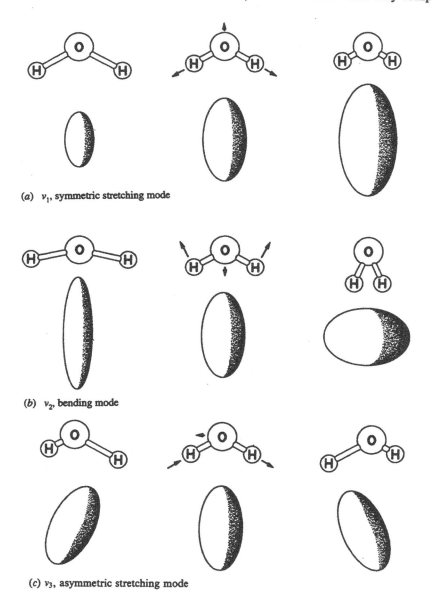

(a) v_1, symmetric stretching mode

(b) v_2, bending mode

(c) v_3, asymmetric stretching mode

Figure 4.6 The change in size, shape, or direction of the polarizability ellipsoid of the water molecule during each of its three vibrational modes. The centre column shows the equilibrium position of the molecule, while to right and left are the (exaggerated) extremes of each vibration.

approximately constant shape. On the other hand, while undergoing the bending motion, in Fig. 4.6(b), it is the shape of the ellipsoid which changes most; thus if we imagine vibrations of very large amplitude, at one extreme (on the left) the molecule approaches the linear configuration with a horizontal axis, while at the other extreme (on the right) it approximates to a diatomic molecule (if the two H atoms are almost coincidental) with a vertical axis. Finally in Fig. 4.6(c) we have the asymmetric stretching motion, v_3, where both the size and shape remain approximately constant, but the direction of the major axis changes markedly. Thus all three vibrations involve obvious changes in at least one aspect of the polarizability ellipsoid, and all are Raman *active*.

Now consider the linear triatomic molecule CO_2, whose three fundamental vibrational modes have been shown in Fig. 3.11; in the first three columns of Fig. 4.7 we illustrate the extreme and equilibrium configurations of the molecule and their approximate polarizability ellipsoids. The question of the Raman behaviour of the symmetric stretching mode, v_1, is easily decided—during the motion the molecule changes size, and so there is a corresponding fluctuation in the size of the ellipsoid; the motion is thus Raman *active*. It might be thought that the v_2 and v_3 vibrations are also Raman active, because the molecule changes shape during each vibration and hence, presumably, so does the ellipsoid; however, both these modes are observed to be Raman *inactive*. We must, then, consider this example rather more carefully.

To do this it is usual to discuss the change of polarizability with some *displacement coordinate*, normally given the symbol ξ. Thus for a stretching motion, ξ is a measure of the extension (positive ξ) or compression (negative ξ) of the bond under consideration, while for a bending mode, ξ measures the displacement of the bond angle from its equilibrium value, positive and negative ξ referring to opposite displacement directions.

Consider, as an example, the v_1 stretch of carbon dioxide sketched in Fig. 4.7(a). If the equilibrium value of the polarizability is α_0 (second picture) then, when the bonds stretch (ξ positive), α increases (remember that the extent of the ellipsoid measures the *reciprocal* of α), while when the bonds contract (negative ξ) α decreases. Thus we can sketch the variation of α with ξ as shown on the right of Fig. 4.7. The details of the curve are not important since we are concerned only with *small* displacements; it is plain that near the equilibrium position ($\xi = 0$) the curve has a distinct slope, that is $d\alpha/d\xi \neq 0$ at $\xi = 0$. Thus for small displacements the motion produces a change in polarizability and is therefore Raman *active*.

If we now consider the situation for v_2, the bending motion of Fig. 4.7(b), we can count a downwards displacement of the oxygen atoms as negative ξ and an upwards displacement as positive. Although it is not clear from the diagrams whether the motion causes an increase or a decrease in polarizability (actually it is an *increase*) it *is* plain that the change is exactly the same for both positive and negative ξ. Thus we can plot α against ξ on the right of Fig. 4.7(b) with, as before, $\alpha = \alpha_0$ at $\xi = 0$. Now for small displacements we evidently have $d\alpha/d\xi = 0$ and hence for *small displacements* there is effectively no change in the polarizability and the motion is Raman *inactive*.

Exactly the same argument applies to the asymmetric stretch, v_3, shown in Fig. 4.7(c). Here the polarizability decreases equally for positive and negative ξ, so the plot of polarizability against ξ has the appearance shown. Again $d\alpha/d\xi = 0$ for small displacements and the motion is Raman *inactive*.

We could have followed the same reasoning for the three vibrations of water discussed previously. In each case we would have discovered that the α versus ξ curve has the general shape of Fig. 4.7(a) or its mirror image; in other words, in each case $d\alpha/d\xi \neq 0$ and the motion is Raman *active*. In general, however, the slopes of the three curves would be different at $\xi = 0$, that is $d\alpha/d\xi$ would have different values. Since we have seen that the Raman spectrum is forbidden for $d\alpha/d\xi = 0$ but allowed for $d\alpha/d\xi \neq 0$ we can imagine that the 'degree of allowedness' varies with $d\alpha/d\xi$. Thus if the polarizability curve has a large slope at $\xi = 0$ the Raman

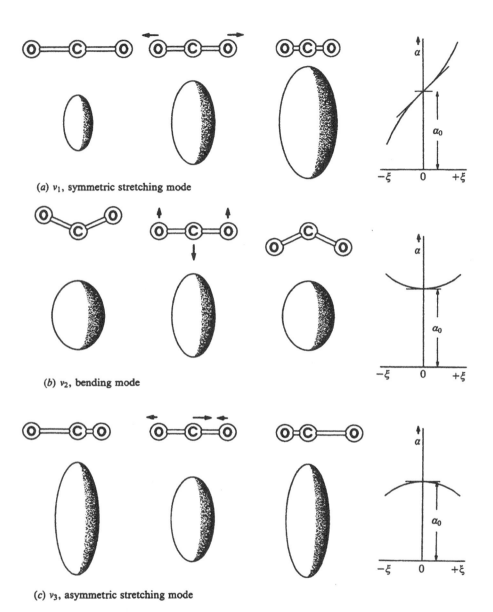

(a) ν_1, symmetric stretching mode

(b) ν_2, bending mode

(c) ν_3, asymmetric stretching mode

Figure 4.7 The changes in the polarizability ellipsoid of carbon dioxide during its vibrations, and a graph showing the variation of the polarizability, α, with the displacement coordinate, ξ, during each vibration.

line will be strong; if the slope is small it will be weak; and if zero, not allowed at all. From this stems the following very useful general rule:

Symmetric vibrations give rise to intense Raman lines; non-symmetric ones are usually weak and sometimes unobservable.

In particular, a bending motion usually yields only a very weak Raman line; e.g. the ν_2 motion of H_2O (Fig. 4.6(b)), although allowed in the Raman, has not been observed, nor has ν_3, for which $d\alpha/d\xi$ is also small.

Table 4.1 Raman and infra-red activities of carbon dioxide

Mode of vibration of CO_2	Raman	Infra-red
v_1 : symmetric stretch	Active	Inactive
v_2 : bend	Inactive	Active
v_3 : asymmetric stretch	Inactive	Active

4.3.2 Rule of Mutual Exclusion

A further extremely important general rule has been established whose operation may be exemplified by carbon dioxide. We can summarize our conclusions about the Raman and infra-red activities of the fundamental vibrations of this molecule in Table 4.1, and we see that, for this molecule, no vibration is simultaneously active in both Raman and infra-red. The corresponding general rule is:

Rule of mutual exclusion. If a molecule has a *centre of symmetry* then Raman active vibrations are infra-red inactive, and vice versa. If there is no centre of symmetry then some (but not necessarily all) vibrations may be both Raman and infra-red active.

The converse of this rule is also true, i.e. the observance of Raman and infra-red spectra showing no common lines implies that the molecule has a centre of symmetry; but here caution is necessary since, as we have already seen, a vibration may be Raman active but too weak to be observed. However, if some vibrations are observed to give coincident Raman and infra-red absorptions it is certain that the molecule has no centre of symmetry. Thus extremely valuable structural information is obtainable by comparison of the Raman and infra-red spectra of a substance; we shall show examples of this in Sec. 4.5.

4.3.3 Overtone and Combination Vibrations

Without detailed consideration of the symmetry of a molecule and of its various modes of vibration, it is no easy matter to predict the activity, either in Raman or infra-red, of its overtone and combination modes. The nature of the problem can be seen by considering v_1 and v_2 of carbon dioxide; the former is Raman active only, the latter infra-red active. What, then, of the activity of $v_1 + v_2$? In fact it is only infra-red active, but this is not at all obvious merely from considering the dipole or polarizability changes during the motions. Again, when discussing Fermi resonance (Sec. 3.5.2) we chose as an example the resonance of v_1 and $2v_2$ of carbon dioxide in the *Raman* effect. Thus $2v_2$ is Raman active although the fundamental v_2 is only infra-red active.

We shall not attempt here to discuss this matter further, being content to leave the reader with a warning that the activity or inactivity of a fundamental in a particular type of spectroscopy does not necessarily imply corresponding behaviour of its overtones or combinations, particularly if the molecule has considerable symmetry. A more detailed discussion is to be found in Herzberg's book *Infra-red and Raman Spectra* and others mentioned in the bibliography.

4.3.4 Vibrational Raman Spectra

The structure of vibrational Raman spectra is easily discussed. For every vibrational mode we can write an expression of the form:

$$\varepsilon = \bar{\omega}_e(v + \tfrac{1}{2}) - \bar{\omega}_e x_e(v + \tfrac{1}{2})^2 \quad \text{cm}^{-1} \qquad (v = 0, 1, 2, \ldots) \qquad (4.15)$$

where, as before (cf. Eq. (3.12)), $\bar{\omega}_e$ is the equilibrium vibrational frequency expressed in wavenumbers and x_e is the anharmonicity constant. Such an expression is perfectly general, whatever the shape of the molecule or the nature of the vibration. Quite general, too, is the selection rule:

$$\Delta v = 0, \pm 1, \pm 2, \ldots \qquad (4.16)$$

which is the same for Raman as for infra-red spectroscopy, the probability of $\Delta v = \pm 2, \pm 3, \ldots$ decreasing rapidly.

Particularizing, now, to Raman active modes, we can apply the selection rule (4.16) to the energy level expression (4.15) and obtain the transition energies (cf. Eq. (3.15)):

$$\left.\begin{array}{lll} v = 0 \rightarrow v = 1: & \Delta\varepsilon_{\text{fundamental}} = \bar{\omega}_e(1 - 2x_e) & \text{cm}^{-1} \\ v = 0 \rightarrow v = 2: & \Delta\varepsilon_{\text{overtone}} \quad = 2\bar{\omega}_e(1 - 3x_e) & \text{cm}^{-1} \\ v = 1 \rightarrow v = 2: & \Delta\varepsilon_{\text{hot}} \quad\;\; = \bar{\omega}_e(1 - 4x_e) & \text{cm}^{-1} \quad \text{etc.} \end{array}\right\} \qquad (4.17)$$

Since the Raman scattered light is, in any case, of low intensity we can ignore completely all the weaker effects such as overtones and 'hot' bands, and restrict our discussion merely to the fundamentals. This is not to say that active overtones and hot bands cannot be observed, but they add little to the discussion here.

We would expect Raman lines to appear at distances from the exciting line corresponding to each active fundamental vibration. In other words we can write:

$$\bar{\nu}_{\text{fundamental}} = \bar{\nu}_{\text{ex.}} \pm \Delta\varepsilon_{\text{fundamental}} \quad \text{cm}^{-1} \qquad (4.18)$$

where the minus sign represents the Stokes' lines (i.e. for which the molecule has gained energy at the expense of the radiation) and the plus sign refers to the anti-Stokes' lines. The latter are often too weak to be observed, since as we saw earlier (cf. Sec. 3.1.3) very few of the molecules exist in the $v = 1$ state at normal temperatures.

The vibrational Raman spectrum of a molecule is, then, basically simple. It will show a series of reasonably intense lines to the low-frequency side of the exciting line with a much weaker, mirror-image series on the high-frequency side. The separation of each line from the centre of the exciting line gives immediately the Raman active fundamental vibration frequencies of the molecule.

As an example we illustrate the Raman spectrum of chloroform, $CHCl_3$, a symmetric top molecule (Fig. 4.8(a)). The exciting line in this case is the 488 nm argon ion laser line (at a power of 100 mW), and a wavenumber scale is drawn from this line as zero. Raman lines appear at 262, 366, 668, 761, 1216, and 3019 cm^{-1} on the low-frequency (Stokes') side of the exciting line while the line at 262 cm^{-1} on the frequency (anti-Stokes') side is included for a comparison of its intensity.

For comparison also we show at Fig. 4.8(b) the *infra-red* spectrum of the same molecule. The range of the instrument used precluded measurements below 600 cm^{-1}, but we see clearly that strong (and hence fundamental) lines appear in the spectrum at wavenumbers corresponding very precisely with those of lines in the Raman spectrum but with very different relative intensities.

Figure 4.8 Comparison between (a) the Raman and (b) the infra-red spectra of chloroform, CHCl₃, showing the coincidence of bands. (*Thanks are due to Miss S. Coulthurst and Mr M. Russell for (a) and to Mr J. Camplin for (b), all of the University of York.*)

For this molecule, cotaining five atoms, nine fundamental vibrations (that is $3N - 6$) are to be expected. The molecule has considerable symmetry, however, and three of these vibrations are doubly degenerate (see Herzberg's *Infra-red and Raman Spectra* for details) leaving six different fundamental absorptions; we see that these are all active in both the infra-red and Raman. The immediate conclusion, not at all surprisingly, is that the molecule has no centre of symmetry.

4.3.5 Rotational Fine Structure

We need not consider in detail the rotational fine structure of Raman spectra in general, if only because such fine structure is rarely resolved, except in the case of diatomic molecules. For the latter we can write the vibration–rotation energy levels (cf. Eq. (3.18)) as:

$$\varepsilon_{J,v} = \bar{\omega}_e(v + \tfrac{1}{2}) - \bar{\omega}_e x_e(v + \tfrac{1}{2})^2 + BJ(J+1) \quad \text{cm}^{-1}$$
$$(v = 0, 1, 2, \ldots; \ J = 0, 1, 2, \ldots) \tag{4.19}$$

where, as before in Raman, we ignore centrifugal distortion. For diatomic molecules the J selection rule is $\Delta J = 0, \pm 2$ (Sec. 2.1) and, combining this with the vibrational change $v = 0 \rightarrow v = 1$, we have:.

$$\Delta J = 0: \qquad \Delta\varepsilon_Q = \bar{v}_o \quad \text{cm}^{-1} \qquad\qquad \text{(for all } J) \tag{4.20}$$

$$\left.\begin{array}{llll} \Delta J = +2: & \Delta\varepsilon_S = \bar{v}_o + B(4J+6) & \text{cm}^{-1} & (J = 0, 1, 2, \ldots) \\[4pt] \Delta J = -2: & \Delta\varepsilon_O = \bar{v}_o - B(4J+6) & \text{cm}^{-1} & (J = 2, 3, 4, \ldots) \end{array}\right\} \tag{4.21}$$

where we write $\bar{v}_o$ for $\bar{\omega}_e(1 - 2x_e)$ and use the subscripts O, Q, and S to refer to the O branch lines ($\Delta J = -2$), Q branch lines ($\Delta J = 0$), and S branch lines ($\Delta J = +2$), respectively.

Stokes' lines (i.e. lines to *low* frequency of the exciting radiation) will occur at wavenumbers given by:

$$\bar{v}_Q = \bar{v}_{\text{ex.}} - \Delta\varepsilon_Q = \bar{v}_{\text{ex.}} - \bar{v}_o \quad \text{cm}^{-1} \qquad\qquad \text{(for all } J)$$
$$\bar{v}_O = \bar{v}_{\text{ex.}} - \Delta\varepsilon_O = \bar{v}_{\text{ex.}} - \bar{v}_o + B(4J+6) \quad \text{cm}^{-1} \qquad (J = 2, 3, 4, \ldots)$$
$$\bar{v}_S = \bar{v}_{\text{ex.}} - \Delta\varepsilon_S = \bar{v}_{\text{ex.}} - \bar{v}_o - B(4J+6) \quad \text{cm}^{-1} \qquad (J = 0, 1, 2, \ldots)$$

The spectrum arising is sketched in Fig. 4.9 where, for completeness, the pure rotation lines in the immediate vicinity of the exciting line are also shown. The presence of the strong Q branch in the Raman spectrum is to be noted and compared with the P and R branches only which occur for a diatomic molecule in the infra-red (cf. the spectrum of carbon monoxide in Fig. 3.7). The analysis of the O and S branches in the Raman spectrum to give a value for B and hence for the moment of inertia and bond length is straightforward.

Much weaker anti-Stokes' lines will occur at the same distance from, but to high frequency of, the exciting line.

The resolution of Raman spectra is not sufficient to warrant the inclusion of finer details such as centrifugal distortion or the breakdown of the Born–Oppenheimer approximation which were discussed in Chapter 3 for the corresponding infra-red spectra.

For larger molecules we can, in fact, ignore the rotational fine structure altogether since it is not resolved. Even the O and S (or O, P, R, and S) band contours are seldom observed since they are very weak compared with the Q branch. While some information is denied us in Raman spectra because of this, it does represent a considerable simplification of the overall appearance of such spectra.

In Table 4.2 we collect together some of the information on bond lengths and vibration frequencies which have been obtained from vibrational–rotational Raman spectra. In the case of CS_2 and CH_4 the symmetrical stretching modes only are given since the wavenumbers of the other modes are determined from infra-red techniques.

Figure 4.9 The pure rotation and the rotation–vibration spectrum of a diatomic molecule having a fundamental frequency of $\bar{\nu}_0$ cm^{-1}. Stokes' lines only are shown.

Table 4.2 Some molecular data determined by Raman spectroscopy

Molecule	Bond length (nm)	Vibration (cm^{-1})
H_2	$0.074\ 13 \pm 0.000\ 01$	4395.2
N_2	$0.109\ 76 \pm 0.000\ 01$	2359.6
F_2	$0.141\ 8 \pm 0.000\ 1$	802.1
CS_2	$0.155\ 3 \pm 0.000\ 5$	656.6 (symmetrical stretch)
CH_4	$0.109\ 4 \pm 0.000\ 1$	2914.2 (symmetrical stretch)

4.4 POLARIZATION OF LIGHT AND THE RAMAN EFFECT

4.4.1 The Nature of Polarized Light

It is well known that when a beam of light is passed through a Nicol prism or a piece of crystal filter (e.g. polaroid) the only light passing has its electric (or magnetic) vector confined to a particular plane; it is *plane polarized light*. Although superficially this light is indistinguishable from ordinary (or unpolarized) light, it has a very important property which can be demonstrated by using a second Nicol prism or crystal filter. When previously polarized light falls on the second polarizing device (now called the 'analyser') it will be passed with undiminished intensity only if the polarizing axes of the two prisms or crystal sheets are parallel to each other. At any other orientation of these axes the intensity passed will decrease until, when the axes are perpendicular, no light at all passes through the analyser. Thus the analyser serves both to detect polarized light and to determine its plane of polarization.

  If the light incident upon the analyser is only partially polarized—i.e. if the majority, but not all, of the rays have their electric vectors parallel to a given plane—then the light will not be *completely* extinguished at any orientation of the analyser; its intensity will merely go through a minimum when the analyser is perpendicular to the plane of maximum polarization. We could, then, measure the degree of polarization in terms of the intensity of light transmitted parallel and perpendicular to this plane; it is more convenient, however, to measure the *degree of depolarization*, ρ, as:

$$\rho = I_\perp / I_\parallel \tag{4.22}$$

where $I_\parallel$ is the maximum and $I_\perp$ the minimum intensity passed by the analyser. Thus for completely plane polarized light $I_\perp = 0$ and hence the degree of depolarization is zero also; for completely unpolarized (i.e. ordinary) light, $I_\perp = I_\parallel$ and $\rho = 1$. For intermediate degrees of polarization ρ lies between 0 and 1.

The relevance of this to Raman spectroscopy is that lines in some Raman spectra are found to be plane polarized to different extents even though the exciting radiation is completely depolarized. The reason for this is most easily seen if we consider the vibrations of spherical top molecules.

4.4.2 Vibrations of Spherical Top Molecules

The tetrahedral molecule methane, CH_4, is a good example of a spherical top and we can see, from Fig. 4.10, that its polarizability ellipsoid is spherical. During the vibration known as the symmetric stretch all four C—H distances increase and diminish in phase so that the polarizability ellipsoid contracts and expands but *remains spherical*. For this reason the motion is often referred to as the 'breathing frequency'; it is plainly Raman active.

Let us now consider a beam of unpolarized radiation falling on this molecule, and let us designate the direction of this exciting radiation as the z axis. Since all diameters of a sphere are equal the molecule is equally polarizable in all directions; hence the induced dipole in the molecule will lie along the direction of greatest electric vector in the exciting radiation, i.e. perpendicular to the direction of propagation. Thus the induced dipole will lie in the xy plane whatever the plane of the incident radiation. This behaviour is illustrated in Fig. 4.11, where we show an incident beam with its electric vector in the vertical (zy) plane (Fig. 4.11(a)) and some other plane making an angle α with the horizontal (Fig. 4.11(b)). In both cases the induced dipole is in the xy plane. A non-polarized incident beam will contain components having all values α.

To an observer studying the scattered radiation at right angles to the incident beam, i.e. along the x axis, the oscillating dipole emitting the radiation is confined to the xy plane—the radiation is plane polarized. When the molecule undergoes the breathing vibration, the polarizability ellipsoid remains spherical and the dipole change remains in the xy plane. Thus for this vibration the Raman line will be completely plane polarized, and $\rho = 0$, quite irrespective of the nature of the exciting radiation.

Figure 4.10 The tetrahedral structure of methane, CH_4, and the spherical polarizability ellipsoid of the molecule.

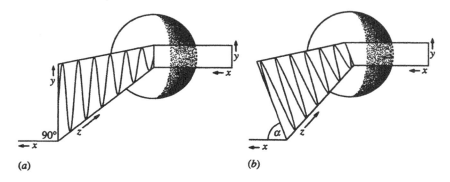

Figure 4.11 To illustrate the plane polarization of Raman scattering from the symmetric vibration ('breathing vibration') of a spherical top molecule.

Now consider a less symmetric vibration of this molecule, e.g. the asymmetric stretching mode where one C—H bond stretches while the other three contract, and vice versa. During this vibration the polarizability surface loses its spherical symmetry and becomes ellipsoid at the extremes of the vibration. When exciting radiation interacts with the molecule the induced dipole moment will be greatest along the direction of easiest polarizability, i.e. along one of the *minor* axes of the ellipsoid. In a sample of molecules these axes will be oriented in random directions to the incident radiation, so now, because of the lack of spherical symmetry in the vibration, the induced dipole will be randomly oriented and the observed Raman line will be depolarized.

Thus we have immediately a method of assigning some observed Raman lines to their appropriate molecular vibrations—in the case of methane the totally symmetric vibration gives rise to a completely polarized Raman line whereas the non-symmetric vibrations give depolarized lines. The degree of polarization of spectral lines can be readily estimated by noting how the intensity of each line varies when a piece of polaroid or other analyser is put into the scattered radiation firstly with its polarizing axis parallel to the xy plane (where z is defined by the direction of the incident beam) and secondly perpendicular to this plane.

4.4.3 Extension to Other Types of Molecule

Precise calculation, rather than the somewhat pictorial argument used above, shows that Raman scattering may be to some extent polarized when emitted by molecules with less symmetry than the tetrahedral ones. In general it can be stated that a completely symmetric vibration gives rise to a polarized or partially polarized Raman line while a non-symmetric vibration gives a depolarized line. Theoretically, if the degree of depolarization, ρ, is less than $\frac{3}{4}$ when laser excitation is used, then the vibration concerned is symmetric and the Raman line is described as 'polarized', while if $\rho \geq \frac{3}{4}$ the line is 'depolarized' and the corresponding vibration non-symmetric. If we can speak loosely of molecules with increasing symmetry—e.g. linear molecules are less symmetric than symmetric tops which, in turn, are less symmetric than spherical tops—then the higher the molecular symmetry the smaller will be the degree of depolarization of the Raman line for a symmetrical vibration.

We can see the usefulness of polarization measurements by considering a simple example. The molecule nitrous oxide has the formula N_2O. Knowing nothing about the structure of this molecule we might turn for help to its infra-red and Raman spectra. The strongest lines in these

Table 4.3 Infra-red and Raman spectra of nitrous oxide

$\bar{\nu}$ (cm^{-1})	Infra-red	Raman
589	Strong; *PQR* contour	—
1285	Very strong; *PR* contour	Very strong; polarized
2224	Very strong; *PR* contour	Strong: depolarized

spectra are collected in Table 4.3 together with their band contours (infra-red) and state of polarization (Raman).

The data tell us immediately that the molecule has no centre of symmetry (Raman and infra-red lines occur at the same wavenumber) and so the structure is *not* N—O—N. The fact that some infra-red bands have *PR* contour indicates that the molecule is linear, however, so we are led to the conclusion that the structure is N—N—O. Such a molecule should have $3N - 5 = 4$ fundamental modes but two of these (the bending modes) will be degenerate; all three different fundamental frequencies should be both infra-red and Raman active but we note that the perpendicular infra-red band (plainly to be associated with the bending mode) does not appear in the Raman. This accords with expectations—bending modes are often weak and even unobservable in the Raman.

We are left with the assignment of the 1285 and 2224 cm^{-1} bands to the symmetric and asymmetric stretching modes. Both infra-red bands have the same *PR* (parallel) contour, but we note that only the 1285 cm^{-1} is strongly Raman polarized. This, then, we assign to the symmetric mode, leaving the 2224 cm^{-1} band as the asymmetric.

The analysis would not normally rest there. The overtone and combination bands would also be studied to ensure that their activities and contours are in agreement with the molecular model proposed; the fine structure of the infra-red bands also supports the structure; and finally isotopic substitution leads to changes in vibrational frequencies in excellent agreement with the model and assignments.

In this rather simple case polarization data were hardly essential to the analysis, but certainly useful. In more complicated molecules it can give very valuable information indeed, particularly for the identification of completely symmetrical vibrations.

4.5 STRUCTURE DETERMINATION FROM RAMAN AND INFRA-RED SPECTROSCOPY

In this section we shall discuss some examples of the combined use of Raman and infra-red spectroscopy to determine the shape of some simple molecules. The discussion must necessarily be limited and the molecules considered (CO_2, N_2O, SO_2, NO_3^-, ClO_3^-, and ClF_3) have been chosen to illustrate the principles used; extension to other molecular types should be obvious.

Dealing first with the triatomic AB_2 molecules, the questions to be decided are whether each molecule is linear or not and, if linear, whether it is symmetrical (B—A—B) or asymmetrical (B—B—A). In the case of carbon dioxide and nitrous oxide, both molecules give rise to some infra-red bands with *PR* contours; they must, therefore, be linear. The mutual exclusion rule (cf. Sec. 4.3.2) shows that CO_2 has a centre of symmetry (O—C—O) while N_2O has not (N—N—O), since only the latter has bands common to both its infra-red and Raman spectra. Thus the structures of these molecules are completely determined.

The infra-red and Raman absorptions of SO_2 are collected in Table 4.4. We see immediately that the molecule has no centre of symmetry, since all three fundamentals are both Raman and infra-red active. In the infra-red all three bands show very complicated rotational fine structure,

Table 4.4 Infra-red and Raman bands of sulphur dioxide

Wavenumber	Infra-red contours	Raman
519	∥ type band	Polarized
1151	∥ type band	Polarized
1361	⊥ type band	Depolarized

and it is evident that the molecule is non-linear—no band shows the simple PR structure of, say, carbon dioxide. the molecule, has, then, the bent shape.

The AB_3 type molecules require rather more discussion. In general we would expect $3N - 6 = 6$ fundamental vibrations for these four-atomic molecules. However, if the molecular shape has some symmetry this number will be reduced by degeneracy. In particular, for the symmetric planar and symmetric pyramidal shapes, one stretching mode and one angle deformation mode are each doubly degenerate and so only four different fundamental frequencies should be observed. These are sketched in Table 4.5 where their various activities and band contours or polarizations are also collected. Both molecular shapes are in fact symmetric tops with the main (threefold) axis passing through atom A perpendicular to the B_3 plane. It is with respect to this axis that the vibrations can be described as ∥ or ⊥. The symmetric modes of

Table 4.5 Activities of vibrations of planar and pyramidal AB_3 molecules

Symmetric planar	Activity (R = Raman, I = infra-red)	Vibration	Pyramidal	Activity (R = Raman, I = infra-red)
(B–A–B₃ planar diagram)	R: active (pol.) strong I: inactive	ν_1 symmetric stretch	(A–B₃ pyramidal diagram)	R: active (pol.) strong I: active ∥
(B–A–B₃ diagram) (⊕ = upwards ⊖ = downwards)	R: inactive I: active ∥	ν_2 out-of-plane symmetric deformation	(A–B₃ pyramidal diagram)	R: active (pol.) medium I: active ∥
(B–A–B₃ diagram)	R: active (depol.) weak I: active ⊥	ν_3 asymmetric stretch	(A–B₃ pyramidal diagram)	R: active (depol.) weak I: active ⊥
(B–A–B₃ diagram)	R: active (depol.) weak I: active ⊥	ν_4 asymmetric deformation	(A–B₃ pyramidal diagram)	R: active (depol.) weak I: active ⊥

Table 4.6 Infra-red and Raman spectra of NO_3^- and ClO_3^-

Nitrate ion (NO_3^-)			Chlorate ion (ClO_3^-)		
Raman (cm^{-1})	Infra-red (cm^{-1})	Assignment	Raman (cm^{-1})	Infra-red (cm^{-1})	Assignment
690 (depol.)	680 $\perp$	ν_4	450 (depol.)	434 $\perp$	ν_4
—	830 ‖	ν_2	610 (pol.)	624 ‖	ν_2
1049 (pol.)	—	ν_1	940 (depol.)	950 $\perp$	ν_3
1355 (depol.)	1350 $\perp$	ν_3	982 (pol.)	994 ‖	ν_1

vibration are parallel and Raman polarized while the asymmetric are perpendicular and depolarized. All the vibrations of the pyramidal molecule change both the dipole moment and the polarizability; hence all are both infra-red and Raman active. The symmetric stretching mode (ν_1) of the planar molecule, however, leaves the dipole moment unchanged (it remains zero throughout) and so is infra-red inactive, while the symmetric bending mode does not change the polarizability (cf. the discussion of the bending mode of CO_2 in Sec. 4.3.1) and so ν_2 is Raman inactive for planar AB_3.

The overall pattern of the spectra, then, should be as follows:

Planar AB_3:   1 vibration Raman active only (ν_1)
                      1 infra-red active only (ν_2)
                      2 vibrations both Raman and infra-red active (ν_3, ν_4).
Pyramidal AB_3:   All four vibrations both Raman and infra-red active.
Non-symmetric AB_3: Possibly more than four different fundamental frequencies.

With this pattern in mind we can consider the spectra of NO_3^- and ClO_3^- ions. The spectroscopic data are summarized in Table 4.6. Without considering any assignment of the various absorption bands to particular vibrations, we can see immediately that the nitrate ion fits the expected pattern for a planar system, while the chlorate ion is pyramidal. Detailed assignments follow by comparison with Table 4.5. Thus for the nitrate ion, the band which is Raman active only is obviously ν_1 while that which appears only in the infra-red is ν_2. If we make the very reasonable assumption that stretching frequencies are larger than bending, then the assignment of ν_3 and ν_4 is self-evident. This same assumption, coupled with polarization and band contour data, gives the assignment shown in the table for the chlorate ion.

Finally we consider the spectroscopic data for ClF_3. This is found to have no less than *six* strong (and hence fundamental) infra-red absorptions, some of which also occur in the Raman. We know immediately, then, that the molecule is neither symmetric planar nor pyramidal. A complete analysis is not possible from the Raman and infra-red spectra alone, but the use of microwave spectroscopy shows that the molecule is T-shaped with bond angles of nearly 90°.

4.6 TECHNIQUES AND INSTRUMENTATION

The invention of the laser and its subsequent use as a source for Raman spectrometers revolutionized that technique. Previously, Raman measurements were taken with a mercury arc as the exciting source, but, because Raman is an inherently weak process, these sources were not intense enough for routine use, many hours sometimes being necessary to record a single spectrum.

The laser is almost ideal as a source for Raman experiments; it gives a very narrow, highly monochromatic, coherent beam, which can be focused very finely into a small sample. In addition, lasers can be extremely powerful, ranging from milliwatts to several watts, concentrated into a small energy spread. The rare gas lasers (Ar^+, Kr^+, etc.), which are most commonly used for Raman measurements, can provide intensities as great as one million times that of sunlight.

Figure 4.12 illustrates a typical Raman spectrometer. The laser beam is passed through a cell, usually a narrow glass or quartz tube filled with the sample. Light scattered sideways from the sample is collected by a lens and passed into a grating monochromator similar to that used in a dispersive infra-red instrument. The signal is measured by a sensitive photomultiplier and, after amplification, it is usually processed by a computer which plots the Raman spectrum.

As we have seen in Sec. 4.4, the use of plane polarized radiation gives information about the symmetries of molecular vibrations. To make these measurements, the laser beam in Fig. 4.12 is plane polarized perpendicularly to the plane of the paper, and a polarizing filter is placed between the sample and the collecting lens. The Raman spectrum is then measured twice, first with the polarizing filter set to pass light polarized perpendicularly to the paper, and then at right angles to this. The ratio of the two signals for each Raman line is a measure of the degree of polarization of that line.

For vibrational measurements the Raman technique has several advantages over infra-red. Firstly, because both the incident and scattered radiation are at ultra-violet or visible frequencies, conventional optics and sample cells (glass, quartz, etc.) can be used, so avoiding the problems inherent in NaCl windows, atmospheric absorption, etc. Secondly, because the beam can be focused extremely finely (diameters as small as 0.1 nm are possible), very small samples can be studied. This, combined with pulsed techniques which can give very short time resolutions, enables very small quantities of transient species to be studied. Thirdly, water, which has strong infra-red absorptions, is a rather weak Raman scatterer and so aqueous solutions can be studied using Raman, because the sample signal is not swamped by that of the solvent.

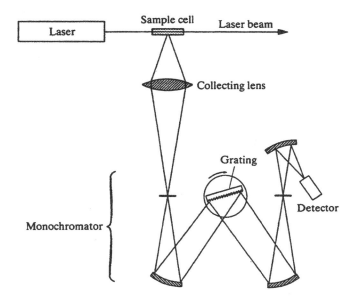

Figure 4.12 Schematic diagram of a Raman spectrometer.

These reasons collectively ensure that Raman spectroscopy is particularly well suited to the study of biological systems. As an example, the Raman spectrum of myoglobin, which is essentially a protein chain surrounding a single iron atom, is shown in Fig. 4.13(a). The spectrum, which was obtained using less than 1 mg of myoglobin, shows that a number of Raman modes are allowed, due mainly to the protein chain. The iron in myoglobin can bind reversibly to gas molecules, such as oxygen and carbon dioxide, and more firmly to carbon monoxide (it is in this way that the molecule assists the transport of gases to muscle in the body

(a)

Frequency (cm^{-1})

(b)

cm^{-1}

Figure 4.13 (a) The Raman spectrum of myoglobin. (b) Detail of the iron–carbon stretching region for carboxymyglobin. (*Thanks are due to Dr D. Biram for spectrum (a); (b) is redrawn from R. E. Hester,* Proceedings of 13th International Conference on Raman Spectroscopy, *John Wiley, 1992, pp. 12–16.*)

and can be so easily poisoned with carbon monoxide). Figure 4.13(b) shows, in more detail, the iron–carbon stretching frequency region of the Raman spectrum of carbon monoxide bound to myoglobin (carboxymyoglobin). Here the full curve is the (smoothed) Raman spectrum, while the dashed curves show the result of using a computer to resolve the spectrum into its component lines. Clearly the spectrum reveals several peaks in this region, which indicates that carbon monoxide takes up slightly different configurations in different myoglobin molecules. The relative intensities of the component peaks vary for carboxymyoglobin formed from myoglobin taken from different animal species.

In addition to liquids and solutions, Raman spectra can be obtained from gas and solid samples. In the case of gases, multiple reflection techniques are sometimes adopted, where the laser beam is reflected several times back and forth through the sample, in order to enhance the signals.

One major problem with some Raman samples—particularly if they are coloured—is that the heat generated by the intense, focused laser beam may cause decomposition. If the sample has an ultra-violet or visible absorption peak at about the same frequency as the incident laser, rapid heating will certainly occur, which can best be overcome by using a different laser. A more general way to reduce sample heating, however, is to spin the sample so that no single spot is continuously irradiated by the laser.

Another problem which sometimes arises is that of sample fluorescence. Fluorescence occurs when an electronically excited molecule decays back to the ground state spontaneously, emitting radiation at a frequency characteristic of the transition between the excited and ground states (see Chapter 6, Sec. 6.3). Such radiation can totally swamp the weak Raman signal. If it results from defects and/or impurities in solid samples, prolonged exposure to the laser beam can sometimes burn out the impurity, provided this treatment does not also destroy the sample. Where the latter may occur, or where the fluorescence is intrinsic to the sample, the simplest approach is to record spectra using a different exciting frequency. The incident beam and Raman signals shift by the same amount, while the fluorescence remains in its original position. Alternatively, pulsed laser excitation and time-gated detection can be used, which can discriminate between the slower fluorescence and the fast spontaneous Raman emission. Finally, a 'quenching agent' can be added to liquids or solutions which allows the excited molecule to decay via a pathway involving the quenching molecule instead of by emission of radiation. However, more recently the development of FT–Raman spectroscopy using near-IR lasers (see the next section) has greatly helped to overcome fluorescence problems.

Computer averaging of multiple scans is commonly used to improve the signal-to-noise ratio, while other developments such as multi-channel detection (effectively an array of signal detectors, each sensing a different small part of the frequency range of the scattered radiation) can also be used to increase the sensitivity of the experiment.

4.7 NEAR-INFRA-RED FT–RAMAN SPECTROSCOPY

For a given energy of the laser beam, the strength of Raman scattering depends on $(\text{frequency})^4$, and it was for this reason that lasers in the high-frequency, visible end of the spectrum were formerly used. More recently, however, near-infra-red laser excitation has been successful, usually using an NdYAG laser operating at 9398 cm^{-1}. Raman scattering is inherently weaker at this lower frequency, and less sensitive indium gallium arsenide (InGaAs) or germanium detectors have to be used, but the application of FT methods, already well developed in the infra-red region, can restore the lost sensitivity. An added economy is that, once the optical path of the FT spectrometer has been properly aligned, it is a fairly simple matter to interchange

Figure 4.14 FT–Raman spectrum of powdered C$_{60}$. (*Reproduced with permission from P. J. Hendra,* J. of Molecular Structure, **266**, *97–114, 1992.*)

sources, beam splitters and detectors, and a single FT instrument can thus be operated either as an infra-red spectrometer or as a Raman spectrometer.

Another advantage of near-infra-red FT–Raman work is that it is less prone to sample heating and sample fluorescence because samples have few absorptions bands in the near-infra-red region. As a result excellent spectra can be obtained from even highly coloured materials. Two disadvantages are that samples heated above 200 °C give a background of thermal emission in this region and that water is no longer a useful solvent, since it has absorption bands from overtones of the — OH stretching fundamentals.

Figure 4.14 shows the near-infra-red FT–Raman spectrum of the highly coloured compound buckminster fullerene (C$_{60}$); this is the spherical allotrope of carbon, which looks rather like a football. The molecule has 174 normal vibrational modes but, because of its high symmetry, many of them are multiply degenerate or forbidden, resulting in this remarkably simple Raman spectrum.

BIBLIOGRAPHY

Andrews, D. L.: *Lasers in Chemistry*, 2nd ed., Springer, 1990.

Carey, P. R.: *Biochemical Applications of Raman and Resonance Raman Spectroscopies*, Academic Press, 1982.

Grasselli, J. G., M. K. Snavely, and B. J. Bulkin: *Chemical Applications of Raman Spectroscopy*, Wiley, 1981.

Hendra, P., C. Jones, and G. Warnes: *FT–Raman Spectroscopy*, Ellis-Horwood, 1991.

Herzberg, G.: *Molecular Structure and Molecular Spectra:* vol. 1, *Spectra of Diatomic Molecules*, 2nd ed., Van Nostrand, 1950.

Herzberg, G.: *Molecular Structure and Molecular Spectra:* vol. 2, *Infra-red and Raman Spectra of Polyatomic Molecules*, Van Nostrand, 1945.

Siesler, H. W., and K. Holland-Moritz: *Infra-red and Raman Spectroscopy of Polymers*, Marcel Dekker Inc., 1980.

Williams, D. H., and I. Fleming: *Spectroscopic Methods in Organic Chemistry*, 4th ed., McGraw-Hill, 1989.

PROBLEMS

(Useful constants: $h = 6.626 \times 10^{-34}$ J s; $c = 2.998 \times 10^{8}$ m s^{-1}; $8\pi^2 = 78.957$; mass of H atom $= 1.673 \times 10^{-27}$ kg.)

4.1 (*a*) What is meant by the term *polarizability*?
(*b*) State the selection rules for Raman scattering.
(*c*) What technological advances have enabled the routine use of Raman spectroscopy?

4.2 With which type of spectroscopy would one observe the pure rotation spectrum of H_2? If the bond length of H_2 is 0.074 17 nm, what would be the spacing of the lines in the spectrum?

4.3 The spin of the hydrogen nucleus is $\frac{1}{2}$; does this make any difference to your answer to Prob. 4.2?

4.4 Which type of vibrational spectroscopy (IR or Raman) would you use to measure the vibrational frequency of the following bonds:

The stretching frequency of $^{14}N - {}^{15}N$
The $C \equiv C$ stretch in ethyne, $HC \equiv CH$
The $C = O$ stretch in acetone, $CH_3.CO.CH_3$
The Re—Re stretch of the inorganic cluster compound, $(CO)_5Re - Re(CO)_5$

4.5 The Re—Re vibration of $(CO)_5Re - Re(CO)_5$ is observed at 122 cm^{-1}, while that of the $Re_2Cl_8{}^{2-}$ ion occurs at 275 cm^{-1}. Without calculating separate force constants, calculate the ratio of the Re—Re bond force constants of the two molecules. Use your result to comment on the bond orders in the two species. (*Note*: Here, as frequently, it is a sufficiently good approximation to treat the vibration of the heavy Re atoms as being independent of the rest of the molecule.)

4.6 A molecule A_2B_2 has infra-red absorptions and Raman spectral lines as in the following table:

cm^{-1}	Infra-red	Raman
3374	—	Strong
3287	Very strong; *PR* contour	—
1973	—	Very strong
729	Very strong; *PQR* contour	—
612	—	Weak

Deduce what you can about the structure of the molecule and assign the observed vibrations to particular molecular modes as far as possible.
Hint: Use data from Table 3.4 to help with your answer.

4.7 A molecule AB_2 has the following infra-red and Raman spectra:

cm^{-1}	Infra-red	Raman
3756	Very strong; perpendicular	—
3652	Strong; parallel	Strong; polarized
1595	Very strong; parallel	—

The rotational fine structure of the infra-red bands is complex and does not show simple *PR* or *PQR* characteristics. Comment on the molecular structure, and assign the observed lines to specific molecular vibrations as far as possible.

4.8 Both N_2O and NO_2 exhibit three different fundamental vibrational frequencies, and for the two molecules some modes are observed in both the infra-red and the Raman. The bands in N_2O show only simple *PR* structure (no *Q* branches) while those in NO_2 show complex rotational structure. What can be deduced about the structure of each molecule?

ELECTRONIC SPECTROSCOPY OF ATOMS

5.1 THE STRUCTURE OF ATOMS

5.1.1 Electronic Wave Functions

It is well known that an atom consists of a central, positively charged nucleus, which contributes nearly all the mass to the system, surrounded by negatively charged electrons in sufficient number to balance the nuclear charge. Hydrogen, the smallest and simplest atom, has a nuclear charge of $+1$ units (where the unit is the electronic charge 1.60×10^{-19} coulomb) and one electron; each succeeding atom increases the nuclear charge and electron total by unity, up to atoms with 100 or more electrons.

Modern theories have long ceased to regard the electron as a particle which obeys the laws of classical mechanics applicable to massive, everyday objects; instead, in common with all entities of subatomic size, we consider that it obeys the laws of quantum mechanics (or wave mechanics) as embodied in the Schrödinger wave equation. In principle, this equation may be used to determine many things: e.g. the way in which electrons group themselves about a nucleus when forming an atom, the energy which each electron may have, the way in which it can undergo transitions between energy states, etc. In practice the application of the Schrödinger equation to these problems presents difficulties which can only be overcome in the case of the simplest atoms or by the use of gross approximations. Here, however, we shall be concerned only with the results obtained—and then only in qualitative terms—rather than the mathematical theory of the process.

The Schrödinger theory can be used to predict the *probability* of an electron with a particular energy being at a particular point in space, and it expresses this probability in terms of a very important algebraic expression called the *wave function* of the electron. The wave function is given the Greek symbol ψ. Quite simply, the probability of finding an electron, whose wave function is ψ, within unit volume at a given point in space, is proportional to the value of ψ^2 at that point.

$$\text{Relative probability density} = \psi^2 \tag{5.1}$$

Let us see what this means. Electronic wave functions consist of three elements: (1) some fundamental *physical constants* (π, h, c, m, e, etc.—where m and e are the mass and charge, respectively, of the electron); (2) *parameters* peculiar to the system under discussion—e.g. for atoms, distance from the nucleus, either radially (r) or along some coordinate axes (x, y, z); and (3) one or more *quantum numbers*. These latter are by no means arbitrarily introduced into the problem in order to make the predictions match experiment; they *belong* to the solution of the Schrödinger equation in the sense that ψ represents a sensible physical situation *only if* the quantum numbers have certain values.

As an example we may quote here the expression for a set of wave functions, ψ_n, which are solutions to the Schrödinger equation for the hydrogen atom:

$$\psi_n = f\left(\frac{r}{a_0}\right) \exp\left(-\frac{r}{na_0}\right) \tag{5.2}$$

where $a_0 = h^2/4\pi^2 me^2$, r is radial distance from the nucleus, $f(r/a_0)$ is a power series of degree $(n-1)$ in r/a_0, and n is the *principal quantum number*, which can have only integral values, 1, 2, 3, . . . , ∞. The constant a_0 has dimensions of length (and is, in fact, about $53\,\text{nm}$) and so the quantity r/a_0 is a pure number. Thus for particular values of r and n, ψ_n and ψ_n^2 are also simply numbers, and ψ_n^2 represents the probability of finding the electron at our chosen distance r from the nucleus when it is in the state represented by the given n value.

It is found that the electronic wave functions of all atoms require the introduction of only four quantum numbers. We shall describe these briefly here, leaving a more thorough discussion to later sections.

5.1.2 The Shape of Atomic Orbitals; Atomic Quantum Numbers

Table 5.1 lists the four quantum numbers, gives the allowed values of each, and states what is the function of each. The *principal quantum number*, as stated earlier, can take integral values from one to infinity. It governs the energy of the electron mainly (although we shall see later that the other quantum numbers also affect this energy to some extent). The table shows that n also governs the size of the electronic *orbital*; this latter is a term used to represent the space within which an electron can move according to the Schrödinger theory—it corresponds approximately to the earlier idea of Bohr that electrons move in circular or elliptical *orbits* like planets round a sun. Energy and size of the orbital are connected in that the smaller the orbital the closer to the nucleus the electron will be and hence the more firmly bound.

The *orbital* (or *azimuthal*) *quantum number* l also has integral values only, but these must be less than n. Thus for $n = 3$, l can be 2, 1, or 0. It governs the shape of the orbital (cf. Fig. 5.1) and the angular momentum of the electron as it circulates about the nucleus in its orbital.

Table 5.1 The atomic quantum numbers

Quantum No.	Allowed values	Function
Principal, n	1, 2, 3, . . .	Governs the energy and size of the orbital
Orbital, l	$(n-1), (n-2), \ldots, 0$	Governs the shape of the orbital and the electronic angular momentum
Magnetic, m	$\pm l, \pm(l-1), \ldots, 0$	Governs the direction of an orbital and the electrons' behaviour in a magnetic field
Spin, s	$+\frac{1}{2}$	Governs the axial angular momentum of the electron

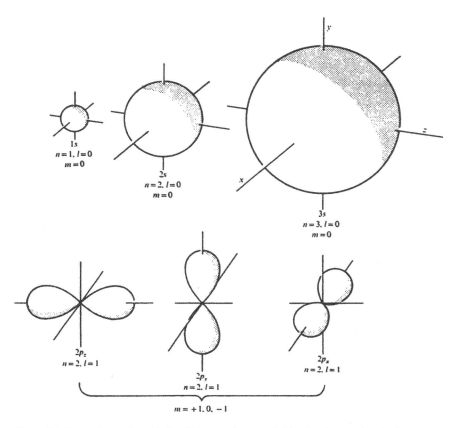

Figure 5.1 Some electronic orbitals which may be occupied by the electron in a hydrogen atom.

The *magnetic quantum number m* takes integral values which depend on l. Thus for $l = 2$, m can be $+2$, $+1$, 0, -1, or -2; in general there are $2l + 1$ values of m. Besides denoting the behaviour of electrons in orbitals when the atom is placed in a magnetic field, the m quantum number can also be used to specify the direction of a particular orbital.

The *spin quantum number s* is of magnitude $+\frac{1}{2}$ only (but cf. Sec. 5.2.2). It measures the spin angular momentum which the electron possesses whether it is present in an atom or in free space.

Since wave functions represent only a probability distribution of an electron it is difficult to define precisely the shape and size of an orbital. From Eq. (5.2) we see that even at very large values of r, ψ_n (and hence ψ_n^2) still has a value, even though small. Thus an orbital tails off to infinity (although, because of the smallness of a_0, 'infinity' on the atomic scale might be taken as 10^{-4} or 10^{-5} cm) in all directions. However, the difficulty can be overcome if we agree to draw a three-dimensional shape *within* which the electron spends, say, 95 per cent or some other fraction of its time. This can be taken as the effective boundary of the electron's domain and it can be called the *orbital*.

Considering still the wave function of Eq. (5.2) we see that the corresponding orbital must be spherical, for at any given distance r from the nucleus ψ_n has the same value irrespective of direction. Thus the 95 per cent boundary will be spherical. For larger n the function tails off less rapidly with distance and so the electron can spend proportionately more of its time further from the nucleus; thus the 95 per cent sphere will increase in size with n. We have drawn the cases $n = 1$, 2, and 3 at the top of Fig. 5.1. These spherical orbitals, it so happens, are associated with an l value of zero (and hence $m = 0$) and they are referred to as *s orbitals*. (Although it is perhaps helpful to connect s with 'spherical'—in fact the label arose historically because of the alleged particular *sharpness* of spectral lines arising from transition of electrons occupying s orbitals—the connection which should be remembered is between s orbitals and $l = 0$.) The s orbitals are labelled according to their n quantum numbers: $1s, 2s, \ldots , ns$.

Orbitals with $l = 1$ (and hence $n \geq 2$) also arise as solutions to the Schrödinger equation for the hydrogen atom. These are twin-lobed and have the approximate shape shown for $n = 2$ in the lower half of Fig. 5.1. Orbitals with $n = 3$, $l = 1$ are larger but have the same shape. Such orbitals are labelled p (historically their transitions were thought to be 'principal') and we see that, for a given n, there are *three* of them, one along each coordinate axis. They can be distinguished as np_x, np_y, and np_z, if necessary. The fact that there are three of them is connected with the three values of m, $m = +1$, 0, and -1, allowed for $l = 1$ states. It is conventional to associate the value $m = 0$ with the np_z orbitals but, for good physical reasons which lie outside the scope of this book, it is not then possible to associate the other m values with either np_x or np_y. Other representations of these orbitals can be drawn, however, in such a way that there is a one-to-one correspondence between each m value and an orbital; these representations are less convenient for the descriptive purposes of this book and we shall not discuss them here.

We can go further: for $l = 2$ (hence $n \geq 3$) we have a set of *d orbitals* (historically 'diffuse') and $l = 3(n \geq 4)$ *f orbitals* (historically 'fundamental'): there are five of the former ($m = \pm 2, \pm 1$ or 0) and seven of the latter ($m = \pm 3, \pm 2, \pm 1$, or 0). Sketches of d orbitals show that they have four lobes, while the f have six, but we shall not attempt to reproduce these here. Orbitals with higher l values, $l = 4, 5, 6, \ldots$, are of less importance and we shall not consider them further; if necessary they are labelled alphabetically after f, that is $l = 4$, g; $l = 5$, h, etc.

5.1.3 The Energies of Atomic Orbitals; Hydrogen Atom Spectrum

However large an atom its electrons take up orbitals of the s, p, d, $\ldots$ type (according to very specific laws which we shall discuss later) and so the overall shape of each electron's domain is unaltered. The *energy* of each orbital, on the other hand, varies considerably from atom to atom. There are two main contributions to this energy: (1) attraction between electrons and nucleus, (2) repulsion between electrons in the same atom.

We consider first the case of hydrogen in some detail: this is the simplest because, having only one electron, factor (2) is completely absent. We shall later see how the picture should be modified for larger atoms.

Because of the absence of interelectronic effects all orbitals with the same n value have the same energy in hydrogen. Thus the $2s$ and $2p$ orbitals, for instance, are degenerate, as are the $3s$, $3p$ and $3d$. However, the energies of the $2s$, $3s$, $4s$, $\ldots$ orbitals differ considerably. For the s orbitals given by Eq. (5.2):

$$\psi_{ns} = f\left(\frac{r}{a_0}\right) \exp\left(-\frac{r}{na_0}\right)$$

the Schrödinger equation shows that the energy is:

$$E_n = -\frac{me^4}{8h^2\varepsilon_o^2 n^2} \quad \text{J}$$

$$\varepsilon_n = -\frac{me^4}{8h^3 c\varepsilon_o^2 n^2} = -\frac{R}{n^2} \quad \text{cm}^{-1} \quad (n = 1, 2, 3, \ldots)$$

(5.3)

where ε_o is the vacuum permittivity, and where the fundamental constants have been collected together and given the symbol R, called the Rydberg constant. Since p, d, ... orbitals have the same energies as the corresponding s (for hydrogen *only*), Eq. (5.3) represents *all* the electronic energy levels of this atom.

The lowest value of ε_n is plainly $\varepsilon_n = -R\,\text{cm}^{-1}$ (when $n = 1$), and so this represents the most stable (or ground) state; ε_n increases with increasing n, reaching a limit, $\varepsilon_n = 0$ for $n = \infty$. This represents complete removal of the electron from the nucleus, i.e. the state of ionization. We sketch these energy levels for $n = 1$ to 5 and $l = 0$, 1, and 2 only in Fig. 5.2. (Some possible transitions, also shown, will be discussed shortly.) The three p states and five d states for each n are degenerate and not shown separately.

Equation (5.3) and Fig. 5.2, then, represent the energy levels of the atom; in order to discuss the spectra which may arise we need the selection rules governing transitions. The Schrödinger equation shows these to be:

$$\Delta n = \text{anything} \quad \text{and} \quad \Delta l = \pm 1 \text{ only} \tag{5.4}$$

From these selection rules we see immediately that an electron in the ground state (the $1s$) can undergo a transition into any p state:

$$1s \rightarrow np \quad (n \geq 2)$$

while a $2p$ electron can have transitions either into an s state or a d state:

$$2p \rightarrow ns \quad \text{or} \quad nd$$

Since s and d orbitals are here degenerate the energy of both these transitions will be identical. These transitions are sketched in Fig. 5.2.

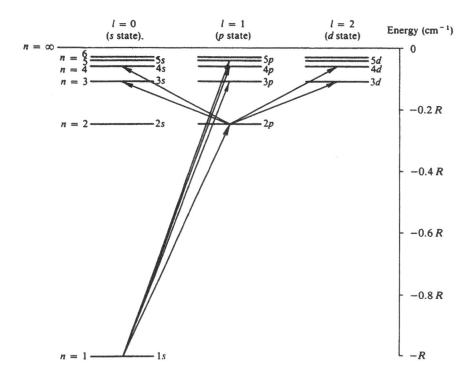

Figure 5.2 Some of the lower electronic energy levels and transitions between them for the single electron of the hydrogen atom.

In general an electron in a lower state n'' can undergo a transition into a higher state n', with absorption of energy:

$$\Delta\varepsilon = \varepsilon_{n'} - \varepsilon_{n''} \quad \text{cm}^{-1}$$

Therefore:

$$\bar{v}_{\text{spect.}} = -\frac{R}{n'^2} - \left(-\frac{R}{n''^2}\right) = R\left\{\frac{1}{n''^2} - \frac{1}{n'^2}\right\} \quad \text{cm}^{-1} \tag{5.5}$$

An identical spectral line will be produced in emission if the electron falls from state n' to state n''. In both cases l must change by unity. Let us consider a few of these transitions, restricting ourselves to absorption for simplicity.

Transitions $1s \rightarrow n'p$, $n' = 2, 3, 4, \ldots$. For these:

$$\bar{v}_{\text{Lyman}} = R\left\{\frac{1}{1} - \frac{1}{n'^2}\right\} = R - \frac{R}{n'^2} \quad \text{cm}^{-1}$$

$$= \frac{3R}{4}, \frac{8R}{9}, \frac{15R}{16}, \frac{24R}{25}, \ldots \quad \text{cm}^{-1} \quad (\text{for } n' = 2, 3, 4, 5, \ldots)$$

Hence we expect a series of lines at the wavenumbers given above. Just such a series is indeed observed in the atomic hydrogen spectrum, and it is called the Lyman series after its discoverer. The appearance of this spectrum is sketched in Fig. 5.3 together with a scale in units of R and in wavenumbers. We can see that the spectrum converges to the point $R\,\text{cm}^{-1}$, and from the observed spectrum the very precise value $R = 109\,677.581\,\text{cm}^{-1}$ is obtained. This convergence limit, which arises when $n' = \infty$, is shown dashed on the figure. It plainly represents complete removal of the electron—i.e. ionization—and the energy required to ionize the atom is given, in cm^{-1}, by the value of R. Using the conversion factor $1\,\text{cm}^{-1} = 1.987 \times 10^{-23}\,\text{J}$, we have a very precise measure of the *ionization potential* from the ground ($1s$) state: $2.1781 \times 10^{-18}\,\text{J}$ (which may be more familiar in non-SI units as $13.595\,\text{eV}$).

Another set of transitions arises from an electron initially in the $2s$ or $2p$ states: $2s \rightarrow n'p$ or $2p \rightarrow n's$, $n'd$. For these we write:

$$\bar{v}_{\text{Balmer}} = R\left\{\frac{1}{4} - \frac{1}{n'^2}\right\} \quad \text{cm}^{-1}$$

$$= \frac{5R}{36}, \frac{3R}{16}, \frac{21R}{100}, \ldots \quad \text{cm}^{-1} \quad (\text{for } n' = 3, 4, 5, \ldots)$$

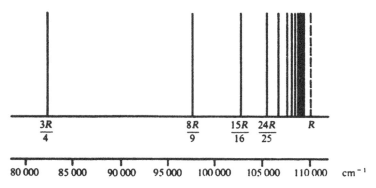

Figure 5.3 Representation of part of the Lyman series of the hydrogen atom, showing the convergence (ionization) point.

Thus we expect another series of lines converging to $\frac{1}{4}R\,\mathrm{cm}^{-1}$ ($n' = \infty$); this series, called the Balmer series after its discoverer, is observed and the value of $\frac{1}{4}R$ obtained from its convergence limit—which represents the ionization potential from the first excited state—is in excellent agreement with the value of R from the Lyman series.

Other similar line series (called the Paschen, Brackett, Pfund, etc., series) are observed for $n'' = 3, 4, 5, \ldots$; indeed these spectra were observed long before the modern theory of atomic structure had been developed. The spectral lines were correlated empirically by Rydberg, and he showed that an equation of the form given in Eq. (5.5) described the wavenumbers of each. It is after him that the Rydberg constant is named.

It should be mentioned that each line series discussed above shows a *continuous absorption* or *emission* to high wavenumbers of the convergence limits. The convergence limit represents the situation where the atomic electron has absorbed just sufficient energy from radiation to escape from the nucleus with zero velocity. It can, however, absorb more energy than this and hence escape with higher velocities and, since the kinetic energy of an electron moving in free space is *not* quantized, *any* energy above the ionization energy can be absorbed. Hence the spectrum in this region is continuous.

This completes our discussion of what might be termed the coarse structure of the hydrogen atom spectrum. In order to consider the fine structure we need to know how the other quantum numbers, besides n, affect the electronic energy levels.

5.2 ELECTRONIC ANGULAR MOMENTUM

5.2.1 Orbital Angular Momentum

An electron moving in its orbital about a nucleus possesses *orbital* angular momentum, a measure of which is given by the l value corresponding to the orbital. This momentum is, of course, quantized, and it is usually expressed in terms of the unit $h/2\pi$, where h is Planck's constant. We may write:

$$\text{Orbital angular momentum} = \sqrt{l(l+1)} \cdot \frac{h}{2\pi} = \sqrt{l(l+1)} \text{ units} \tag{5.6}$$

Angular momentum is a *vector* quantity, by which we mean that its *direction* is important as well as its magnitude—the axis of a spinning top, for instance, points in a particular direction. Conventionally, vectors may be represented by arrows, and the angular momentum vector is represented by an arrow based at the centre of the top, along the top axis, and of length proportional to the magnitude of the angular momentum. Such an arrow can lie in two different directions, at 180° to each other; these directions are associated, depending on the sign convention used, with clockwise and anticlockwise rotations of the top. Mathematically we can ignore the spinning body and deal merely with the properties of the arrow.

It is usual to distinguish vector quantities by the use of *bold-face* type and we shall accordingly represent orbital angular momentum by the symbol **l** where:

$$\mathbf{l} = \sqrt{l(l+1)} \text{ units} \tag{5.7}$$

In this equation l is always zero or positive and hence so is **l**. Since **l** and l are so closely connected they are often loosely used interchangeably: thus we speak of an electron having 'an angular momentum of 2' when we strictly mean that $l = 2$ and $\mathbf{l} = \sqrt{2 \times 3} = 2.44$ units.

We might at first think that the angular momentum vector of an electron could point in an infinite number of different directions. This, however, would be to reckon without the quantum theory. In fact, once a reference direction has been specified (and this may be done in many ways, either externally, such as by applying an electric or magnetic field, or internally, perhaps in terms of the angular momentum vector of one particular electron), the angular momentum vector can point *only so that its components along the reference direction are integral multiples of* $h/2\pi$. Figure 5.4(a) and (b) shows the situation for an electron with $l = 1$ and $l = 2$, respectively (i.e. a p and a d electron). The reference direction, here taken to be vertical in the figure, is conventionally used to define the z axis, and so we can write the *components* of $\mathbf{l}$ in this direction as $\mathbf{l}_z$ (note the use of bold-face type for $\mathbf{l}_z$, since the components of angular momentum are clearly vectors, having both magnitude and direction). Alternatively, since we know that the $\mathbf{l}_z$ are integral multiples of $h/2\pi$, we can represent the components in terms of an integral number l_z (not bold-face), where:

$$\mathbf{l}_z = l_z \cdot \frac{h}{2\pi} \tag{5.8}$$

This latter notation is used in Fig. 5.4, and we see there that for $l = 1$, l_z takes values of $+1$, 0, and -1, while for $l = 2$ the values are $+2$, $+1$, 0, -1, and -2. In general we see that l_z has values:

$$l_z = l, \; l-1, \; \ldots, \; 0, \; \ldots, -(l-1), l \tag{5.9}$$

and that there are $2l + 1$ values of l_z for a given l. Plainly l_z is to be identified with the magnetic quantum number m introduced in Sec. 5.1.2:

$$l_z \equiv m$$

and this justifies our previous assertion that m governs essentially the *direction* of an orbital.

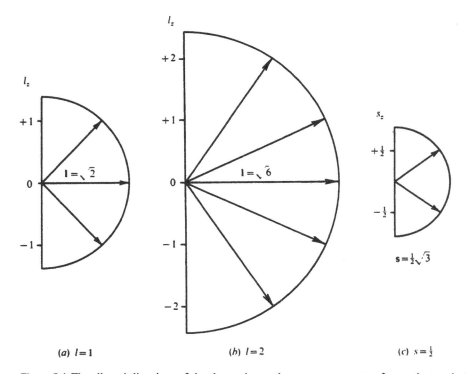

Figure 5.4 The allowed directions of the electronic angular momentum vector for an electron in (a) a p state ($l = 1$), (b) a d state ($l = 2$), and (c) the allowed directions of the electronic spin angular momentum vector. The reference direction is taken arbitrarily as upwards in the plane of the paper.

Before proceeding further let us reiterate the distinction between l, **l**, l_z, and l_z. The quantum number l is an integer, positive or zero, representing the state of an electron in an atom and determining its orbital angular momentum. The vector **l** designates the magnitude and direction of this momentum as shown by the vector arrows of Fig. 5.4. When expressed in units of $h/2\pi$, **l** is numerically equal to $\sqrt{l(l+1)}$. Once a reference direction is specified (and this is often arbitrary) **l** can point only so as to have components $l_z = l_z h/2\pi$ (with l_z an integer or zero) along that direction.

Usually the orbital energy of the electron depends only on the magnitude and not the direction of its angular momentum; thus the $2l+1$ values of l_z are all *degenerate*. However, we should note that it is possible to *lift the degeneracy* (cf. Sec. 5.6) so that levels with different l_z have different energy.

5.2.2 Electron Spin Angular Momentum

Every electron in an atom can be considered to be spinning about an axis as well as orbiting about the nucleus. Its spin motion is designated by the *spin quantum number s*, which can be shown to have a value of $\frac{1}{2}$ only. Thus the spin angular momentum is given by:

$$\mathbf{s} = \sqrt{s(s+1)} \, \frac{h}{2\pi} = \sqrt{\frac{1}{2} \times \frac{3}{2}} \text{ units}$$

$$= \tfrac{1}{2}\sqrt{3} \text{ units} \qquad (5.10)$$

The quantization law for spin momentum is that the vector can point so as to have components in the reference direction which are *half-integral* multiples of $h/2\pi$, i.e. so that $\mathbf{s}_z = s_z h/2\pi$ with s_z taking the values $+\frac{1}{2}$ or $-\frac{1}{2}$ only. The two (that is $2s+1$) allowed directions are shown in Fig. 5.4(c); they are normally degenerate.

5.2.3 Total Electronic Angular Momentum

We now need to discover some means whereby the orbital and spin contributions to the electronic angular momentum may be combined. Formally we can write:

$$\mathbf{j} = \mathbf{l} + \mathbf{s} \qquad (5.11)$$

where **j** is the *total angular momentum*. Since **l** and **s** are vectors, Eq. (5.11) must be taken to imply *vector addition*. Also formally, we can express **j** in terms of a total angular momentum quantum number j:

$$\mathbf{j} = \sqrt{j(j+1)} \, \frac{h}{2\pi} = \sqrt{j(j+1)} \text{ units} \qquad (5.12)$$

where j is *half-integral* (since s is half-integral for a one-electron atom), and a quantal law applies equally to **j** as to **l** and **s** : **j** *can have z-components which are half-integral only*, i.e.:

$$j_z = \pm j, \pm(j-1), \pm(j-2), \ldots, \tfrac{1}{2} \qquad (5.13)$$

There are two methods by which we can deduce the various allowed values of **j** for particular **l** and **s** values. We shall consider them both briefly.

1. *Vector summation.* In ordinary mechanics two forces in different directions may be added by a graphical method in which vector arrows are drawn to represent the magnitude and direction of the forces, the 'parallelogram is completed', and the magnitude and direction of the resultant given by the diagonal of the parallelogram. Exactly the same method can be used to find the resultant (**j**) of the vectors **l** and **s**. The important difference is that quantum

mechanical laws restrict the angle between **l** and **s** to values such that **j** is given by Eq. (5.12) with *half-integral j*. Thus **j** can take values

$$\tfrac{1}{2}\sqrt{3}, \tfrac{1}{2}\sqrt{15}, \tfrac{1}{2}\sqrt{35}, \ldots \quad \text{corresponding to} \quad j = \tfrac{1}{2}, \tfrac{3}{2}, \tfrac{5}{2}, \ldots$$

The method is illustrated in Fig. 5.5(*a*) and (*b*) for the case $l = 1$ (that is $\mathbf{l} = \sqrt{2}$) and $s = \tfrac{1}{2}(\mathbf{s} = \tfrac{1}{2}\sqrt{3})$. In Fig. 5.5(*a*) the summation yields $\mathbf{j} = \tfrac{1}{2}\sqrt{15}$, which corresponds to a *j* value of $\tfrac{3}{2}$, while in Fig. 5.5(*b*) $\mathbf{j} = \tfrac{1}{2}\sqrt{3}$ or $j = \tfrac{1}{2}$. Construction or calculation shows that **l** and **s** may not be combined in any other way to give an allowed value of **j**.

Note that we can get exactly the same answer by summing the *quantum numbers l* and *s* to get the *quantum number j*. In this example $l = 1$, $s = \tfrac{1}{2}$, and hence:

$$j = l + s = \tfrac{3}{2} \quad \text{or} \quad j = l - s = \tfrac{1}{2}$$

This simple approach, although adequate for systems with one electron only, is not readily extended to multi-electron systems. For these we must use the rather more fundamental method outlined below.

2. *Summation of z components.* If the components along a common direction of two vectors are added, the summation yields the component in that direction of their resultant. We have seen (cf. Eq. (5.9)) that the *z* components of $l = 1$ are ± 1 and 0, while those of $s = \tfrac{1}{2}$ are $\pm\tfrac{1}{2}$ only. Taking all possible sums of these quantities we have:

$$j_z = l_z + s_z$$

Therefore:

$$j_z = 1 + \tfrac{1}{2}, \ 1 - \tfrac{1}{2}, \ 0 + \tfrac{1}{2}, \ 0 - \tfrac{1}{2}, \ -1 + \tfrac{1}{2}, \ -1 - \tfrac{1}{2}$$
$$= \tfrac{3}{2}, \tfrac{1}{2}, \tfrac{1}{2}, -\tfrac{1}{2}, -\tfrac{1}{2}, -\tfrac{3}{2}$$

In this list of six j_z components, the maximum value is $\tfrac{3}{2}$, which we know (cf. Eq. (5.13)) must belong to $j = \tfrac{3}{2}$. Other components of $j = \tfrac{3}{2}$ are $\tfrac{1}{2}, -\tfrac{1}{2}$, and $-\tfrac{3}{2}$ and, striking these from the above six, we are left with $j_z = +\tfrac{1}{2}$ and $-\tfrac{1}{2}$. These values are plainly consistent with $j = \tfrac{1}{2}$.

Thus all the six components are accounted for if we say that the states $j = \tfrac{3}{2}$ and $j = \tfrac{1}{2}$ may be formed from $l = 1$ and $s = \tfrac{1}{2}$. This is, of course, in agreement with the vector summation method.

Both these methods show that for a *p* electron (that is $l = 1$), the orbital and spin momenta may be combined to produce a total momentum of $\mathbf{j} = \tfrac{1}{2}\sqrt{15}$ when **l** and **s** reinforce (physically

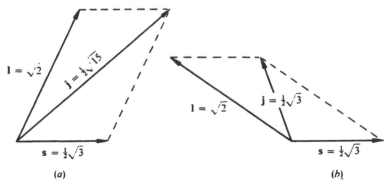

(a)   (b)

Figure 5.5 The two energy states having different total angular momentum which can arise as a result of the vector addition of $\mathbf{l} = \sqrt{2}$ and $\mathbf{s} = \tfrac{1}{2}\sqrt{3}$.

we would say that the angular momenta have the same *direction*) or to give $\mathbf{j} = \frac{1}{2}\sqrt{3}$ when $\mathbf{l}$ and $\mathbf{s}$ oppose each other. Thus the total momentum is different in *magnitude* in the two cases and hence we have arrived at two *different energy states* depending on whether $\mathbf{l}$ and $\mathbf{s}$ reinforce or oppose. Both energy states are p states, however (since l is 1 for both), and they may be distinguished by writing the j quantum number value as a subscript to the *state symbol P*, thus: $P_{3/2}$ or $P_{1/2}$. (We here use a capital letter for the state of a whole atom and a small letter for the state of an individual electron; in the hydrogen atom, which contains only one electron, the distinction is trivial.) States such as these, split into two energies, are termed *doublet states*; their doublet nature is usually indicated by writing a superscript 2 to the state symbol, thus: $^2P_{3/2}$, $^2P_{1/2}$. The state (or term) symbols produced are to be read 'doublet P three halves' or 'doublet P one half', respectively.

All other higher l values for the electron will obviously produce doublet states when combined with $s = \frac{1}{2}$; for instance, $l = 2, 3, 4, \ldots$ will yield $^2D_{5/2,3/2}$, $^2F_{7/2,5/2}$, $^2G_{9/2,7/2}$, etc. The student should be satisfied of this, preferably by using the z-component summation method outlined above. There is, however, a slight difficulty with s states ($l = 0$). Here, since $l = 0$, it can make no contribution to the vector sum, and the only possible resultant is $\mathbf{s} = \frac{1}{2}\sqrt{3}$ or $s = \frac{1}{2}$. Remember that $s = -\frac{1}{2}$ is not allowed, since the *quantum number* cannot be negative; it is only the z component of the vector which can have negative values. Thus for an s electron we would have the state symbol $S_{1/2}$ only. This is nonetheless formally written as a doublet state ($^2S_{1/2}$) for reasons which should become clear during the discussion of multiplicity in Sec. 5.4.3.

We can now consider the relevance of this discussion to atomic spectroscopy.

5.2.4 The Fine Structure of the Hydrogen Atom Spectrum

The hydrogen atom contains but one electron and so the coupling of orbital and spin momenta and consequent splitting of energy levels will be exactly as described above. We summarize the essential details of the energy levels in Fig. 5.6. Each level is labelled with its n quantum number on the extreme left and its j value on the right; the l value is indicated by the state symbols S, P, D, $\ldots$ at the top of each column. There is no attempt to show the energy-level splitting of the P and D states to scale in this diagram—the separation between levels differing only in j is many thousands of times smaller than the separation between levels of different n. However, we do indicate that the j-splitting *decreases* with increasing n and with increasing l. The F, G, $\ldots$ states, not shown on the diagram, follow the same pattern.

The selection rules for n and l are the same as before:

$$\Delta n = \text{anything} \qquad \Delta l = \pm 1 \text{ only} \tag{5.14}$$

but now there is a selection rule for j:

$$\Delta j = 0, \pm 1 \tag{5.15}$$

These selection rules indicate that transitions are allowed between any S level and any P level:

$$^2S_{1/2} \rightarrow {}^2P_{1/2} \quad (\Delta j = 0)$$
$$^2S_{1/2} \rightarrow {}^2P_{3/2} \quad (\Delta j = +1)$$

Thus the spectrum to be expected from the ground ($1s$) state will be identical with the Lyman series (cf. Sec. 5.1.3) except that *every line will be a doublet*. In fact the separation between the lines is too small to be readily resolved but we shall shortly consider the spectrum of sodium in which this splitting is easily observed.

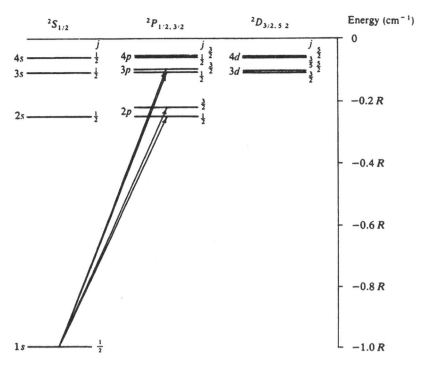

Figure 5.6 Some of the lower energy levels of the hydrogen atom, showing the inclusion of *j*-splitting. The splitting is greatly exaggerated for clarity.

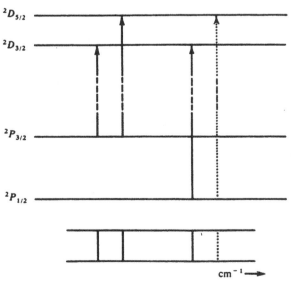

Figure 5.7 The 'compound doublet' spectrum arising as the result of transitions between ^2P and ^2D levels in the hydrogen atom.

Transitions between the ^2P and ^2D states are rather more complex; Fig. 5.7 shows four of the energy levels involved. Plainly the transition at lowest frequency will be that between the closest pair of levels, the $^2P_{3/2}$ and $^2D_{3/2}$. This, corresponding to $\Delta j = 0$, is allowed. The next transition, $^2P_{3/2} \rightarrow {}^2D_{5/2}$ ($\Delta j = +1$), is also allowed and will occur close to the first because the separation between the doublet D states is very small. Thirdly, and more widely spaced, will be $^2P_{1/2} \rightarrow {}^2D_{3/2}$ ($\Delta j = +1$), but the fourth transition (shown dotted), $^2P_{1/2} \rightarrow {}^2D_{5/2}$, is not allowed since for this $\Delta j = +2$.

Thus the spectrum will consist of the three lines shown at the foot of the figure. This, arising from transitions between doublet levels, is usually referred to as a 'compound doublet' spectrum.

We see, then, that the inclusion of coupling between orbital and spin momenta has led to a slight increase in the complexity of the hydrogen spectrum. In practice, the complexity will be observed only in the spectra of heavier atoms, since for them the j-splitting is larger than for hydrogen. In principle, however, all the lines in the hydrogen spectrum should be close doublets if the transitions involve s levels, or 'compound doublets' if s electrons are not involved.

5.3 MANY-ELECTRON ATOMS

5.3.1 The Building-Up Principle

The Schrödinger equation shows that electrons in atoms occupy orbitals of the same type and shape as the s, p, d, . . . orbitals discussed for the hydrogen atom, but that the energies of these electrons differ markedly from atom to atom. There is no general expression for the energy levels of a many-electron atom comparable to Eq. (5.3) for hydrogen; each atom must be treated as a special case and its energy levels either tabulated or shown on a diagram similar to Fig. 5.2 or Fig. 5.6.

There are three basic rules, known as the building-up rules, which determine how electrons in large atoms occupy orbitals. These may be summarized as:

1. Pauli's principle: no two electrons in an atom may have the same set of values for n, l, $l_z (\equiv m)$, and s_z.
2. Electrons tend to occupy the orbital with lowest energy available.
3. Hund's principle: electrons tend to occupy degenerate orbitals singly with their spins parallel.

Rule 1 effectively limits to two the number of electrons in each orbital. An example may make this clear: we may characterize both an orbital and an electron occupying it by specifying the n, l, and m quantum numbers. Thus a $1s$ orbital or $1s$ electron has $n = 1$, $l = 0$, and $m = l_z = 0$; the electron (but not the orbital) is further characterized by a statement of its spin direction, i.e. by specifying $s_z = +\frac{1}{2}$ or $s_z = -\frac{1}{2}$. Two electrons can together occupy the $1s$ orbital provided, according to rule 1, that one has the set of values $n = 1$, $l = 0$, $l_z = 0$, $s_z = +\frac{1}{2}$, and the other $n = 1$, $l = 0$, $l_z = 0$, $s_z = -\frac{1}{2}$. We talk, rather loosely, of two electrons occupying the same orbital only if their spins are *paired* (or *opposed*). A third electron cannot exist in the same orbital without repeating a set of values for n, l, l_z, and s_z already taken up. It would have to be placed into some other orbital and the choice is determined by rule 2: it would go into the next higher vacant or half-vacant orbital. In general, orbital energies in many-electron atoms increase with increasing n, as they do for hydrogen, but they also increase with increasing l, whereas we noted for hydrogen that all s, p, d . . . orbitals with the same n were degenerate. In fact the order of the energy levels for most atoms is as follows:

$$1s < 2s < 2p < 3s < 3p < 4s < 3d < 4p < 5s < 4d \ldots \tag{5.16}$$

Thus when the $1s$ orbital is full (i.e. contains two electrons) the next available orbital is the $2s$, and after this the $2p$. Now we remember that there are *three* $2p$ orbitals, one along each coordinate axis, and *each* of these can contain two electrons. We may write the n, l, l_z, and s_z values as:

$$\left.\begin{array}{llll} n=2 & l=1 & l_z=1 & s_z=\pm\tfrac{1}{2} \\ n=2 & l=1 & l_z=0 & s_z=\pm\tfrac{1}{2} \\ n=2 & l=1 & l_z=-1 & s_z=\pm\tfrac{1}{2} \end{array}\right\} \text{ total six electrons}$$

All three p orbitals remain degenerate (as do the five d orbitals, seven f, etc.) for a given n. It is rule 3 which tells us how electrons occupy these degenerate orbitals. Hund's rule states that when, for example, the $2p_x$ orbital contains an electron, the next electron will go into a *different* $2p$, say $2p_y$, orbital, and a third into the $2p_z$. This may be looked upon as a consequence of repulsion between electrons. A fourth electron has no choice but to pair its spin with an electron already in one $2p$ orbital, while a fifth and sixth will complete the filling of the three $2p$'s.

On this basis we can build up the *electronic configurations* of the 10 smallest atoms, from hydrogen to neon. This is shown in Table 5.2 where each box represents an orbital occupied by one or two electrons with spin directions shown by the arrows. A convenient notation for the electronic configuration is also shown in the table.

When a set of orbitals of given n and l is filled it is referred to as a *closed shell*. Thus the $1s^2$ set of helium, the $2s^2$ set of beryllium, and the $2p^6$ set of neon are all closed shells. The convenience of this is that closed shells make no contribution to the orbital or spin angular momentum of the whole atom and hence they may be ignored when discussing atomic spectra. This represents a considerable simplification.

Table 5.2 Electronic structure of some atoms

	$1s$	$2s$	$2p$	
Hydrogen	↑			$1s^1$
Helium	↑↓			$1s^2$
Lithium	↑↓	↑		$1s^2 2s^1$
Beryllium	↑↓	↑↓		$1s^2 2s^2$
Boron	↑↓	↑↓	↑	$1s^2 2s^2 2p_x^1$
Carbon	↑↓	↑↓	↑ ↑	$1s^2 2s^2 2p_x^1 2p_y^1$
Nitrogen	↑↓	↑↓	↑ ↑ ↑	$1s^2 2s^2 2p_x^1 2p_y^1 2p_z^1$
Oxygen	↑↓	↑↓	↑↓ ↑ ↑	$1s^2 2s^2 2p_x^2 2p_y^1 2p_z^1$
Fluorine	↑↓	↑↓	↑↓ ↑↓ ↑	$1s^2 2s^2 2p_x^2 2p_y^2 2p_z^1$
Neon	↑↓	↑↓	↑↓ ↑↓ ↑↓	$1s^2 2s^2 2p_x^2 2p_y^2 2p_z^2$

5.3.2 The Spectrum of Lithium and Other Hydrogen-like Species

The alkali metals, lithium, sodium, potassium, rubidium, and cesium, all have a single electron outside a closed-shell core (cf. lithium in Table 5.2). Superficially, then, they resemble hydrogen and this resemblance is augmented by the fact that we can ignore the angular momentum of the core and deal merely with the spin and orbital momentum of the outer electron. Thus we immediately expect the p, d, ... levels to be split into doublets because of coupling between l and s.

The energy levels of lithium are sketched in Fig. 5.8, which should be compared with the corresponding Fig. 5.6 for hydrogen. The two diagrams are similar except for the energy difference between the s, p, and d orbitals of given n in the case of lithium and the fact that,

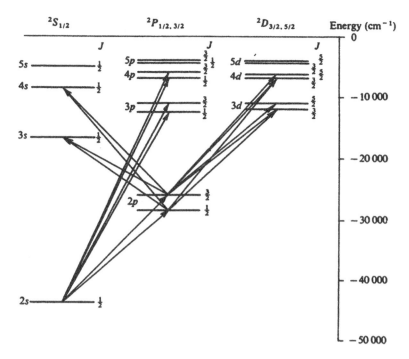

Figure 5.8 Some of the lower energy levels of the lithium atom, showing the difference in energy of s, p, and d states with the same value of n. Some allowed transitions are also shown. The j-splitting is greatly exaggerated.

for this metal, the $1s$ state is filled with electrons which do not generally take part in spectroscopic transitions, as it requires much less energy to induce the $2s$ electron to undergo a transition. Under high energy conditions, however, one or both of the $1s$ electrons may be promoted.

The selection rules for alkali metals are the same as for hydrogen, that is Δn = anything, $\Delta l = \pm 1$, $\Delta j = 0, \pm 1$, and so the spectra will be similar also. Thus transitions from the ground state ($1s^2 2s$) can occur to p levels: $2S_{1/2} \rightarrow nP_{1/2,3/2}$, and a series of doublets similar to the Lyman series will be formed, converging to some point from which the ionization potential can be found. From the $2p$ state, however, two separate series of lines will be seen:

$$2\,^2P_{1/2,3/2} \rightarrow n\,^2S_{1/2}$$

and

$$2\,^2P_{1/2,3/2} \rightarrow n\,^2D_{3/2,5/2}$$

The former will be doublets, the latter compound doublets, but their frequencies will differ because the s and d orbital energies are no longer the same.

The same remarks apply to the other alkali metals, the differences between their spectra and that of lithium being a matter of scale only. For instance the j-splitting due to coupling between **l** and **s** increases markedly with the atomic number. Thus the doublet separation of lines in the spectral series, which is scarcely observable for hydrogen, is less than $1\,\mathrm{cm}^{-1}$ for the $2p$ level of lithium, about $17\,\mathrm{cm}^{-1}$ for sodium, and over $5000\,\mathrm{cm}^{-1}$ for cesium.

Any atom which has a single electron moving outside a closed shell will exhibit a spectrum of the type discussed above. Thus ions of the type He^+, Be^+, B^{2+}, etc., should, and indeed do, show what are termed 'hydrogen-like spectra'.

5.4 THE ANGULAR MOMENTUM OF MANY-ELECTRON ATOMS

We turn now to consider the contribution of two or more electrons in the outer shell to the total angular momentum of the atom. There are two different ways in which we might sum the orbital and spin momentum of several electrons:

1. First sum the orbital contributions, then the spin contributions separately, and finally add the total orbital and total spin contributions to reach the grand total. Symbolically:

$$\sum l_i = L \quad \sum s_i = S \quad L + S = J$$

   where we use bold-face capital letters to designate total momentum.
2. Sum the orbital and spin momenta of each electron separately, finally summing the individual totals to form the grand total:

$$l_i + s_i = j_i \quad \sum j_i = J$$

The first method, known as Russell–Saunders coupling, gives results in accordance with the spectra of small and medium-sized atoms, while the second (called j–j coupling, since individual j's are summed) applies better to large atoms. We shall consider only the former in detail.

5.4.1 Summation of Orbital Contributions

The orbital momenta l_1, l_2, . . . of several electrons may be added by the same methods as were discussed in Sec. 5.2.3 for the summation of the orbital and spin momenta of a single electron. Thus we could:

1. Add the vectors l_1, l_2, . . . graphically, remembering that their resultant L must be expressible by:

$$L = \sqrt{L(L+1)} \qquad (L = 0, 1, 2, \ldots) \tag{5.17}$$

   where L is the total orbital momentum quantum number. Thus L can have values 0, $\sqrt{2}$, $\sqrt{6}$, $\sqrt{12}$, . . . only. Figure 5.9 illustrates the method for a p and a d electron, $l_1 = 1$, $l_2 = 2$; hence $l_1 = \sqrt{2}$, $l_2 = \sqrt{6}$. There are three, and only three, ways in which the two vectors may be combined to give L consistent with Eq. (5.17). The three values of L are

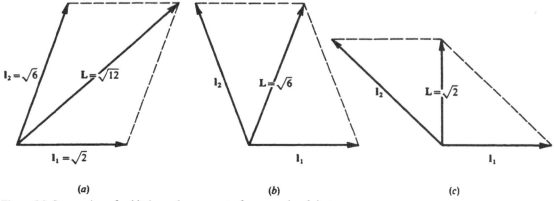

(a) (b) (c)

Figure 5.9 Summation of orbital angular momenta for a p and a d electron.

seen to be $\sqrt{12}$, $\sqrt{6}$, and $\sqrt{2}$, corresponding to the quantum numbers $L = 3$, 2, and 1, respectively.

2. Alternatively we could add the individual *quantum numbers* l_1 and l_2 to obtain the total quantum number L according to:

$$L = l_1 + l_2, \quad l_1 + l_2 - 1, \ldots \quad |l_1 - l_2| \tag{5.18}$$

where the modulus sign $|\ldots|$ indicates that we are to take $l_1 - l_2$ or $l_2 - l_1$, whichever is positive. For two electrons, there will plainly be $2l_i + 1$ different values of L, where l_i is the smaller of the two l values.

3. Finally we could add the z components of the individual vectors, picking out from the result sets of components corresponding to the various allowed **L** values. Symbolically this process is:

$$L_z = \sum l_{i_z}$$

Of these, method 2 is the simplest but it is only applicable when the individual electrons concerned have different n or different l values (these are termed *non-equivalent* electrons). If n and l are the same for two or more electrons they are termed *equivalent* and method 3 must be used. Examples will be given later.

5.4.2 Summation of Spin Contributions

The same methods may be used here as in Sec. 5.4.1. Briefly, if we write the total spin angular momentum as **S** and the total spin quantum number as S (which is often simply called the total spin), we can have:

1. Graphical summation, provided the resultant is

$$\mathbf{S} = \sqrt{S(S+1)} \tag{5.19}$$

where S is either *integral or zero* only, if the number of contributing spins is *even*, or *half-integral* only, if the number is *odd*.

2. Summation of individual quantum numbers. For N spins we have:

$$
\begin{aligned}
S &= \sum s_i, \sum s_i - 1, \sum s_i - 2, \ldots \\
&= \frac{N}{2}, \frac{N}{2} - 1, \ldots, \tfrac{1}{2} \quad \text{(for } N \text{ odd)} \\
&= \frac{N}{2}, \frac{N}{2} - 1, \ldots, 0 \quad \text{(for } N \text{ even)}
\end{aligned}
\tag{5.20}
$$

3. Summation of individual s_z to give S_z.

Method 2, which is *always* applicable, is the simplest. Thus for two electrons we have the two possible spin states:

$$S = \tfrac{1}{2} + \tfrac{1}{2} = 1 \quad \text{or} \quad S = \tfrac{1}{2} + \tfrac{1}{2} - 1 = 0$$

In the former the spins are called *parallel* and the state may be written ($\uparrow\uparrow$), while in the latter they are *paired* or *opposed* and written ($\uparrow\downarrow$).

Again, for three electrons we may have:

$$S = \tfrac{1}{2} + \tfrac{1}{2} + \tfrac{1}{2} = \tfrac{3}{2} \quad (\uparrow\uparrow\uparrow)$$
$$S = \tfrac{1}{2} + \tfrac{1}{2} + \tfrac{1}{2} - 1 = \tfrac{1}{2} \quad (\uparrow\uparrow\downarrow, \uparrow\downarrow\uparrow, \text{ or } \downarrow\uparrow\uparrow)$$

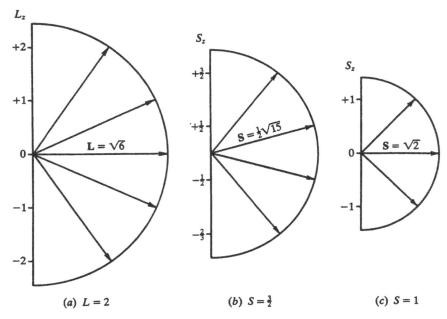

Figure 5.10 The z components of (a) an orbital angular momentum vector, (b) a spin vector for which S is half-integral, and (c) a spin vector for which S is integral.

where we see that there are three ways in which the $S = \frac{1}{2}$ state may be realized and one in which $S = \frac{3}{2}$.

As we have implied above, both **L** and **S** have z components along a reference direction. For **L** these components are limited to integral values by quantum laws and, as we can see from Fig. 5.10(a), there are, in general, $2L + 1$ of them, while for **S** the S_z will be integral only or half-integral only, depending on whether S is integral or half-integral. We show examples in Fig. 5.10(b) and (c). In both cases there are $2S + 1$ components.

5.4.3 Total Angular Momentum

The addition of the total orbital momentum **L** and the total spin momentum **S** to give the grand total momentum **J** can be carried out in the same ways as the addition of **l** and **s** to give **j**, for a single electron. The only additional point is that the quantum number J in the expression:

$$\mathbf{J} = \sqrt{J(J+1)} \cdot \frac{h}{2\pi} \tag{5.21}$$

must be *integral* if S is integral and *half-integral* if S is half-integral. In terms of the quantum numbers we can write immediately:

$$J = L + S, \ L + S - 1, \ \ldots, \ |L - S| \tag{5.22}$$

where, as before, the *positive* value of $L - S$ is the lowest limit of the series of values.

For example, if $L = 2$, $S = \frac{3}{2}$, we would have:

$$J = \tfrac{7}{2}, \tfrac{5}{2}, \tfrac{3}{2}, \text{ and } \tfrac{1}{2}$$

while if $L = 2$, $S = 1$, the J values are:

$$J = 3, 2, \text{ or } 1 \text{ only}$$

In general we see that there are $2S + 1$ different values of J and hence $2S + 1$ states with different total momentum. The energy of a state depends on its total momentum, so we arrive at $2S + 1$ different energy levels, the energy of each depending on the way in which **L** and **S** are combined. The quantity $2S + 1$, which occupies a special place in atomic spectroscopy, is called the *multiplicity* of the system.

We recall that, when discussing the total angular momentum of a single electron (Sec. 5.2.3), we found that each state, except those with $l = 0$, consisted of two very slightly different energy levels owing to j-splitting; we called these 'doublet states'. We also called the $l = 0$ states doublets, although each has but one energy level since $j = \frac{1}{2}$ only. We now see that the concept of multiplicity justifies us in labelling *all* one-electron states as doublets, since for all of them $S = \frac{1}{2}$, and hence $2S + 1 = 2$, and they all have a multiplicity of two.

It is a general rule that in states with $L \geq S$, whether consisting of one electron or of many, the multiplicity is equal to the actual number of levels with different J, whereas if $L < S$, then there are only $2L + 1$ different J values, which is less than the multiplicity. As an example of the latter, if $L = 1$ and $S = 2$, there are only three different J values, $J = 3$, 2, or 1, whereas the multiplicity is $2S + 1 = 5$.

It is in fact highly convenient to label states with their multiplicities rather than to show the number of different J values because, as we shall see shortly, there is a selection rule which forbids transitions between states with different multiplicities; thus designating the multiplicity of a particular state immediately indicates to which other levels in the system a transition may take place.

5.4.4 Term Symbols

In the whole of this section we have been describing the way in which the total angular momentum of an atom is built up from its various components. Using one sort of coupling only (the Russell–Saunders coupling) we arrive at vector quantities **L**, **S**, and **J** for a system which may be expressed in terms of quantum numbers L, S, and J:

$$\mathbf{L} = \sqrt{L(L + 1)} \quad \mathbf{S} = \sqrt{S(S + 1)} \quad \mathbf{J} = \sqrt{J(J + 1)} \tag{5.23}$$

where the integral L and integral or half-integral S and J are themselves combinations of individual electronic quantum numbers.

In any particular atom, then, we see that the individual electronic angular momenta may be combined in various ways to give different states each having a different total angular momentum (**J**) and hence a different energy (unless some states happen to be degenerate). Before discussing the effect of these states on the spectrum of an atom we require some symbolism which we may use to describe states conveniently. We have already introduced such *state symbols* or *term symbols* in Sec. 5.2.3, but we now consider them rather more fully.

The term symbol for a particular atomic state is written as follows:

$$\text{Term symbol} = {}^{2S+1}L_J \tag{5.24}$$

where the numerical superscript gives the *multiplicity* of the state, the numerical subscript gives the total angular momentum quantum number J, and the value of the orbital quantum number L is expressed by a *letter*:

For $\quad L = 0$, 1, 2, 3, 4, . . .
Symbol $\quad = S, P, D, F, G, . . .$

a symbolism which is comparable with the $s, p, d, . . .$ already used for single-electron states with $l = 0$, 1, 2, ...

Let us now see some examples.

1. $S = \frac{1}{2}$, $L = 2$; hence $J = \frac{5}{2}$ or $\frac{3}{2}$ and $2S + 1 = 2$. Term symbols: $^2D_{5/2}$ and $^2D_{3/2}$, which are to be read 'doublet D five halves' and 'doublet D three halves', respectively.
2. $S = 1$, $L = 1$; hence $J = 2$, 1, or 0 and $2S + 1 = 3$. Term symbols: $^3P_2$, $^3P_1$, or $^3P_0$ (read 'triplet P two', etc.).
   In both these examples we see that (since $L \geq S$), the multiplicity is the same as the number of different energy states.
3. $S = \frac{3}{2}$, $L = 1$; hence $J = \frac{5}{2}$, $\frac{3}{2}$, or $\frac{1}{2}$ and $2S + 1 = 4$. Term symbols: $^4P_{5/2}$, $^4P_{3/2}$, $^4P_{1/2}$ (read 'quartet P five halves', etc.) where, since $L < S$, there are only three different energy states but each is nonetheless described as *quartet* since $2S + 1 = 4$.

The reverse process is equally easy; given a term symbol for a particular atomic state we can immediately deduce the various total angular momenta of that state. Some examples:

4. $^3S_1$: we read immediately that $2S + 1 = 3$, and hence $S = 1$, and that $L = 0$ and $J = 1$.
5. $^2P_{3/2}$: $L = 1$, $J = \frac{3}{2}$, $2S + 1 = 2$; hence $S = \frac{1}{2}$.

Note, however, that the term symbol tells us only the total spin, total orbital, and grand total momenta of the whole atom—it tells us nothing of the states of the individual electrons in the atoms, nor even how many electrons contribute to the total. Thus in example 5 above, the fact that $S = \frac{1}{2}$ implies that the atom has an odd number of contributing electrons, all except one of which have their spins paired. Thus a single electron ($\uparrow$), three electrons ($\uparrow\uparrow\downarrow$), five electrons ($\uparrow\downarrow\uparrow\downarrow\uparrow$), etc., all form a doublet state. Similarly, the value $L = 1$ implies, perhaps, one p electron, or perhaps one p and two s electrons, or one of many other possible combinations.

Normally this is not important; the spectroscopist is interested only in the energy state of the atom as a whole. Should we wish to specify the energy states of individual electrons, however, we can do so by including them in the term symbol as a prefix. Thus in example 5 we might have $2p\,^2P_{3/2}$, or $1s2p3s\,^2P_{3/2}$, etc.

We can now apply our knowledge of atomic states to the discussion of the spectra of some atoms with two or more electrons. We start with the simplest, that of helium.

5.4.5 The Spectrum of Helium and the Alkaline Earths

Helium, atomic number two, consists of a central nucleus and two outer electrons. Clearly there are only two possibilities for the relative spins of the two electrons:

1. Their spins are paired; in which case if s_{1_z} is $+\frac{1}{2}$, s_{2_z} must be $-\frac{1}{2}$; hence $S_z = s_{1_z} + s_{2_z} = 0$, and so $S = 0$ and we have *singlet* states.
2. Their spins are parallel; now $s_{1_z} = s_{2_z} = +\frac{1}{2}$, say, so that $S_z = 1$ and the states are *triplet*.

The lowest possible energy state of this atom is when both electrons occupy the $1s$ orbital; this, by Pauli's principle, is possible only if their spins are paired, so the ground state of helium must be a singlet state. Further, $L = l_1 + l_2 = 0$, and hence J can only be zero. The ground state of helium, therefore, is $^1S_0$.

The relevant selection rules for many-electron systems are:

$$\Delta S = 0 \quad \Delta L = \pm 1 \quad \Delta J = 0, \pm 1 \tag{5.25}$$

(There is a further rule that a state with $J = 0$ cannot make a transition to another $J = 0$ state, but this will not concern us here.) We see immediately that, since S cannot change during a transition, the singlet ground state can undergo transitions only to other singlet states. The selection rules for L and J are the same as those for l and j considered earlier.

For the moment we shall imagine that only one electron undergoes transitions, leaving the other in the $1s$ orbital, and the left-hand side of Fig. 5.11 shows the energy levels for the various singlet states which arise.

Initially the $1s^2\ {}^1S_0$ state can undergo a transition only to $1s^1\ np^1$ states (abbreviated to $1snp$); in the latter $L = 1$, $S = 0$, and hence $J = 1$ only, so the transition may be symbolized:

$$1s^2\ {}^1S_0 \rightarrow 1snp\ {}^1P_1$$

or, briefly:

$$^1S_0 \rightarrow {}^1P_1$$

From the ${}^1P_1$ state the system could either revert to ${}^1S_0$ states, as shown in the figure, or undergo transitions to the higher ${}^1D_2$ states (for these $S = 0$, $L = 2$, and hence $J = 2$ only). In general, then, all these transitions will give rise to spectral series very similar to those of lithium except that here transitions are between *singlet* states only and all the spectral lines will be single.

Returning now to the situation in which the electron spins are parallel (case 2, the triplet states) we see that, since the electrons are now forbidden by Pauli's principle from occupying the same orbital, the lowest energy state is $1s2s$. This and other triplet energy levels are shown on the right of Fig. 5.11. The $1s2s$ state has $S = 1$, $L = 0$, and hence $J = 1$ only, and so it is ${}^3S_1$; by the selection rules of Eq. (5.25) it can only undergo transitions into the $1snp$ triplet states. These, with $S = 1$, $L = 1$, have $J = 2$, 1, or 0, and so the transitions may be written.

$$^3S_1 \rightarrow {}^3P_2,\ {}^3P_1,\ {}^3P_0$$

All three transitions are allowed, since $\Delta J = 0$ or ± 1, so the resulting spectral lines will be *triplets*.

Transitions from the ${}^3P$ states may take place either to ${}^3S$ states (spectral series of triples) or to ${}^3D$ states. In the latter case the spectral series may be very complex if completely resolved. For ${}^3D$ we have $S = 1$, $L = 2$, and hence $J = 3$, 2, or 1, and we show in Fig. 5.12 a transition

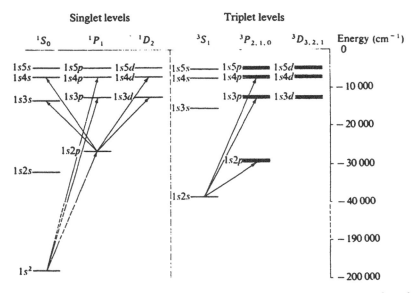

Figure 5.11 Some of the energy levels of the electrons in the helium atom, together with a few allowed transitions.

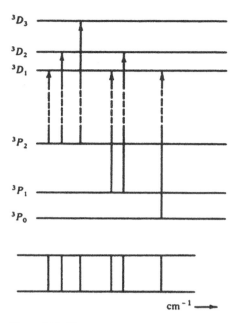

Figure 5.12 The 'compound triplet' spectrum arising from transitions between ^3P and ^3D levels in the helium atom. The separation between levels of different J is much exaggerated.

between ^3P and ^3D states, bearing in mind the selection rule $\Delta J = 0, \pm 1$. We note that $^3P_2$ can go to each of $^3D_{3,2,1}$, $^3P_1$ can go only to $^3D_{2,1}$, and $^3P_0$ can go only to $^3D_1$. Thus the complete spectrum (shown at the foot of the figure) should consist of six lines. Normally, however, the very close spacing is not resolved, and only three lines are seen; for this reason the spectrum is referred to as a *compound triplet*.

We might note in passing that Fig. 5.12 shows that levels with higher J have a higher energy in helium, and that the separation decreases from top to bottom. This is not the case with all atoms, however. If higher J is equivalent to lower energy then the separation *increases* from top to bottom and the multiplet is described as *inverted*. In helium, and other atoms with similar behaviour, the multiplet is *normal* or *regular*.

We see, then, that the spectrum of helium consists of spectral series grouped into two types which overlap each other in frequency. In one type, involving transitions between singlet levels, all the spectral lines are themselves singlets, while in the other the transitions are between triplet states and each 'line' is at least a close triplet and possibly even more complex. Because of the selection rule $\Delta S = 0$ there is a strong prohibition on transitions between singlet and triplet states, and transitions *cannot occur* between the right- and left-hand sides of Fig. 5.11. Early experimenters, noting the difference between the two types of spectral series, suggested that helium exists in two modifications, *ortho-* and *para*-helium. This is not far from the truth, although we know now that the difference between the two forms is very subtle: it is merely that one has its electron spins always opposed and the other always parallel.

Other atoms containing two outer electrons exhibit spectra similar to that of helium. Thus the alkaline earths, beryllium, magnesium, calcium, etc., fall into this category, as do ionized species with just two remaining electrons, for example, B^+, C^{2+}, etc.

We should remind the reader at this point that the above discussion on helium has been carried through on the assumption that one electron remains in the $1s$ orbital all the time. This is a reasonable assumption since a great deal of energy would be required to excite *two* electrons simultaneously, and this would not happen under normal spectroscopic conditions. However, not all atoms have only s electrons in their ground state configuration, and we consider next some of the consequences.

5.4.6 Equivalent and Non-Equivalent Electrons; Energy Levels of Carbon

The ground state electronic configuration of carbon is $1s^2 2s^2 2p^2$, which indicates that both the $1s$ and $2s$ orbitals are filled (and hence contribute nothing to the angular momentum of the atom), while the $2p$ orbitals are only partially filled. The $2p$ electrons, also, are most easily removed, so it is these which normally undergo spectroscopic transitions.

Two or more electrons are referred to as *equivalent* if they have the same value of n and of l. Thus the two $2p$ electrons in the ground state of carbon are equivalent ($n_1 = n_2 = 2$, $l_1 = l_2 = 1$) while the set of $1s2s$ are non-equivalent ($n_1 \neq n_2$ although $l_1 = l_2$), as are, for example, $2s2p$ ($n_1 = n_2$ but $l_1 \neq l_2$). Special care is necessary when considering the total angular momentum of equivalent electrons since restrictions are placed on the values of the quantum numbers which each may have. Let us consider the case of $2p^2$ in some detail.

The first restriction arises from Pauli's principle (Sec. 5.3.1). Since we have $n_1 = n_2$ and $l_1 = l_2$ we cannot simultaneously choose $l_{1_z} = l_{2_z}$ and $s_{1_z} = s_{2_z}$.

Further restrictions follow from physical considerations. The basic principle is that electrons cannot be distinguished from each other and so if the energies of two electrons are exchanged we have no way of discovering experimentally that such exchange has taken place. The implication is that if the values of all four numbers n, l, l_z, and s_z for each of two electrons are exchanged, the initial situation is identical in every way with the final. When considering total momentum and the term symbols of atoms we are interested only in different situations and we must not count twice those systems which are interconvertible merely by an exchange of all four numbers, n, l, l_z, and s_z. Consider some examples chosen from the $2p^2$ case:

1. We have $n_1 = n_2$, $l_1 = l_2$ and if we also choose $l_{1_z} = l_{2_z}$ then we know (Pauli's principle) that if $s_{1_z} = +\frac{1}{2}$, then $s_{2_z} = -\frac{1}{2}$; alternatively, if $s_{1_z} = -\frac{1}{2}$ then $s_{2_z} = +\frac{1}{2}$. Now these two cases are completely identical because one can be reached from the other by exchanging n, l, l_z, and s_z. Thus, while we can consider either set alone as typical of the state, we must not consider both together.

2. Similarly, if we assume $s_{1_z} = s_{2_z}$ then we know $l_{1_z} \neq l_{2_z}$. For p electrons $l = 1$ and hence $l_z = 1$, 0, or -1. So we might have:

$$s_{1_z} = s_{2_z}: \quad l_{1_z} = 1 \qquad l_{2_z} = 0 \text{ or } -1$$
$$l_{1_z} = 0 \qquad l_{2_z} = 1 \text{ or } -1$$
$$l_{1_z} = -1 \qquad l_{2_z} = 1 \text{ or } 0$$

Note, however, that the system represented by the pair of values (1, 0) for (l_{1_z}, l_{2_z}) is identical with that for (0, 1); (0, -1) is identical with (-1, 0) and (1, -1) with (-1, 1). Thus we reduce the above six pairs to only three *different* sets:

$$s_{1_z} = s_{2_z}: \quad l_{1_z} = 1 \qquad l_{2_z} = 0$$
$$l_{1_z} = 1 \qquad l_{2_z} = -1$$
$$l_{1_z} = 0 \qquad l_{2_z} = -1$$

3. Finally we note that if $l_{1_z} \neq l_{2_z}$ and $s_{1_z} \neq s_{2_z}$, then interchange of only *one* pair of values (for example $s_{1_z} = +\frac{1}{2}$, $s_{2_z} = -\frac{1}{2}$, $\rightarrow s_{1_z} = -\frac{1}{2}$, $s_{2_z} = +\frac{1}{2}$) does produce a *different* situation; physically an electron already distinguishable by its l_z value is being reversed in spin. All four numbers n, l, l_z, and s_z must be exchanged to produce an indistinguishable state.

Table 5.3 Substates of two equivalent p electrons ($n_1 = n_2 = 2$; $l_1 = l_2 = 1$)

l_{1_z}	l_{2_z}	s_{1_z}	s_{2_z}	L_z $(l_{1_z} + l_{2_z})$	S_z $(s_{1_z} + s_{2_z})$	Substate
+1	+1	$+\frac{1}{2}$	$-\frac{1}{2}$	+2	0	(a)
+1	0	$+\frac{1}{2}$	$+\frac{1}{2}$	+1	+1	(b)
+1	0	$+\frac{1}{2}$	$-\frac{1}{2}$	+1	0	(c)
+1	0	$-\frac{1}{2}$	$+\frac{1}{2}$	+1	0	(d)
+1	0	$-\frac{1}{2}$	$-\frac{1}{2}$	+1	−1	(e)
+1	−1	$+\frac{1}{2}$	$+\frac{1}{2}$	0	+1	(f)
+1	−1	$+\frac{1}{2}$	$-\frac{1}{2}$	0	0	(g)
+1	−1	$-\frac{1}{2}$	$+\frac{1}{2}$	0	0	(h)
+1	−1	$-\frac{1}{2}$	$-\frac{1}{2}$	0	−1	(i)
0	0	$+\frac{1}{2}$	$-\frac{1}{2}$	0	0	(j)
0	−1	$+\frac{1}{2}$	$+\frac{1}{2}$	−1	+1	(k)
0	−1	$+\frac{1}{2}$	$-\frac{1}{2}$	−1	0	(l)
0	−1	$-\frac{1}{2}$	$+\frac{1}{2}$	−1	0	(m)
0	−1	$-\frac{1}{2}$	$-\frac{1}{2}$	−1	−1	(n)
−1	−1	$+\frac{1}{2}$	$-\frac{1}{2}$	−2	0	(o)

Keeping these rules in mind we can construct Table 5.3, in which the first four columns list combinations of l_{1_z}, l_{2_z}, s_{1_z}, and s_{2_z} leading to different energy states (let us call them substates). We are interested in the *total* energy and so we show in the next two columns the values of $l_{1_z} + l_{2_z} = L_z$ and $s_{1_z} + s_{2_z} = S_z$, respectively. The final column merely supplies a convenient label to each substate for the following discussion. The 15 substates in the table constitute the z components of the various **L** and **S** vectors which may be formed from two equivalent p electrons. We can find the term symbols in the following way:

1. Note first that the largest L_z value in the table is $L_z = +2$ (substate (a)), and this is associated with $S_z = 0$. $L_z = +2$ must be a z component of the state $L = 2$, i.e. one component of a D state, the other components of which are $L_z = +1$, 0, −1, and −2. In the table we can find several substates of the requisite L_z values, all associated with $S_z = 0$; it is immaterial which of the alternatives we choose, so let us take substates (c), (g), (l), and (o). Since $S_z = 0$ the state must be singlet; hence we have:

$$^1D = \text{substates } (a), (c), (g), (l), \text{ and } (o)$$

   With $L = 2$, $S = 0$ this can only be a $^1D_2$ state.
2. Of the remaining substates the largest L_z is $+1$ associated with $S_z = +1$ (substate (b)). This is plainly one component of a ^3P state, the other components of which may be selected as:

$$^3P \text{ state:} \quad S_z = +1: \quad L_z = 1, 0, \text{ and } -1, \text{ that is } (b), (f), (k)$$
$$S_z = 0: \quad L_z = 1, 0, \text{ and } -1, \text{ that is } (d), (h), (m)$$
$$S_z = -1: \quad L_z = 1, 0, \text{ and } -1, \text{ that is } (e), (i), (n)$$

Here we have considered the three components of $S = 1$, $S_z = +1$, 0, and −1, to be associated in turn with the components of $L = 1$. The three states listed correspond to term symbols $^3P_2$, $^3P_1$, and $^3P_0$.

3. Finally the one remaining substate, (j), has $L_z = S_z = 0$ and this plainly comprises a $^1S_0$ state.

Overall, then, two equivalent p electrons give rise to the three different energy states ^1D, ^1S, and ^3P, of which the latter, being a triplet state, has three close energy levels $^3P_2$, $^3P_1$, and $^3P_0$. Hund's rule, which we quoted in Sec. 5.3.1, may be expressed for equivalent electrons as:

> The state of lowest energy for a given electronic configuration is that having the *greatest multiplicity*. If more than one state has the same multiplicity then the lowest of these is that with the greatest L value.

Thus for carbon the ground state is the ^3P, the next in energy is the ^1D, and finally the ^1S. We note that this new expression of Hund's rule implies that electrons in degenerate orbitals tend to have their spins parallel (since this gives the greatest multiplicity, and hence lowest energy); this in turn means that electrons tend to go into separate orbitals since in the same orbital they must have paired spins. Thus we are justified in writing the electronic structures of carbon, nitrogen, and oxygen as in Table 5.2.

If now one of the $2p$ electrons of carbon is promoted to the $3p$ state we have the configuration $1s^2 2s^2 2p3p$. This is an excited state in which the p electrons are non-equivalent. The interested student should show, by the method of Table 5.3, that six different term symbols can be found for this configuration, that is ^1S, ^1P, ^1D, ^3S, ^3P, and ^3D. In this case, since $n_1 \neq n_2$, neither the Pauli principle nor the principle of indistinguishability offers restrictions to l_z and s_z values, and hence more terms result.

For non-equivalent electrons, however, it is simpler to deal directly with L and S values. Thus we have $s_1 + s_2 = 1$ or 0 depending upon whether the electron spins are parallel or opposed, while for $l_1 = l_2 = 1$ we can have $L = 2$, 1 or 0. We can then tabulate L, S, and J directly, together with their term symbols:

L	S	J	Term symbol
2	1	3, 2, or 1	$^3D_{3,\,2,\,1}$
2	0	2	$^1D_2$
1	1	2, 1, or 0	$^3P_{2,\,1,\,0}$
1	0	1	$^1P_1$
0	1	1	$^3S_1$
0	0	0	$^1S_0$

and we arrive at the six states listed previously. Note that this direct method is not applicable to equivalent electrons because summation of l to give L implies that all l_z are allowed: this, we have seen, is not true when the electrons are equivalent.

Many other electronic configurations occur, both for carbon and other atoms, in which two or more equivalent electrons contribute to the total energy. We shall not discuss these further, however, except to state that their total energies and term symbols may be discovered by the same process as exemplified above for $2p^2$ electrons.

We can, however, now conveniently discuss rather more fully the operation of the helium–neon laser which was mentioned briefly in Sec. 1.10. This is an example of a continuous laser; by means of an electric discharge the helium atoms in a mixture of helium and neon are excited and ionized. Those which are excited into singlet states decay by emitting radiation until they arrive in the ground state once more, ready for re-excitation; those excited to triplet states, however (see Fig. 5.11), can decay only as far as the $1s2s\ ^3S_1$ state, which is metastable, since the selection rule $\Delta S = 0$ prevents its reversion to the ground $1s^2\ ^1S_0$ state. The $1s2s\ ^3S_1$ state is about $160\,000\,\text{cm}^{-1}$ above the ground state.

Turning now to the other component of the mixture, neon, this has a ground state configuration $1s^2 2s^2 2p^6\ ^1S_0$; it happens that one of its excited states, the $1s^2 2s^2 2p^5 4s^1$, where one of the $2p$ electrons has been promoted to the $4s$ orbital, is very nearly $160\,000\ cm^{-1}$ above the ground state, so collisions between excited helium atoms and ground state neon atoms can result in a resonance exchange of energy:

$$He^* + Ne \rightarrow He + Ne^*$$

Thus the electric discharge essentially pumps *neon* atoms into an excited state. This state can undergo spontaneous decay to lower singlet states, but here the induced decay is quite important and, if radiation of about $8700\ cm^{-1}$ is present, the decay to $1s^2 2s^2 2p^5 3p^1$ is induced, while radiation of $15\,800\ cm^{-1}$ results in decay to $1s^2 2s^2 2p^5 3s^1$—it is this latter which gives the usual 632.8 nm radiation from this laser.

The presence of radiation of the appropriate frequency is ensured by keeping the helium–neon mixture at low pressure in a tube placed between a pair of highly efficient mirrors. Thus the majority of the radiation is repeatedly reflected up and down the tube, and it is only the one per cent or so which 'escapes' through the mirrors that constitutes the useful output from the laser. Nonetheless, because all the available power is concentrated into a very narrow, highly monochromatic, and coherent beam, these lasers are increasingly used as sources of light and power.

5.5 PHOTOELECTRON SPECTROSCOPY

Photoelectron spectroscopy (PES) is an excellent technique for probing atomic and molecular electronic energy levels. When an atom or molecule is subjected to high-energy radiation, photons in the radiation collide with and eject electrons from atoms, leaving behind ions. Ejected electrons depart with different velocities and photoelectron spectroscopy measures the velocity distribution of the released electrons. In this section we describe the technique generally and concentrate on its effect on individual atoms (either alone or in a molecule); we shall consider in Chapter 6 (Sec. 6.5) features arising from the vibration of molecules undergoing ionization by PES and the effect of chemical bonding.

The process involved in photoelectron spectroscopy is sketched in Fig. 5.13, where we depict a few electrons (shown as circles) in some of the energy levels of an atom. Each electron is held in place by the nucleus with a characteristic *binding energy*—a term used in this context in preference to *ionization potential* which is usually taken to mean the *least* energy required to remove an electron from the atom, i.e. the energy needed to remove an electron from the *highest* occupied atomic orbital. For PES we must consider removing an electron from *any* orbital—hence the use of the term binding energy. A photon (depicted as a wave with a particular energy $h\nu$) is seen arriving from the left of Fig. 5.13 and colliding with an electron. The energy of the photon is imparted to the electron and, if this energy is greater than the binding energy, the electron will leave the atom and carry with it the excess energy—thus it will have a certain kinetic energy (and velocity). Clearly the total energy must be conserved, so we have:

$$h\nu = \text{binding energy} + \text{kinetic energy}$$

or
$$\text{Binding energy} = h\nu - \text{kinetic energy} \qquad (5.26)$$

and it follows that, provided we know the energy of the monochromatic exciting radiation, we can measure the binding energies of electrons in the atom under examination by observing the kinetic energies with which they leave.

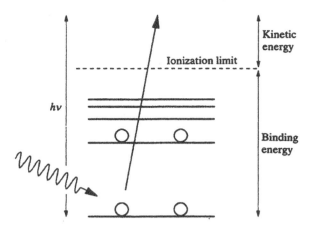

Figure 5.13 The principle of photoelectron spectroscopy.

Electrons can be ejected from either the core or the valence levels of the atom, depending on the energy of the exciting radiation, and their kinetic energies are characteristic of the atom from which they have been emitted. If the atom is part of a molecule, the energy of its valence electrons (and, to a much lesser extent, its core electrons) will be modified by the bonding, but are nonetheless characteristic of the orbital in question.

When ionization occurs the kinetic energy acquired by the emitted electrons is not quantized, so that any incident energy higher than the ionization energy can be used for excitation. Experimentally a monochromatic radiation source is ideal or one with only a few sharp emission frequencies. A helium lamp (emission wavelength 58.4 nm in the ultra-violet region, which has an equivalent photon energy of 21.2 eV) is commonly used for excitation of valence electrons, and the technique is then referred to as *ultra-violet photoelectron spectroscopy*, or UPES. A higher energy is required to ionize inner core electrons, however, for which an X-ray source is suitable; the technique is then called *X-ray photoelectron spectroscopy* (or XPES). In this case the X-ray beam is produced by electron bombardment of a clean metal target, such as Al or Mg, resulting in the emission of radiation at very specific energies; e.g. the k_α line for Al occurs at 1486.6 eV. If a very high intensity source is required, synchrotron radiation can be used (see Chapter 1, Sec. 1.11), which has the further advantage that its emission frequency can be tuned over a range.

Experimentally, the detection of electrons must be carried out in a high vacuum, as electrons are chemically active. There are no known window materials for He lamp sources, and so the lamp is usually mounted in a chamber which has a small hole into the sample compartment through which the emitted radiation can pass; continuous pumping maintains the high vacuum as helium leaks slowly through the hole. Obviously, the requirement for high vacuum means that measurements on solid samples are far easier than those on liquids and gases. However, modifications such as the use of flow techniques have ensured that a useful body of data has been obtained for the latter.

Figure 5.14 shows the binding energies of electrons in the energy levels of three different elements. The levels are labelled with their term symbols (1s, 2p, etc.) and with the X-ray notation which is sometimes used in PES. In this notation the levels are given letters corresponding to their shell:

$$\text{For } n = \qquad 1 \quad 2 \quad 3 \quad 4 \quad \ldots, \text{ etc.}$$
$$\text{X-ray notation} \quad K \quad L \quad M \quad N \quad \ldots, \text{ etc.}$$

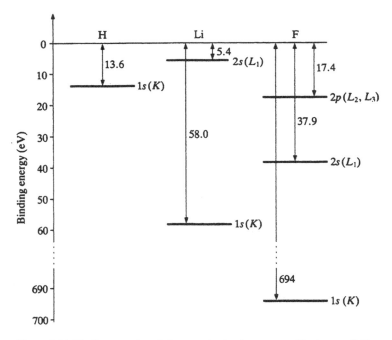

Figure 5.14 Binding energies of electrons in the free atoms of hydrogen, lithium, and fluorine; the X-ray notation for each level is also shown.

If necessary, a subscript denotes the sublevel from which the electron is ejected. The following examples show both notations for a few selected states:

Term symbol	$^1S_0$	$^2S_0$	$^2P_{1/2}$	$^2P_{3/2}$
X-ray notation	K	L_1	L_2	L_3

In both XPES and UPES the kinetic energy of the ejected electrons is measured using a hemispherical analyser such as that shown in Fig. 5.15. Monochromatic X-ray or ultra-violet

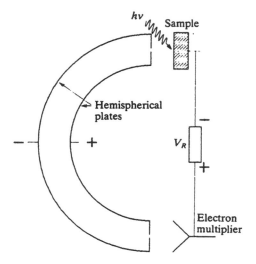

Figure 5.15 Schematic diagram of a photoelectron spectrometer.

radiation falls on the sample and ejected electrons pass between a pair of electrically charged hemispherical plates which act as an energy filter, allowing electrons of only a particular kinetic energy to pass through—the pass energy, E_{pass}. The resulting electron current, measured by an electron multiplier, indicates the number of electrons ejected from the surface with that kinetic energy. E_{pass} can be systematically varied by changing the retarding voltage (V_R) applied to the analyser. Thus a photoelectron spectrum is a plot of the number of electrons emitted against their kinetic energy.

Figure 5.16 shows part of the XPES and UPES spectra of argon with an illustration of the

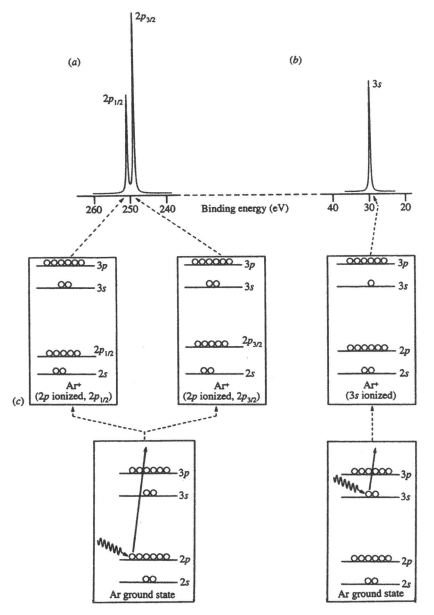

Figure 5.16 (*a*) The X-ray and (*b*) the ultra-violet photoelectron spectra of gaseous argon. (*c*) The energy levels of the states from which the spectra in (*a*) and (*b*) arise.

associated levels. A number of peaks are observed which are characteristic of the Ar^+ resulting from the removal of a single electron from each of the different levels. For example, an electron removed from the $2p$ orbital leaves behind a $1s^2 2s^2 2p^5 3s^2 3p^6$ configuration which, because of the possible combinations of l and s values, results in two slightly different energy states, $^2P_{3/2}$ and $^2P_{1/2}$. Of these, the $^2P_{3/2}$ is slightly higher in energy and so, when this state is formed, the electron leaving carries with it slightly *less* kinetic energy. This j-splitting can be seen to give rise to two well-resolved peaks in the spectrum.

The probability of ejecting an electron from a particular energy level will obviously depend on the number of electrons originally in that level, i.e. on its degeneracy. Under ideal operating conditions, the area under a PES peak is a measure of that degeneracy. Comparison of the total areas under the peaks resulting from excitation of the $3s$ and $2p$ electrons in Fig. 5.16 shows a ratio of about 1:3, as expected (two electrons in $3s$ and six in $2p$). It is also evident from the 250 eV peak that there is twice the chance of forming an ion in the $^2P_{3/2}$ state compared to ion formation in the $^2P_{1/2}$ state; this is because each level has a degeneracy of $2J + 1$, which is four for the $^2P_{3/2}$ state and two for $^2P_{1/2}$.

5.6 THE ZEEMAN EFFECT

We have been concerned in this chapter with two sorts of electronic energy. Firstly, there is energy of *position*—energy arising by virtue of interaction between electrons and the nucleus and between electrons and other electrons in the same atom. This energy can be described in terms of the n and l quantum numbers, although we have discussed it, rather less precisely, by drawing energy-level diagrams. Secondly, there is energy of *motion*—energy arising from the summed orbital and spin momenta of the electrons in the atom which depends on the l_z and s_z values of each electron and the way in which these are coupled. This gives rise to the fine structure of spectroscopic lines discussed earlier.

Angular momentum can be considered as arising from a physical movement of electrons about the nucleus and, since electrons are charged, such motion constitutes a circulating electric current and hence a magnetic field. This field can, indeed, be detected, and it is its interaction with exterior fields which is the subject of this section.

We can represent the angular momentum field by a vector μ—the magnetic dipole of the atom—and it is readily shown that μ is directly proportional to the angular momentum **J** and has the same direction. If the electron is considered as a point of mass m and charge e, then we have:

$$\mu = -\frac{e}{2m} \mathbf{J} \quad \mathrm{J\,T^{-1}}$$

(Here we use the SI unit of magnetic field, the tesla (T), which is equivalent to 10 000 gauss in electromagnetic units.) However, quantum mechanics indicates that the electron is not a point charge and a more exact expression for μ is:

$$\mu = -\frac{ge}{2m} \mathbf{J} = -\frac{ge}{2m} \sqrt{J(J+1)} \frac{h}{2\pi} \quad \mathrm{J\,T^{-1}} \tag{5.27}$$

where g is a purely numerical factor, called the Landé splitting factor. This factor depends on the state of the electrons in the atom and is given by:

$$g = \frac{3}{2} + \frac{S(S+1) - L(L+1)}{2J(J+1)} \tag{5.28}$$

In general g lies between 0 and 2.

We now recall (cf. Eq. (5.13) for one electron) that **J** can have either integral or half-integral components J_z along a reference direction, depending upon whether the quantum number J is integral or half-integral. Figure 5.17(a) shows this for a state with $J = \frac{3}{2}$, the $2J + 1$ components being given in general by:

$$J_z = J, J - 1, \ldots, \tfrac{1}{2} \text{ or } 0, \ldots, -J \tag{5.29}$$

Further, since μ is proportional to **J**, μ will also have components in the z direction which are given by:

$$\mu_z = -\frac{ge}{2m}\frac{h}{2\pi} J_z \tag{5.30}$$

These are shown diagrammatically in Fig. 5.17(b). If now an external field is applied to the atom, thus specifying the previously arbitrary z direction, the atomic dipole μ will interact with the applied field to an extent depending on its component in the field direction. If the strength of the applied field is B_z then the extent of the interaction is simply $\mu_z B_z$:

$$\text{Interaction} = \Delta E = \mu_z B_z = -\frac{heg}{4\pi m} B_z \quad \text{J} \tag{5.31}$$

In this equation we have expressed the interaction as ΔE since the application of the field splits the originally degenerate energy levels corresponding to the $2J + 1$ values of J_z into $2J + 1$ *different* energy levels. This is shown for $J = \frac{3}{2}$ in Fig. 5.17(c). It is this splitting, or lifting of the degeneracy on the application of an external magnetic field, which is called the *Zeeman effect* after its discoverer.

The energy splitting is very small; the factor $he/4\pi m$ in Eq. (5.31), known as the *Bohr magneton*, has a value of $9.27 \times 10^{-24}\,\mathrm{J\,T^{-1}}$; thus for $g = 1$, and for an applied field B_z of one tesla (that is 10 000 gauss), the interaction energy is only some 10^{-23} joules, which in turn is of the order of $0.5\,\mathrm{cm^{-1}}$. This small splitting is, of course, reflected in a splitting of the spectral transitions observed when a magnetic field is applied to an atom. In order to discuss the effect on the spectrum we need one further selection rule:

$$\Delta J_z = 0, \pm 1$$

Let us consider the doublet lines in the sodium spectrum produced, as we have discussed in Sec. 5.3.2, by transitions between the $^2S_{1/2}$ states and the $^2P_{1/2}$ and $^2P_{3/2}$ states. When a field B_z is applied to the atom, the $^2S_{1/2}$ and $^2P_{1/2}$ states are both split into two (since $J = \frac{1}{2}$, $2J + 1 = 2$), while the $^2P_{3/2}$ is split into four. The extent of the splitting (Eq. (5.31)) is proportional to the g factor in each state and, from Eq. (5.28), we can easily calculate:

$$^2S_{1/2}: \quad S = \tfrac{1}{2},\ L = 0,\ J = \tfrac{1}{2}, \text{ hence } g = 2$$
$$^2P_{1/2}: \quad S = \tfrac{1}{2},\ L = 1,\ J = \tfrac{1}{2}, \text{ hence } g = \tfrac{2}{3}$$
$$^2P_{3/2}: \quad S = \tfrac{1}{2},\ L = 1,\ J = \tfrac{3}{2}, \text{ hence } g = 1\tfrac{1}{3}$$

and we see that the $^2S_{1/2}$, $^2P_{1/2}$, and $^2P_{3/2}$ levels are split in the ratio of $3 : 1 : 2$. We show the situation in Fig. 5.18. On the left of the figure we see the energy levels and transitions *before* the field B_z is applied; the levels are unsplit and the spectrum is a simple doublet. On the right we see the effect of the applied field. The spectrum shows that the original line due to the $^2S_{1/2} \to {}^2P_{1/2}$ transition disappears and is replaced by *four* new lines, while the $^2S_{1/2} \to {}^2P_{3/2}$ transition is replaced by *six* new lines.

The effect described above is usually referred to as the *anomalous Zeeman effect*—although, in fact, most atoms show the effect in this form. The *normal Zeeman effect* applies to transitions

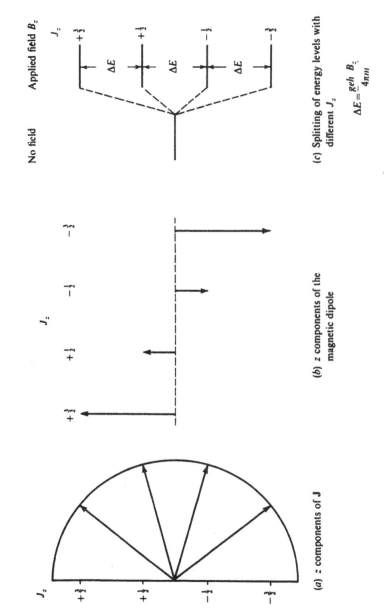

(a) z components of **J**

(b) z components of the
magnetic dipole

(c) Splitting of energy levels with
different J_z

$$\Delta E = \frac{g^{eh}}{4\pi m} B_z$$

Figure 5.17 The effect of an applied magnetic field on the energy levels of an electron with $J = \frac{3}{2}$. In (a) the $2J + 1 = 4$ components of J are shown and in (b) their corresponding magnetic moments in the reference direction. An external field splits the originally degenerate levels into four separate levels as in (c).

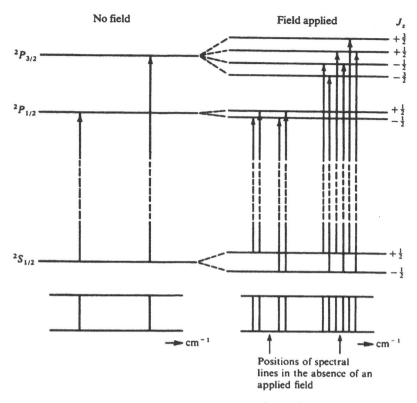

Figure 5.18 The Zeeman effect on transitions between ^2S and ^2P states. The situation before the field is applied is shown on the left, that after on the right.

between singlet states only (e.g. the transitions of electrons in the helium atom shown on the left of Fig. 5.11). For singlet states we have:

$$2S + 1 = 1 \qquad \text{hence } S = 0$$

Therefore:

$$J = L \quad \text{and} \quad g = 1 \quad \text{(cf. Eq. (5.28))}$$

Thus the splitting between all singlet levels is identical for a given applied field and the corresponding Zeeman spectrum is considerably simplified.

In general, the Zeeman effect can give very useful information about the electronic states of atoms. In the first place, the number of lines into which each transition becomes split when a field is applied depends on the J value of the states between which transitions arise. Next the g value, deduced from the splitting for a known applied field, gives information about the L and S values of the electron undergoing transitions. Overall, then, the term symbols for various atomic states can be deduced by Zeeman experiments. In this way all the details of atomic states, term symbols, etc., discussed above, have been amply confirmed experimentally.

5.7 THE INFLUENCE OF NUCLEAR SPIN

The nuclei of many atoms are known to be spinning about an axis. We shall discuss at some length in Chapter 7 the spectrum which this spin may give rise to in the radiofrequency region, but it is pertinent here to consider very briefly what effect such spin may have on the electronic spectra of atoms.

The nuclear spin quantum number I may be zero, integral, or half-integral depending on the particular nucleus considered. Thus the nuclear angular momentum, given by

$$\mathbf{I} = \sqrt{I(I+1)} \, \frac{h}{2\pi} = \sqrt{I(I+1)} \text{ units} \tag{5.32}$$

can have values 0, $\sqrt{3/2}$, $\sqrt{2}$, $\sqrt{15/2}$, etc.

The effect of $\mathbf{I}$ on the spectrum can be understood if we define the total momentum (electronic + nuclear) of an atom by $\mathbf{F}$:

$$\mathbf{F} = \sqrt{F(F+1)} \, \frac{h}{2\pi} = \sqrt{F(F+1)} \text{ units} \tag{5.33}$$

where F is the *total momentum quantum number*. If, as before, J is the total electronic quantum number, then we may write

$$F = J + I, \ J + I - 1, \ldots |J - I| \tag{5.34}$$

thus giving $2J + 1$ or $2I + 1$ different energy states, whichever is the less.

The energy-level splitting due to nuclear spin is of the order of 10^{-3} that due to electron spin; thus extremely fine resolving power is necessary for its observation and it is normally referred to as *hyperfine structure*.

5.8 CONCLUSION

This completes all we have to say about atomic spectroscopy. In the next chapter we extend the ideas introduced here to cover the electronic spectra of simple molecules, and we shall briefly discuss the techniques of electronic spectroscopy.

BIBLIOGRAPHY

See the Bibliography to Chapter 6, with the following additions:

Harris, D. C., and M. D. Bertolucci: *Symmetry and Spectroscopy*, Dover Publications, 1989.
Herzberg, G.: *Atomic Spectra and Atomic Structure*, Dover, 1944.
Kuhn, M. G.: *Atomic Spectra*, Academic Press, 1962.
Shore, B. W., and D. H. Menzel: *Principles of Atomic Spectra*, Wiley, 1968.

PROBLEMS

(Useful constants: $R = 109\,677.581 \text{ cm}^{-1}$; $1 \text{ cm}^{-1} \equiv 11.958 \text{ J mol}^{-1}$.)

5.1 Calculate the first three lines in the absorption spectrum arising from transitions from the $3s$ level of the hydrogen atom. What is the ionization energy of this level?

5.2 The term symbol for a particular atomic state is quoted as $^4D_{5/2}$. What are the values of L, S, and J for this state? What is the minimum number of electrons which could give rise to this? Suggest a possible electronic configuration.

5.3 What are the term symbols for the following pairs of non-equivalent electrons: (a) ss, (b) pp, (c) sd, and (d) pd?

5.4   What are the term symbols for the following pairs of equivalent electrons: (a) s^2, (b) p^2, and (c) d^2?

5.5   The term symbols for particular states of three different atoms are quoted as $^4S_1$, $^2D_{7/2}$, and $^0P_1$; explain why these are erroneous.

5.6   Figure 5.7 shows the three transitions arising between ^2P and ^2D states. Into how many lines would each of these transitions split if a magnetic field were applied? (Assume that the g-value is different for each energy level.)

5.7   The ground state electron configuration of phosphorus is $1s^2 2s^2 2p^6 3s^2 3p^3$. Ignoring spin–orbit coupling, find the terms for this state and identify that which is lowest in energy.

5.8   Work out the lowest energy terms for the following atoms, including spin–orbit coupling (the essential part of the electron configuration is given in brackets):

Ru $(4d^7 5s)$; La $(5d6s^2)$; Nb $(4d^4 5s)$; Ca $(4s^2)$;
Eu $(4f^7 6s^2)$; Gd $(4f^7 5d6s^2)$; Ag $(4d^{10} 5s)$;
W $(5d^4 6s^2)$; Pt $(5d^9 6s)$.

5.9   What are the electron configurations for the ground state and first excited state of the Tl$^+$ ion? List the terms which arise from each of these configurations, including spin–orbit coupling. Place the levels in order of increasing energy. The observed energies of the lowest four levels are:

(a)   zero (by convention);   (b) $49\,451\,\text{cm}^{-1}$;
(c) $52\,393\,\text{cm}^{-1}$ (d) $61\,725\,\text{cm}^{-1}$.

According to the Russell–Saunders coupling scheme, the energy separation between adjacent spin–orbit coupling levels from a given term should be l times the larger value of J (e.g. $3l$ between levels with $J = 2$ and $J = 3$), where l is the spin–orbit coupling constant. On this basis, what would you expect the relative sizes of the energy gaps (b)–(c) and (c)–(d) to be, and why?

ELECTRONIC SPECTROSCOPY OF MOLECULES

In the first section of this chapter we shall discuss, in some detail, the electronic spectra of diatomic molecules. We shall find that the overall appearance of such spectra can be considered without assuming any knowledge of molecular structure, without reference to any particular electronic transition, and, indeed, with little more than a formal understanding of the nature of electronic transitions within molecules. In Sec. 6.2 we shall summarize modern ideas of molecular structure and show how these lead to a classification of electronic states analogous to the classification of atomic states discussed in the previous chapter. Section 6.3 will extend the ideas of Secs 6.1 and 6.2 to polyatomic molecules and Sec. 6.4 will deal briefly with experimental techniques. Section 6.5 discusses the application of photoelectron spectroscopy to molecular species.

6.1 ELECTRONIC SPECTRA OF DIATOMIC MOLECULES

6.1.1 The Born–Oppenheimer Approximation

As a first approach to the electronic spectra of diatomic molecules we may use the Born–Oppenheimer approximation previously mentioned in Sec. 3.2; in the present context this may be written:

$$E_{\text{total}} = E_{\text{electronic}} + E_{\text{vibration}} + E_{\text{rotation}} \tag{6.1}$$

which implies that the electronic, vibrational, and rotational energies of a molecule are completely independent of each other. We shall see later to what extent this approximation is invalid. A change in the total energy of a molecule may then be written:

$$\Delta E_{\text{total}} = \Delta E_{\text{elec.}} + \Delta E_{\text{vib.}} + \Delta E_{\text{rot.}} \quad \text{J}$$

or

$$\Delta \varepsilon_{\text{total}} = \Delta \varepsilon_{\text{elec.}} + \Delta \varepsilon_{\text{vib.}} + \Delta \varepsilon_{\text{rot.}} \quad \text{cm}^{-1} \tag{6.2}$$

The approximate orders of magnitude of these changes are:

$$\Delta\varepsilon_{\text{elec.}} \approx \Delta\varepsilon_{\text{vib.}} \times 10^3 \approx \Delta\varepsilon_{\text{rot.}} \times 10^6 \tag{6.3}$$

and so we see that vibrational changes will produce a 'coarse structure' and rotational changes a 'fine structure' on the spectra of electronic transitions. We should also note that whereas pure rotation spectra (Chapter 2) are shown only by molecules possessing a permanent electric dipole moment and vibrational spectra (Chapter 3) require a change of dipole during the motion, electronic spectra are given by *all* molecules since changes in the electron distribution in a molecule are always accompanied by a dipole change. This means that homonuclear molecules (for example H_2 or N_2), which show no rotation or vibration–rotation spectra, *do* give an electronic spectrum and show vibrational and rotational structure in their spectra from which rotational constants and bond vibration frequencies may be derived.

Initially we shall ignore rotational fine structure and discuss the appearance of the vibrational coarse structure of spectra.

6.1.2 Vibrational Coarse Structure: Progressions

Ignoring rotational changes means that we rewrite Eq. (6.1) as

$$E_{\text{total}} = E_{\text{elec.}} + E_{\text{vib.}} \quad \text{J}$$

or

$$\varepsilon_{\text{total}} = \varepsilon_{\text{elec.}} + \varepsilon_{\text{vib.}} \quad \text{cm}^{-1} \tag{6.4}$$

From Eq. (3.12) we can write immediately:

$$\varepsilon_{\text{total}} = \varepsilon_{\text{elec.}} + (v + \tfrac{1}{2})\bar{\omega}_e - x_e(v + \tfrac{1}{2})^2\bar{\omega}_e \quad \text{cm}^{-1} \qquad (v = 0, 1, 2, \ldots) \tag{6.5}$$

The energy levels of this equation are shown in Fig. 6.1 for two arbitrary values of $\varepsilon_{\text{elec.}}$. As in previous chapters the lower states are distinguished by a *double* prime (v'', $\varepsilon_{\text{elec.}}''$), while the upper states carry only a *single* prime (v', $\varepsilon_{\text{elec.}}'$). Note that such a diagram cannot show correctly the relative separations between levels of different $\varepsilon_{\text{elec.}}$, on the one hand, and those with different v' or v'', on the other (cf. Eq. (6.3)), but that the spacing between the upper vibrational levels is deliberately shown to be rather smaller than that between the lower; this is the normal situation since an excited electronic state usually corresponds to a weaker bond in the molecule and hence a smaller vibrational wavenumber $\bar{\omega}_e$.

There is essentially no selection rule for v when a molecule undergoes an electronic transition, i.e. every transition $v'' \to v'$ has some probability, and a great many spectral lines would, therefore, be expected. However, the situation is considerably simplified if the *absorption* spectrum is considered from the electronic ground state. In this case, as we have seen in Sec. 3.1.3, virtually all the molecules exist in the lowest vibrational state, that is $v'' = 0$, and so the only transitions to be observed with appreciable intensity are those indicated in Fig. 6.1. These are conventionally labelled according to their (v', v'') numbers (note: upper state *first*), that is (0, 0), (1, 0), (2, 0), etc. Such a set of transitions is called a *band* since, under low resolution, each line of the set appears somewhat broad and diffuse, and is more particularly called a v' *progression*, since the value of v' increases by unity for each line in the set. The diagram shows that the lines in a band crowd together more closely at high frequencies; this is a direct consequence of the anharmonicity of the upper state vibration which causes the excited vibrational levels to converge.

An analytical expression can easily be written for this spectrum. From Eq. (6.5) we have immediately:

$$\Delta\varepsilon_{\text{total}} = \Delta\varepsilon_{\text{elec.}} + \Delta\varepsilon_{\text{vib.}}$$

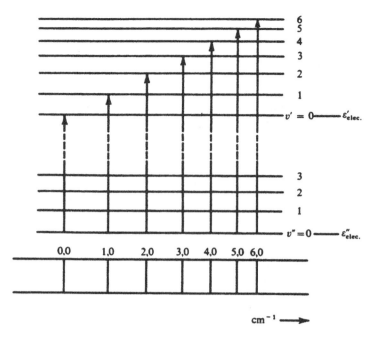

Figure 6.1 The vibrational 'coarse' structure of the band formed during electronic absorption from the ground $(v'' = 0)$ state to a higher state.

Therefore

$$\bar{v}_{\text{spec.}} = (\varepsilon' - \varepsilon'') + \{(v' + \tfrac{1}{2})\bar{\omega}'_e - x'_e(v' + \tfrac{1}{2})^2\bar{\omega}'_e\}$$
$$- \{(v'' + \tfrac{1}{2})\bar{\omega}''_e - x''_e(v'' + \tfrac{1}{2})^2\omega''_e\} \quad \text{cm}^{-1} \tag{6.6}$$

and, provided some half-dozen lines can be observed in the band, values for $\bar{\omega}'_e$, x'_e, $\bar{\omega}''_e$, and x''_e, as well as the separation between electronic states, $(\varepsilon' - \varepsilon'')$, can be calculated. Thus the observation of a band spectrum leads not only to values of the vibrational frequency and anharmonicity constant in the ground state ($\bar{\omega}''_e$ and x''_e), but also to these parameters in the excited electronic state ($\bar{\omega}'_e$ and x'_e). This latter information is particularly valuable since such excited states may be extremely unstable and the molecule may exist in them for very short times; nonetheless, the band spectrum can tell us a great deal about the bond strength of such species.

We shall see later that molecules normally have many excited electronic energy levels, so that the whole absorption spectrum of a diatomic molecule will be more complicated than Fig. 6.1 suggests: the ground state can usually undergo a transition to several excited states, and each such transition will be accompanied by a band spectrum similar to Fig. 6.1.

Further, in *emission spectra* the previously excited molecule may be in one of a large number of available (ε', v') states, and has a similar multitude of (ε'', v'') states to which it may revert. Thus emission spectra are usually extremely complicated, and a great deal of care and patience is needed for a complete analysis.

6.1.3 Intensity of Vibrational–Electronic Spectra: the Franck–Condon Principle

Although quantum mechanics imposes no restrictions on the change in the vibrational quantum number during an electronic transition, the vibrational lines in a progression are not all observed to be of the same intensity. In some spectra the (0, 0) transition is the strongest, in others the intensity increases to a maximum at some value of v', while in yet others only a few vibrational lines with high v' are seen, followed by a continuum. All these types of spectrum are readily explicable in terms of the *Franck–Condon principle* which states that *an electronic transition takes place so rapidly that a vibrating molecule does not change its internuclear distance appreciably during the transition.*

We have already seen in Chapter 3 how the energy of a diatomic molecule varies with internuclear distance (cf. Fig. 3.3). We recall that this figure, the Morse curve, represents the energy when one atom is considered fixed on the $r = 0$ axis and the other is allowed to oscillate between the limits of the curve. Classical theory would suggest that the oscillating atom would spend most of its time *on* the curve at the turning point of its motion, since it is moving most slowly there; quantum theory, while agreeing with this view for high values of the vibrational quantum number, shows that for $v = 0$ the atom is most likely to be found at the *centre* of its motion, i.e. at the equilibrium internuclear distance $r_{eq.}$. For $v = 1, 2, 3, \ldots$ the most probable positions steadily approach the extremities until, for high v, the quantal and classical pictures merge. This behaviour is shown in Fig. 6.2 where we plot the probability distribution in each vibrational state against internuclear distance. Those who have studied quantum mechanics will realize that Fig. 6.2 shows the variation of ψ^2 with internuclear distance, where ψ is the vibrational wave function.

If a diatomic molecule undergoes a transition into an upper electronic state in which the excited molecule is stable with respect to dissociation into its atoms, then we can represent the upper state by a Morse curve similar in outline to that of the ground electronic state. There will

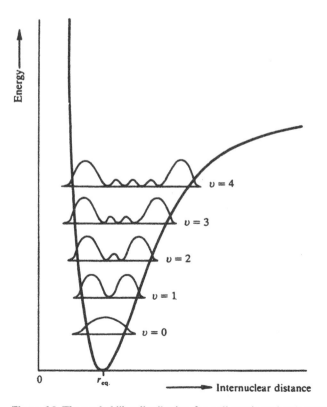

Figure 6.2 The probability distribution for a diatomic molecule according to the quantum theory. The nuclei are most likely to be found at distances apart given by the maxima of the curve for each vibrational state.

probably (but not necessarily) be differences in such parameters as vibrational frequency, equilibrium internuclear distance, or dissociation energy between the two states, but this simply means that we should consider each excited molecule as a new, but rather similar, molecule with a different, but also rather similar, Morse curve.

Figure 6.3 shows four possibilities. In Fig. 6.3(a) we show the upper electronic state having the same equilibrium internuclear distance as the lower. Now the Franck–Condon principle suggests that a transition occurs *vertically* on this diagram, since the internuclear distance does not change, and so if we consider the molecule to be initially in the ground state both electronically (ε'') and vibrationally ($v'' = 0$), then the most probable transition is that indicated by the vertical line in Fig. 6.3(a). Thus the strongest spectral line of the $v'' = 0$ progression will be the (0, 0). However, the quantum theory only says that the *probability* of finding the oscillating atom is greatest at the equilibrium distance in the $v = 0$ state—it allows some, although small, chance of the atom being near the extremities of its vibrational motion. Hence there is some chance of the transition starting from the ends of the $v'' = 0$ state and finishing in the $v' = 1, 2$, etc., states. The (1, 0), (2, 0), etc., lines diminish rapidly in intensity, however, as shown at the foot of Fig. 6.3(a).

In Fig. 6.3(b) we show the case where the excited electronic state has a *slightly* smaller internuclear separation than the ground state. A vertical transition from the $v'' = 0$ level will be most likely to occur into the upper vibrational state $v' = 2$, transitions to lower and higher v' states being less likely; in general the upper state most probably reached will depend on the difference between the equilibrium separations in the lower and upper states. In Fig. 6.3(c), the excited electronic state has a slightly larger internuclear separation than the ground state, but the resulting transitions and spectrum are similar (however, see Sec. 6.5).

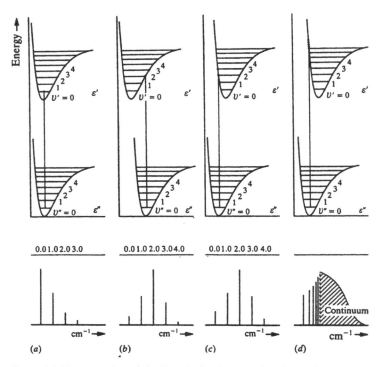

Figure 6.3 The operation of the Franck–Condon principle for (a) internuclear distances equal in the upper and lower states, (b) upper state internuclear distance a little less than that in the lower state, (c) upper state distance a little greater than in the lower, and (d) upper state distance considerably greater.

In Fig. 6.3(*d*) the upper state separation is drawn as *considerably* greater than that in the lower state and we see that, firstly, the vibrational level to which a transition takes place has a high v' value. Further, transitions can now occur to a state where the excited molecule has energy in excess of its own dissociation energy. From such states the molecule will dissociate without any vibrations and, since the atoms which are formed may take up any value of kinetic energy, the transitions are not quantized and a continuum results. This is shown at the foot of the figure. We consider the phenomenon of dissociation more fully in the next section.

The situation is rather more complex for emission spectra and for absorption from an excited vibrational state, for now transitions take place from *both ends* of the vibrational limits with equal probability; hence each progression will show two maxima which will coincide only if the equilibrium separations are the same in both states.

6.1.4 Dissociation Energy and Dissociation Products

Figure 6.4(*a*) and (*b*) shows two of the ways in which electronic excitation can lead to dissociation (a third way called *predissociation* will be considered in Sec. 6.1.7). Figure 6.4(*a*) represents the case, previously discussed, where the equilibrium nuclear separation in the upper state is considerably greater than that in the lower. The dashed line limits of the Morse curves represent the dissociation of the normal and excited molecule into atoms, the dissociation energies being D_0'' and D_0' from the $v = 0$ state in each case. We see that the total energy of the dissociation products (i.e. atoms) from the upper state is greater by an amount called $E_{ex.}$ than that of the products of dissociation in the lower state. This energy is the *excitation energy* of one (or rarely both) of the atoms produced on dissociation.

We saw in the previous section that the spectrum of this system consists of some vibrational transitions (quantized) followed by a continuum (non-quantized transitions) representing dissociation. The lower wavenumber limit of this continuum must represent just sufficient energy to cause dissociation and no more (i.e. the dissociation products separate with virtually zero kinetic energy) and thus we have

$$\bar{v}_{\text{continuum limit}} = D_0'' + E_{ex.} \quad \text{cm}^{-1} \tag{6.7}$$

Figure 6.4 Illustrating dissociation by excitation into (*a*) a stable upper state and (*b*) a continuous upper state.

and we see that we can measure D_0'', the dissociation energy, if we know $E_{ex.}$, the excitation energy of the products, whatever they may be. Now, although the excitation energy of atoms to various electronic states is readily measurable by atomic spectroscopy (cf. Chapter 5), the precise *state* of dissociation products is not always obvious. There are several ways in which the total energy $D_0'' + E_{ex.}$ may be separated into its components, however; here we shall mention just two.

Firstly, thermochemical studies often lead to an approximate value of D_0'' and hence, since $D_0'' + E_{ex.}$ is accurately measurable spectroscopically, a rough value for $E_{ex.}$ is obtained. When the spectrum of the atomic products is studied, it usually happens that only one value of excitation energy corresponds at all well with $E_{ex.}$. Thus the state of the products is known, $E_{ex.}$ measured accurately, and a precise value of D_0'' deduced.

Secondly, if more than one spectroscopic dissociation limit is found, corresponding to dissociation into two or more different states of products with different excitation energies, the separations between the excitation energies are often found to correspond closely with the separations between only one set of excited states of the atoms observed spectroscopically. Thus the nature of the excited products and their energies are immediately known.

In Fig. 6.4(b) we illustrate the case in which the upper electronic state is *unstable*: there is no minimum in the energy curve and, as soon as a molecule is raised to this state by excitation, the molecule dissociates into products with total excitation energy $E_{ex.}$. The products fly apart with kinetic energy $E_{kinetic}$ which represents (as shown on the figure) the excess energy in the final state above that needed just to dissociate the molecule. Since $E_{kinetic}$ is not quantized the whole spectrum for this system will exhibit a continuum, the lower limit of which (if observable) will be precisely the energy $D_0'' + E_{ex.}$. As before, if $E_{ex.}$ can be found from a knowledge of the dissociation products, D_0'' can be measured with great accuracy.

We shall see in Sec. 6.2.2 what sort of circumstances lead to the minimum in the upper state (Fig. 6.4(a)), on the one hand, or the *continuous* upper state (Fig. 6.4(b)), on the other.

In many electronic spectra no continua appear at all—the internuclear distances in the upper and lower states are such that transitions near to the dissociation limit are of negligible probability—but it is still possible to derive a value for the dissociation energy by noting how the vibrational lines converge. We have already seen in Chapter 3 (cf. Eq. (3.12)) that the vibrational energy levels may be written:

$$\varepsilon_v = (v + \tfrac{1}{2})\bar{\omega}_e - x_e(v + \tfrac{1}{2})^2\bar{\omega}_e \quad cm^{-1} \tag{6.8}$$

And so the separation between neighbouring levels, $\Delta\varepsilon$, is plainly:

$$\Delta\varepsilon = \varepsilon_{v+1} - \varepsilon_v$$
$$= \bar{\omega}_e\{1 - 2x_e(v + 1)\} \quad cm^{-1} \tag{6.9}$$

This separation obviously decreases linearly with increasing v and the dissociation limit is reached when $\Delta\varepsilon \to 0$. Thus the maximum value of v is given by $v_{max.}$, where:

$$\bar{\omega}_e\{1 - 2x_e(v_{max.} + 1)\} = 0$$

i.e.

$$v_{max.} = \frac{1}{2x_e} - 1 \tag{6.10}$$

We recall that the anharmonicity constant, x_e, is of the order of 10^{-2}; hence $v_{max.}$ is about 50.

We saw in Sec. 3.1.3 that two vibrational transitions (in the infra-red) were sufficient to determine x_e and $\bar{\omega}_e$. Thus, an example given there for HCl yielded $\bar{\omega}_e = 2990\,cm^{-1}$,

$x_e = 0.0174$. From Eq. (6.10) we calculate $v_{max.} = 27.74$ and the next lowest integer is $v = 27$. Replacing $v = 27$, $\bar{\omega}_e = 2990\,cm^{-1}$, and $x_e = 0.0174$ into Eq. (6.8) gives the maximum value of the vibrational energy as $42\,890\,cm^{-1}$ or $513.0\,kJ\,mol^{-1}$. This is to be compared with a more accurate value of $427.2\,kJ\,mol^{-1}$ evaluated thermochemically.

The discrepancy between these two figures arises from two causes. Firstly, the infra-red data only allow us to consider two or three vibrational transitions (the fundamental plus the first and second overtones). The electronic spectrum, as we have seen, shows many more vibrational lines (in fact the number is limited not by quantum restrictions but by the Franck–Condon principle) and we shall get a better value of D_0'' if we make use of these extra data. Secondly, we have assumed that Eq. (6.8) applies exactly even at high values of v; this is not true because cubic and even quartic terms become important at this stage. Because of these, $\Delta\varepsilon$ decreases more rapidly than Eq. (6.9) suggests.

Both these points may be met if we plot the *separation* between vibrational transitions, $\Delta\varepsilon$, as observed in the electronic spectrum, against the vibrational quantum number. Initially, Eq. (6.9) will apply quite accurately and the graph will be a straight line which may be extrapolated either to find $v_{max.}$ or, since the dissociation energy itself is simply the sum of all the increments $\Delta\varepsilon$ from $v = 0$ to $v = v_{max.}$, the *area* under the $\Delta\varepsilon$ versus v graph gives this energy directly. Such a linear extrapolation was first suggested by Birge and Sponer and is usually given their name.

On the other hand, if extensive data are available about a set of electronic–vibration transitions, the graph of $\Delta\varepsilon$ versus v will, at high v, begin to fall off more sharply as cubic and quartic terms become significant. In this case the most accurate determination of dissociation energy is obtained by extrapolating the smooth curve and finding the area beneath it. Figure 6.5 shows this process for data on iodine vapour given by R. D. Verma.

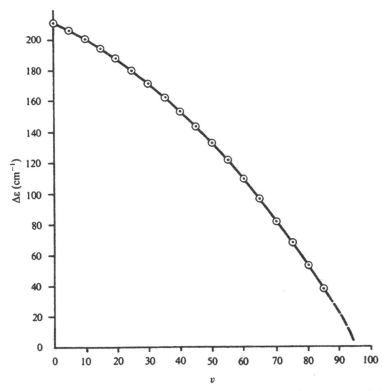

Figure 6.5 Birge–Sponer extrapolation to determine the dissociation energy of the iodine molecule, I_2. (*Taken from the data of R. D. Varma, J. Chem. Phys., **32**, p. 738, 1960, by kind permission of the author.*)

In absorption spectra it is normally the series of lines originating at $v'' = 0$ which is observed (cf. Fig. 6.1). Thus the convergence of the levels in the upper state and hence the dissociation energy of that state is normally found. While this in itself is of great interest, particularly since molecules in excited states usually revert to the ground state within fractions of a microsecond, the dissociation energy in the ground state can be found quite easily provided that, as before, the dissociation products and their excitation energy are known. Thus, in Fig. 6.4(a), if we know $E_{ex.}$ (from atomic spectroscopy) and D_0' (from Birge–Sponer extrapolation), and if we can measure the energy of the (0, 0) transition either directly or by calculation from the observed energy levels, we have:

$$D_0'' = \text{energy of (0, 0)} + D_0' - E_{ex.} \quad \text{cm}^{-1} \tag{6.11}$$

6.1.5 Rotational Fine Structure of Electronic–Vibration Transitions

So far we have seen that the electronic spectrum of a diatomic molecule consists of one or more series of convergent lines constituting the vibrational coarse structure on each electronic transition. Normally each of these 'lines' is observed to be broad and diffuse or, if the resolution is sufficiently good, each appears as a cluster of many very close lines. This is, of course, the rotational fine structure.

To a very good approximation we can ignore centrifugal distortion and we have the energy levels of a rotating diatomic molecule (cf. Eqs (2.11) and (2.12)) as:

$$\varepsilon_{rot.} = \frac{h}{8\pi^2 Ic} J(J+1) = BJ(J+1) \quad \text{cm}^{-1} \quad (J = 0, 1, 2, \dots) \tag{6.12}$$

where I is the moment of inertia, B the rotational constant, and J the rotational quantum number. Thus, by the Born–Oppenheimer approximation, the total energy (excluding kinetic of translation) of a diatomic molecule is:

$$\varepsilon_{total} = \varepsilon_{elec.} + \varepsilon_{vib.} + BJ(J+1) \quad \text{cm}^{-1} \tag{6.13}$$

Changes in the total energy may be written:

$$\Delta\varepsilon_{total} = \Delta\{\varepsilon_{elect.} + \varepsilon_{vib.}\} + \Delta\{BJ(J+1)\} \quad \text{cm}^{-1} \tag{6.14}$$

and the wavenumber of a spectroscopic line corresponding to such a change becomes simply:

$$\bar{v}_{spect.} = \bar{v}_{(v',v'')} + \Delta\{BJ(J+1)\} \quad \text{cm}^{-1} \tag{6.15}$$

where we write $\bar{v}_{(v'',v'')}$ to represent the wavenumber of an electronic–vibrational transition. This plainly corresponds to any *one* of the transitions, for example (0, 0) or (1, 0), etc., considered in previous sections. Here we are mainly concerned with $\Delta\{BJ(J+1)\}$.

The selection rule for J depends upon the type of electronic transition undergone by the molecule. We shall discuss these in more detail in Sec. 6.2.2; for the moment we must simply state that if both the upper and lower electronic states are $^1\Sigma$ states (i.e. states in which there is no electronic angular momentum about the internuclear axis), this selection rule is:

$$\Delta J = \pm 1 \text{ only} \quad \text{for } ^1\Sigma \rightarrow {}^1\Sigma \text{ transitions} \tag{6.16}$$

whereas for all other transitions (i.e. provided either the upper or the lower states (or both) have angular momentum about the bond axis) the selection rule becomes:

$$\Delta J = 0 \text{ or } \pm 1 \tag{6.17}$$

For this latter case there is the added restriction that a state with $J = 0$ cannot undergo a transition to another $J = 0$ state:

$$J = 0 \nleftrightarrow J = 0 \qquad (6.18)$$

Thus we see that for transitions between $^1\Sigma$ states, only P and R branches will occur, while for other transitions Q branches will appear in addition.

We can expand Eq. (6.15) as follows:

$$\bar{\nu}_{\text{spect.}} = \bar{\nu}_{(v',v'')} + B'J'(J' + 1) - B''(J'' + 1) \quad \text{cm}^{-1} \qquad (6.19)$$

where B' and J' refer to the upper electronic state, B'' and J'' to the lower. When we considered vibration–rotational spectra in Chapter 3, we saw (cf. Sec. 3.4) that the difference between B values in different vibrational levels was very small and could be ignored except in explaining finer details of the spectra. However, this is by no means the case in electronic spectroscopy: here we have seen, when discussing the Franck–Condon principle in Sec. 6.1.3, that equilibrium internuclear distances in the lower and upper electronic states may differ considerably, in which case the moments of inertia, and hence B values, in the two states will also differ. We cannot say *a priori* which of the two B values will be greater. Quite often the electron excited is one of those forming the bond between the nuclei; if this is so, the bond in the upper state will be weaker and probably longer (cf. Fig. 6.3(b) or (c)) so that the equilibrium moment of inertia increases during the transition and B decreases. Thus $B' < B''$. The reverse is sometimes true, however, e.g. when the electron is excited from an antibonding orbital (see Sec. 6.2.2).

We can discuss the rotational fine structure quite generally by applying the selection rules of Eqs (6.16), (6.17), and (6.18) to the expression for spectral lines, Eq. (6.19). We may note, in passing, that the treatment given here for the P and R branch lines is identical with that given in Sec. 3.4 for the vibration–rotation spectrum, except that there we were concerned with B_0 and B_1 i.e. B values in lower and upper *vibrational* states. Here our concern is with B values in lower and upper *electronic* states, B'' and B', and we also consider the formation of a Q branch.

Taking the P, R, and Q branches in turn:

1. P branch: $\Delta J = -1$, $J'' = J' + 1$

$$\Delta\varepsilon = \bar{\nu}_P = \bar{\nu}_{(v',v'')} - (B' + B'')(J' + 1) + (B' - B'')(J' + 1)^2 \quad \text{cm}^{-1}$$
$$\text{where } J' = 0, 1, 2, \ldots \qquad (6.20a)$$

2. R branch: $\Delta J = +1$, $J' = J'' + 1$

$$\Delta\varepsilon = \bar{\nu}_R = \bar{\nu}_{(v',v'')} + (B' + B'')(J'' + 1) + (B' - B'')(J'' + 1)^2 \quad \text{cm}^{-1}$$
$$\text{where } J'' = 0, 1, 2, \ldots \qquad (6.20b)$$

These two equations can be combined into:

$$\bar{\nu}_{P,R} = \bar{\nu}_{(v',v'')} + (B' + B'')m + (B' - B'')m^2 \quad \text{cm}^{-1}$$
$$\text{where } m = \pm 1, \pm 2, \ldots \qquad (6.20c)$$

with positive m values comprising the R branch (i.e. corresponding to $\Delta J = +1$) and negative values the P branch ($\Delta J = -1$). Note that m cannot be zero (this would correspond in, for example, the P branch to $J' = -1$ which is impossible) so no line from the P and R branch appears at the *band origin* $\bar{\nu}_{(v',v'')}$. We draw the appearance of the R and P branches separately in Fig. 6.6(a) and (b), respectively, taking a 10 per cent difference between the upper and lower B values and choosing $B' < B''$. Note that, *with this choice*, P branch lines occur on the *low* wavenumber side of the band origin and the spacing between the lines increases

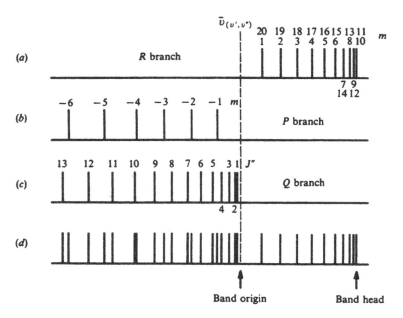

Figure 6.6 The rotational fine structure of a particular vibration–electronic transition for a diatomic molecule. The R, P, and Q branches are shown separately at (a), (b), and (c), respectively, with the complete spectrum at (d).

with m. On the other hand the R branch appears on the *high* wavenumber of the origin and the line spacing decreases rapidly with m—so rapidly that the lines eventually reach a maximum wavenumber and then begin to return to low wavenumbers with increasing spacing.† It will be remembered that in Sec. 3.4, a similar decrease in spacing was observed in the R branch but this was much too slow for a convergence limit to be reached; the rapid convergence here is due simply to the magnitude of $B' - B''$. The point at which the R branch separation decreases to zero is termed the *band head*.

3. Q branch: $\Delta J = 0$, $J' = J''$

$$\Delta \varepsilon = \bar{v}_Q = \bar{v}_{(v',v'')} + (B' - B'')J'' + (B' - B'')J''^2 \quad \text{cm}^{-1}$$

$$\text{where } J'' = 1, 2, 3, \ldots \tag{6.21}$$

Note that here $J'' = J' \neq 0$ since we have the restriction shown in Eq. (6.18). Thus again *no line will appear at the band origin*. We sketch the Q branch in Fig. 6.6(c), again for $B' < B''$ and a 10 per cent difference between the two. We see that the lines lie on the *low* wavenumber side of the origin and their spacing increases. The first few lines of this branch are not usually resolved.

The complete rotational spectrum is shown in Fig. 6.6(d). We have seen in Sec. 2.3.2 that many rotational levels are populated even at room temperature; consequently a large number of the P and R (and Q, where appropriate) lines will appear in the spectrum with comparable intensity. The spectrum is usually dominated by the band head, since here several of the R branch lines crowd together; for this reason, the Q branch is not very apparent if it occurs.

† The returning lines of the R branch coincide with earlier lines if Eq. (6.20(b)) is obeyed exactly. For real molecules cubic and quartic terms become important at high values of m.

In the situation we have been discussing ($B' < B''$), the band head appears in the R branch on the high wavenumber side of the origin; such a band is said to be *degraded* (or *shaded*) *towards the red*—i.e. the tail of the band where the intensity falls off points towards the red (low-frequency) end of the spectrum. If, on the other hand, $B' > B''$, then all our previous arguments are reversed. Briefly: (1) the Q branch spreads to *high* wavenumber, (2) the R branch (still, of course, on the *high* wavenumber side) consists of a series of lines with *increasing* separation, and (3) the band head appears in the P branch to *low* frequency of the origin. Such a band is *shaded to the violet*.

Normally, all the vibrational bands in any one electronic transition (e.g. the set of bands shown as a line spectrum in Fig. 6.1) are shaded in the same direction, while different electronic transitions in the same molecule may well show different shadings. Thus, observation of the shading may assist in the analysis of a complete spectrum. However, it may happen that different shadings are observed in bands belonging to the same electronic transition. This is because the B' and B'' values are not altogether independent of the vibrational state (as we have already seen in Sec. 3.4) so that, if $B' - B''$ is small, it may reverse sign for some higher vibrational levels. This behaviour is observed, for example, in the molecular fragment AlF, but is rare.

6.1.6 The Fortrat Diagram

We may rewrite the expressions for the P, R, and Q lines, Eqs (6.20c) and (6.21), with *continuously variable* parameters p and q:

$$\bar{\nu}_{P,R} = \bar{\nu}_{(v',v'')} + (B' + B'')p + (B' - B'')p^2 \qquad (6.22a)$$

$$\bar{\nu}_Q = \bar{\nu}_{(v',v'')} + (B' - B'')q + (B' - B'')q^2 \qquad (6.22b)$$

when we see that they each represent a parabola, p taking both positive and negative values, while q is positive only. We sketch these parabolae in Fig. 6.7 choosing, as before, $B' < B''$ and a difference of 10 per cent between them, and labelling regions of positive p with $\bar{\nu}_R$ and negative p with $\bar{\nu}_P$. These parabolae are usually referred to as the Fortrat parabolae. If we now illustrate the fact that p and q may in fact take only integral values (but not zero) by drawing circles round the allowed points on the parabolae, we can then read off the $\bar{\nu}$ values of the spectral lines directly from the graph. We show at the foot of the figure the first few lines of each branch with dotted leader lines connecting the spectrum and Fortrat diagram at intervals.

A useful property of the Fortrat diagram is that the band head is plainly at the vertex of the P, R parabola. We may calculate the position of the vertex by differentiation of Eq. (6.22a):

$$\frac{d\bar{\nu}_{P,R}}{dp} = B' + B'' + 2(B' - B'')p = 0$$

or

$$p = -\frac{B' + B''}{2(B' - B'')} \quad \text{for band head} \qquad (6.23)$$

Thus if $B' < B''$ (upper state has longer equilibrium bond length) the band head occurs at *positive* p values (i.e. in the R branch), the line at maximum wavenumber being given by the nearest positive integer to p. Conversely, for $B' > B''$ the band head occurs in the region of p negative, i.e. in the P branch. A simple calculation shows that for a 10 per cent difference between B' and B'' the band head occurs at $p \approx 10$.

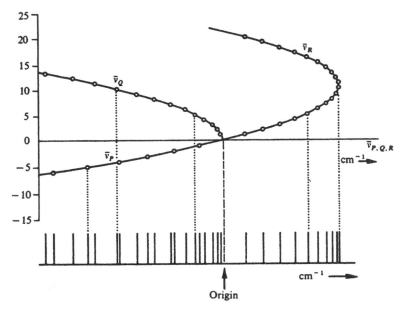

Figure 6.7 The Fortrat diagram sketched for a 10 per cent difference between B' and B'' (with $B' < B''$). The spectrum illustrated at the foot is identical with that of Fig. 6.6(d).

6.1.7 Predissociation

If a large number of vibrational transitions is observed for a particular molecule, it sometimes happens that the vibrational and rotational structures are quite distinct within a progression for large and small changes in the vibrational quantum number, but either the rotational structure is blurred or a complete continuum is observed for intermediate changes. A diagram showing the appearance of such a band is sketched in Fig. 6.8. A continuum at high wavenumber would correspond to ordinary dissociation (cf. Sec. 6.1.4) but the central continuum, occurring at energies well below the true dissociation limit, is referred to as *predissociation*.

Predissociation can arise when the Morse curves of a particular molecule in two different excited states intersect; one such possibility is shown in Fig. 6.9. One of the excited states is stable, since it has a minimum in the curve, and the other is continuous. Some of the vibrational levels are also shown, and let us suppose a transition takes places from some lower state into the vibrational levels shown bracketed on the left. Now if a transition takes place into the levels labelled a, b, or c, a normal vibrational–electronic spectrum occurs complete with rotational fine

Figure 6.8 Diagrammatic illustration of the appearance of predissociation. The rotational fine structure is clearly defined for vibrational transitions both above and below the predissociation region, but in this region the fine structure becomes blurred and lost.

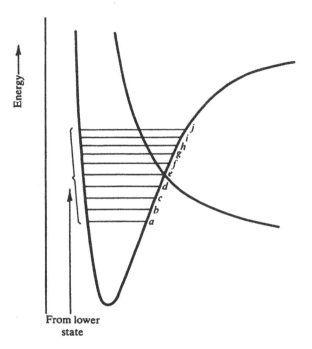

Figure 6.9 Showing the occurrence of predissociation during transitions into a stable upper state intersected by a continuous state.

structure; two such bands appear at the left of Fig. 6.8. If the transition is to levels *d*, *e*, or *f* there is a possibility that the molecule will 'cross over' on to the continuous curve and thus dissociate. In general, transition from one curve to another in this way (a so-called *radiationless transfer* since no energy is absorbed or emitted in the process) is faster than the time taken by the molecule to rotate ($\sim 10^{-10}$ s) but usually slower than the vibrational time ($\sim 10^{-13}$ s). Thus predissociation will occur before the molecule rotates (and thus all rotational fine structure will be destroyed in the spectrum), while the vibrational structure is usually not destroyed. If the cross-over is faster than the vibrational time, then a complete continuum will occur in the spectrum as shown in Fig. 6.8.

On the other hand, transitions into levels *g*, *h*, . . . will give rise to a normal vibrational–electronic spectrum including rotational fine structure once more. As we have seen previously (Sec. 6.1.3) the molecule spends most time at the extreme ends of its vibrational motion when *v* is large, and very little time in between. When moving in the vibrational states *g*, *h*, . . . , the molecule spends insufficient time near the cross-over point for appreciable dissociation to occur and a normal spectrum results.

6.1.8 Diatomic Molecules: A Summary

When the rotational fine structure of electronic spectra can be resolved—as it normally can for diatomic molecules—we see that a great deal of useful information becomes available. We can immediately determine the rotational constant, and hence calculate the moment of inertia and bond length, for both the lower and the upper electronic states. Isotopic species in the molecule will cause a slight difference in the rotational constant, so such isotopes may be detected, and their concentrations measured from the band intensity. Equally, the vibrational levels of the electronic states can be determined from the position of band origins; these lead to the evaluation of fundamental vibration frequencies, of bond force constants, and, perhaps, of dissociation energies too. The latter, however, are more accurately determined if a continuum is observed at the end of a band spectrum.

Where data obtained from such spectra can be checked independently, e.g. by microwave or infra-red spectroscopy, by X-ray or neutron diffraction, or by thermochemical methods, perfectly satisfactory agreement is found. Thus we can use electronic spectroscopic methods with great confidence to determine bond lengths and strengths in those molecules to which such independent methods are not applicable.

Probably the most important application of this type of spectroscopy is to the study of excited states and unstable radicals. Thus we have seen that B values and dissociation energies are obtained for both the upper and lower electronic states—and data for the upper state are not obtainable by other means. Further, considerable amounts of energy are involved in the production of electronic spectra, and complex molecules are frequently disrupted into fragments during the process, the fragments, or free radicals, normally being very short-lived. Examples are legion, a few of the more important diatomic ones being CH, NH, C_2, OH, CN, etc. Spectra arising from these radicals can be recognized and studied, leading to the determination of bond lengths, force constants, dissociation energies, etc. Further, if the variation of the intensity of such spectra over short periods of time is studied—as in the techniques of flash photolysis—information can be obtained about the rate at which the radicals are produced and destroyed. Since the length of time during which some radicals have an independent existence is measured in microseconds or less, it is remarkable that many such 'diatomic molecules' are as well characterized as, for example, the rather more stable H_2 or CO.

6.2 ELECTRONIC STRUCTURE OF DIATOMIC MOLECULES

6.2.1 Molecular Orbital Theory

Several theories have been suggested to account for the formation of molecules from atoms. All, if taken to a sufficiently high degree of approximation, seem to agree with observed data, but the calculation involved is so extensive that complete agreement is seldom reached and then only in the simplest examples. Here we shall discuss just one of these theories—the *molecular orbital theory*; we choose this, not because it is better or simpler than others (such considerations depend upon the particular problem in hand and are, in any case, largely subjective), but because it gives a convenient pictorial representation of molecule formation which is particularly suited to the discussion of electronic transitions and because the ideas it uses are entirely analogous to those of atomic structure which we have discussed in the previous chapter.

Thus we have seen that electrons in atoms do not occupy space haphazardly or have arbitrary energies, but that their distribution and energy are governed by well-defined natural laws. These characteristics may be calculated from the Schrödinger equation and expressed in terms of a three-dimensional wave function, or orbital, ψ, which depends on the values of three quantum numbers, n, l, and m (or l_z); the spin of the electron also contributes to the energy. Definite rules determine which orbitals are occupied in the ground state and what transitions may take place between orbitals.

Molecular orbital theory supposes orbitals to extend about, and embrace, *two or more nuclei*, the shape and energy of these orbitals being calculable from the Schrödinger equation in terms of three quantum numbers. Essentially the same rules (i.e. lowest energy first, maximum of two (paired) electrons per orbital, parallel spins in degenerate orbitals) apply to their filling as to the filling of atomic orbitals.

The situation is relatively simple for diatomic molecules where the molecular orbital embraces two nuclei only and we shall discuss these molecules in some detail first. The extension to polyatomic molecules will be outlined in Sec. 6.3.

6.2.2 The Shapes of Some Molecular Orbitals

As in atomic orbital theory (cf. Sec. 5.1.2) the shape of a molecular orbital is the space within which an electron belonging to that orbital will spend 95 per cent (or some other arbitrary fraction) of its time. While detailed computation of these shapes from the Schrödinger theory may be extremely difficult, a very good qualitative idea of their approximate shape may be obtained by considering molecular orbitals to be made up of sums and differences of the atomic orbitals of the constituent atoms—the so-called *linear combination of atomic orbitals* (LCAO) approximation. Thus for a diatomic molecule we could imagine the formation of two different molecular orbitals whose wave functions would be:

$$\psi_{\text{mo.}} = \psi_1 + \psi_2 \quad \text{or} \quad \psi_{\text{mo.}} = \psi_1 - \psi_2 \qquad (6.24)$$

where ψ_1 and ψ_2 are the relevant atomic orbitals of the two atoms. Note that the function $\psi_2 - \psi_1$ is identical with $\psi_1 - \psi_2$, since it is $\psi_{\text{mo.}}^2$ which represents the probability of finding an electron in a particular place.

Let us consider the hydrogen molecule, H_2, as an example; the obvious atomic orbitals to use are the $1s$ orbitals of each atom. Figure 6.10(a) shows the situation:

$$\psi_{H_2} = \psi_{1s} + \psi_{1s} \qquad (6.25)$$

We recall from the previous chapter (Eq. (5.2)) that ψ_{1s} is everywhere positive in value and so, where the atomic orbitals overlap, the value of ψ_{H_2} will be increased. This suggests (and detailed calculation confirms) that the molecular orbital of Eq. (6.25) is a simple ellipsoid, symmetrical in shape. The concentration of electronic charge between the nuclei acts as a sort of cement keeping the nuclei together and thus this orbital represents the formation of a bond between the atoms. It is called a *bonding orbital* and given the label $1s\sigma$ since it is produced from two $1s$ orbitals (we shall see later that σ has a similar significance regarding the orbital angular momenta of molecular electrons as s has for atoms).

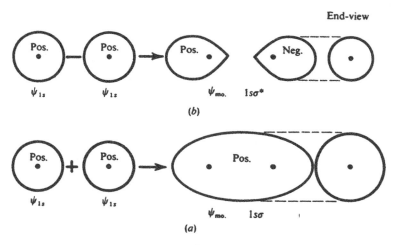

Figure 6.10 The formation of (a) a bonding $1s\sigma$ orbital and (b) an antibonding $1s\sigma^*$ orbital from two atomic $1s$ orbitals.

On the other hand, Fig. 6.10(*b*) shows the situation for

$$\psi_{H_2} = \psi_{1s} - \psi_{1s} \tag{6.26}$$

Since, again, ψ_{1s} is everywhere positive, where the two separate ψ_{1s} orbitals overlap, they will cancel each other. Thus between the nuclei $\psi_{mo.}$ will be zero, while it will be positive near one nucleus and negative near the other (remember that it is ψ^2 which determines probability and this is, in either case, positive). Now, however, the shape of the molecular orbital shows that electronic charge, far from being concentrated *between* the nuclei, is greatest *outside* them; thus the nuclear repulsion is enhanced and the orbital is described as *antibonding*. It leads to a state of higher energy than two separate atoms and is labelled $1s\sigma^*$, the asterisk representing high energy. Figure 6.10 also shows, at the extreme right, an end-view of these orbitals. They are both seen to have cylindrical symmetry about the bond axis; it is this property which leads to their both being described as σ orbitals although in other respects they have quite different appearances.

Another facet of orbital symmetry should be mentioned here. If the molecule considered is homonuclear (i.e. made of two *identical* atoms) then the midpoint of the bond between them is a *centre of symmetry*—starting from any point in the molecule, on or off the internuclear axis, exactly similar surroundings are encountered by proceeding to the point diagonally opposite the centre. Such a process is known as *inversion*, and such molecular properties as electron density, force fields, etc., are quite unchanged by inversion. However, we note that ψ (as opposed to ψ^2) may or may not be changed *in sign* by inversion. Thus inversion of the $1s\sigma$ molecular orbital of Fig. 6.10(*a*) plainly causes no change in ψ since it is everywhere positive; this orbital, in which ψ is completely symmetrical, is described as *even* and usually given the symbol *g* (German: *gerade* = even) as a suffix: $1s\sigma_g$. On the other hand, the $1s\sigma^*$ orbital in Fig. 6.10(*b*) is anti-symmetrical, since inversion reverses the sign of ψ. This orbital is thus *odd* and given the subscript *u*, $1s\sigma_u$ or $1s\sigma_u^*$ (German: *ungerade* = odd). In the case of molecular hydrogen, then, the bonding orbital is even, the antibonding is odd; this situation may be reversed for other molecular orbitals, as we shall see.

If the molecule is heteronuclear (for example CO, HCl, etc.) then there is no centre of symmetry and the odd–even classification of orbitals does not arise.

Before turning to the shapes of other molecular orbitals, it is instructive to consider how the energy of the $1s\sigma_g$ and $1s\sigma_u^*$ orbitals varies with the distance between the nuclei. This variation may be calculated from the Schrödinger equation and the result is shown in Fig. 6.11. The $1s\sigma_g$, the bonding orbital, shows a typical Morse curve for a diatomic molecule, the minimum in the curve showing that a bond is formed between the atoms. The $1s\sigma_u^*$, on the other hand, shows no minimum, but is one of the 'continuous' curves already discussed in Sec. 6.1.4. In this case the dissociation limits of the two curves coincide since the dissociation products are identical—two hydrogen atoms. The relationship of the orbitals sketched in Fig. 6.10 to the energy curve is shown by superimposing the molecular orbitals at their appropriate equilibrium internuclear distance and the separate atomic orbitals at a large distance.

Two 2s atomic orbitals can form $2s\sigma_g$ and $2s\sigma_u^*$ bonding and antibonding orbitals with identical shape to (but larger and with higher energy than) the $1s\sigma_g$ and $1s\sigma_u^*$ orbitals. Two 2*p*-type orbitals can overlap in two different ways depending on their relative orientation. If we label the internuclear axis the *z* direction, then we may consider first the 2*p* orbital which lies along this axis from each atom, i.e. the $2p_z$ orbitals. Now the expression for the wave function of a $2p_z$ orbital has the form $\psi_{2p_z} = zf(r)$, where $f(r)$ is a positive function of distance from the nucleus. We see, then, that for $+z$ directions ψ is also positive, while it is negative for $-z$. The two lobes of a 2*p* orbital thus have opposite signs of ψ (although, of course, ψ^2 is everywhere positive).

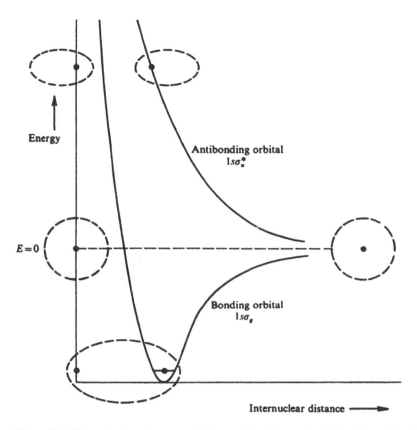

Figure 6.11 The variation of energy with internuclear distance in the bonding and antibonding orbitals, $1s\sigma_g$ and $1s\sigma_u^*$.

We draw these orbitals and indicate their signs on the left of Fig. 6.12 and it is evident that in the molecular orbital

$$\psi_{\text{mo.}} = \psi_{2p_z} + \psi_{2p_z} \tag{6.27}$$

the electron density between the nuclei is cancelled to zero and the orbital will have the shape shown at the top right of the figure. Similarly, in

$$\psi_{\text{mo.}} = \psi_{2p_z} - \psi_{2p_z} \tag{6.28}$$

the electron density increases between the nuclei and the shape is shown at the bottom right of the figure. Plainly the latter is *bonding*, and consideration of its symmetry shows that it is *even* (*g*), while the former is plainly *antibonding* and *odd* (*u*) in character. The end-view of both is the same, however, and shows symmetry about the bond axis; for this reason both are referred to as σ orbitals and they may be labelled $2p\sigma_g$ and $2p\sigma_u^*$, respectively, to indicate their origin.

The overlap of two $2p_y$ orbitals is shown in Fig. 6.13 ($2p_x$ are exactly similar but rotated out of the plane of the paper through 90°). The summed orbitals, which are *bonding*, are sketched at bottom right and we see that the molecular orbital formed consists of two streamers, one above and one below the nuclei. The end-view of this orbital is shown at the extreme right; it has the appearance of an atomic *p* orbital and hence it is labelled π. In this case the bonding orbital is evidently *odd*, so we have a $2p\pi_u$ state.

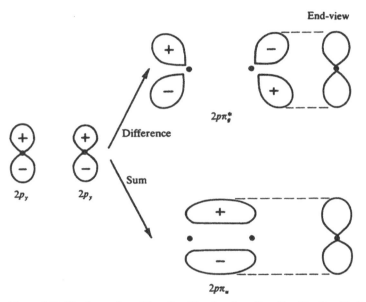

Figure 6.12 The formation of bonding ($2p\sigma_g$) and antibonding ($2p\sigma_u^*$) orbitals from two atomic $2p_z$ orbitals, where the z axis is taken as the internuclear axis.

Figure 6.13 The formation of bonding ($2p\pi_u$) and antibonding ($2p\pi_g^*$) orbitals from two atomic $2p_y$ orbitals, the z axis being the internuclear axis. Atomic $2p_x$ orbitals would form identical molecular orbitals except that all lobes would be rotated through a right angle about the z axis.

On the other hand, if the atomic orbitals are subtracted the orbital picture shown at top right of Fig. 6.13 is produced. This has a similar end-view to $2p\pi_u$, is *antibonding* and *even*, and hence it is labelled π_g^*.

More complex orbitals exist—δ, ϕ, etc., formed by interaction between d, f, etc., atomic orbitals—but they need not concern us; the simple molecules with which we shall deal use σ and π orbitals only. We do need to know, however, the order of increasing energy of the molecular orbitals so far discussed so that we can consider the ground states of some atoms. This we show diagrammatically (and by no means to scale) in Fig. 6.14(a), where we also indicate, to the right and left, the atomic orbitals which combine to form each molecular state. This diagram should be taken to be schematic only, since there is considerable overlap and interaction between the $2p\pi_u$, $2p\sigma_g$, and $2s\sigma_u^*$ orbitals for some lighter diatomic molecules such as Li_2; it will be sufficient for our purposes, however. For H_2 this overlap does not occur and the situation is as shown in Fig. 6.14(b). Using these diagrams and the Pauli principle (not more than two electrons to each

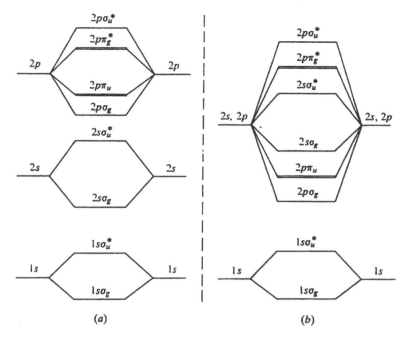

Figure 6.14 Schematic molecular orbital energy level diagrams for (a) simple diatomic molecules other than hydrogen and (b) the hydrogen molecule.

orbital) we can build up the electronic configurations of simple molecules. Some examples follow:

1. Hydrogen, H_2. In the ground state of this molecule the two $1s$ atomic electrons, one from each atom, can both occupy the molecular $1s\sigma_g$ provided their spins are opposed. The energy of the electrons in this state, as we can see from Figs 6.11 and 6.14(b), is lower than that of two separate atoms; hence the molecule is stable. We can write its configuration as $1s\sigma_g^2$.

2. Helium, He_2. If this molecule were to form it would have a total of four electrons to place into orbitals, two from each atom. Only two of these could go into the $1s\sigma_g$; the other two would have to go into $1s\sigma_u^*$. However, we can see from Figs 6.11 and 6.14(a) that more energy would be absorbed by the latter than evolved by the former; hence the molecule is unstable with respect to the atoms.

3. Nitrogen, N_2. The two atoms each have an electronic configuration $1s^2 2s^2 2p^3$ (cf. Table 5.2), so we have a total of 14 electrons to dispose of, in pairs, into the molecular orbitals of Fig. 6.14(a); clearly the configuration will be $1s\sigma_g^2 1s\sigma_u^2 2s\sigma_g^2 2s\sigma_u^2 2p\sigma_g^2 2p\pi_u^4$ (where we omit the asterisks to avoid confusion). As a good approximation we can allow the bonding and antibonding contributions of the $1s\sigma_g^2$ and $1s\sigma_u^2$ to cancel, and similarly with the $2s\sigma_g^2$ and $2s\sigma_u^2$. We are left, then, with three pairs of electrons in bonding orbitals, the $2p\sigma_g$ and the $2p\pi_u$, so we conclude that the molecule has a triple bond.

4. Oxygen, O_2. Each atom has one electron more than the nitrogen atom (Table 5.2), and clearly the lowest available orbital in Fig. 6.14(a) is the *antibonding* $2p\pi_g$. Two electrons in this orbital effectively cancel the bonding contribution of a pair in the $2p\pi_u$, and we are left with a net bonding of four electrons, i.e. a double bond. One unusual characteristic of the structure of O_2 concerns the two electrons in the antibonding $2p\pi_g$. Since there are here two degenerate orbitals, the electrons will occupy one each to satisfy electron repulsion; however, according to Hund's rule (Sec. 5.4.6) which applies to molecules just as to atoms, they will

preferentially have their spins *parallel* rather than paired. Thus O_2 is a molecule containing some unpaired electrons—a structural characteristic which gives it its most unusual magnetic properties.

6.2.3 Electronic Angular Momentum in Diatomic Molecules; Classification of States

We found in Sec. 5.2 that the total energy of an electron, while depending mainly on its average distance from the nucleus (represented by the quantum number n), also depends on its orbital and spin angular momenta (quantum numbers l and s) and on the way in which these are coupled together (quantum number j). For several electrons in an atom we found that their separate energies can be combined in different ways to produce a variety of states; simple rules allow the ground state to be predicted in any particular case.

Much the same comments apply to electrons in molecules. Thus a single electron in a molecule has a quantum number n specifying the size of its orbital and mainly determining its energy, and a number l specifying its orbital angular momentum. Small letters $s, p, d, \ldots$ are used, as before, to designate l values of $0, 1, 2, \ldots$. However, it will be remembered that in order to discuss the components of l we required to invoke some reference direction called the z direction; in a diatomic molecule a reference direction is already quite obviously specified—the internuclear axis, or bond—and it would be perverse (not to say *wrong*) to discuss the components of l along any other direction. Furthermore, a force field exists along this direction due to the presence of two nuclear charges; therefore different l components are not degenerate but represent *different energies*.

The axial component of orbital angular momentum is of more importance in molecules than the momentum itself and for this reason it is given the special symbol λ. Formally $\lambda v |l_z|$, so that λ takes *positive* integral values or is zero, and we designate the λ state of an electron in a molecule by using the small Greek letters corresponding to the $s, p, d, \ldots$ of atomic nomenclature. Thus we have, for

$$l_z = 0, \pm 1, \pm 2, \pm 3, \ldots$$
$$\lambda = 0, \quad 1, \quad 2, \quad 3, \ldots$$

and the symbols are:
$$\sigma, \quad \pi, \quad \delta, \quad \phi, \ldots$$

Since λ has positive values only, each λ state with $\lambda > 0$ is *doubly degenerate*, because it corresponds to l_z being both positive and negative. The significance of λ is that the *axial component of orbital angular momentum* $= \lambda h/2\pi$ or λ *units*.

The total orbital angular momentum of several electrons in a molecule can be discussed, as for atoms, in terms of the quantum number $L = \Sigma l, \Sigma l - 1$, etc., with $\mathbf{L} = \sqrt{L(L+1)}h/2\pi$, but again the axial component, denoted by Λ, is of greatest significance. Since, by definition, all individual λ_i lie along the internuclear axis, their summation is particularly simple. We have:

$$\Lambda = |\Sigma \lambda_i| \tag{6.29}$$

and states are designated by capital Greek letters Σ, Π, Δ, etc., for $\Lambda = 0, 1, 2, \ldots$. We must take into account, when using Eq. (6.29), that the individual λ_i may have the same or opposite directions and all possible combinations which give a positive Λ should be considered. Thus for a π and a δ electron ($\lambda_1 = 1$, $\lambda_2 = 2$) we could have $\Lambda = 1$ or 3 (but not -1), i.e. a Π or a Φ state.

Electron *spin* momentum, on the other hand, is not greatly affected by the electric field of the two nuclei—we say the spin–axial coupling is weak, whereas the orbital–axial coupling is usually strong. Normally, therefore, we use the same notation for electronic spin in molecules as

Table 6.1 Comparison of symbols used for electronic angular momenta

	Orbital momentum	Spin momentum	Total momentum
For atoms			
Single electron	l (symbol s, p, d for $l = 0, 1, 2, \ldots$)	s	j
Single electron (z component)	l_z	s_z	j_z
Several electrons	L (symbol S, P, D for $L = 0, 1, 2, \ldots$)	S	J
Several electrons (z component)	L_z	S_z	J_z
For molecules			
Single electron	l	s	j_a (seldom used)
Single electron (axial component)	λ (symbol σ, π, δ for $\lambda = 0, 1, 2, \ldots$)	σ	ω
Several electrons	L	S	J_a (seldom used)
Several electrons (axial component)	Λ (symbol Σ, Π, Δ for $\Lambda = 0, 1, 2, \ldots$)	Σ	Ω

in atoms; the total spin momentum **S** is given by $\sqrt{S(S+1)}$ where the total spin quantum number S is:

$$S = \Sigma S_i, \ \Sigma S_i - 1, \ \Sigma S_i - 2, \ \ldots, \ \tfrac{1}{2} \text{ or } 0 \qquad (6.30)$$

The multiplicity of a molecular state is, as for atoms, $2S + 1$ and this is usually indicated as an upper prefix to the state symbol. Thus for the Π and Φ states discussed in the previous paragraph, the states would be written $^3\Pi$ or $^3\Phi$ if the individual π and δ electron spins are parallel ($S = \tfrac{1}{2} + \tfrac{1}{2} = 1$, $2S + 1 = 3$), or as $^1\Pi$ or $^1\Phi$ if the spins are paired.

When the *axial* component of a spin is required, however, it is often designated by σ for a single electron or Σ for several (corresponding to s and S for the atomic case). In this case the multiplicity is $2\Sigma + 1$.

Finally, we consider the axial component of the *total* electronic angular momentum, i.e. the sum of the axial components of spin and orbital motion. In general the total momentum is strongly coupled to the axis and its axial component is more significant than the momentum itself. If we write the axial component as Ω we have simply:

$$\Omega = |\Lambda + \Sigma| \qquad (6.31)$$

but we must remember that Λ and Σ may have the same or opposite directions along the internuclear axis. Thus in the $^3\Pi$ state described above we have $\Lambda = 1$, $\Sigma = 1$; hence $\Omega = 2$ or 0. The $^1\Pi$ state has $\Lambda = 1$, $\Sigma = 0$; hence we have $\Omega = 1$ only. These states would be indicated by writing their Ω values as subscripts: $^3\Pi_2$, $^3\Pi_0$, $^3\Pi_1$.

Perhaps it will be of assistance to the student if we draw up a table (Table 6.1) showing the symbols used to designate the various sorts of angular momentum in atoms and molecules, together with their axial components, for one or more electrons.

6.2.4 An Example: The Spectrum of Molecular Hydrogen

Before turning to polyatomic molecules, let us see how the above ideas may be applied to the simplest molecule, H_2. We shall consider first the nature of the ground state and some excited states of the molecule and how these relate to occupancy of the molecular orbitals of Fig.

6.14(b), then the energy of these states and, finally, what transitions may arise between them. The student may find it helpful to discover the many points of similarity between the discussion which follows and that given in Sec. 5.4.5 on helium.

The hydrogen molecule contains two electrons, one contributed to by each of the atoms. We would thus expect to find singlet and triplet states, depending on whether the electron spins are paired or parallel. In the *ground state* both electrons will occupy the same lowest orbital, i.e. the $1s\sigma_g$ of Fig. 6.14(b), and, by Pauli's principle, they must then form a singlet state. Both electrons are σ electrons (since both are in a σ orbital); hence $\lambda_1 = \lambda_2 = 0$ and $\Lambda = 0$ also. The state is thus $^1\Sigma$. We could indicate the value of Ω as a subscript ($\Omega = \Lambda + \Sigma = 0 + 0 = 0$, since $\Sigma = 0$ for singlet states) but it is more informative to specify the symmetry (g or u) of the orbital. In this case both electrons are in the same g orbital; hence the total state is $^1\Sigma_g$.

A further subdivision of Σ states is normally made, representing another facet of molecular symmetry. In any diatomic molecule (whether homo- or heteronuclear) any plane drawn through both nuclei is a *plane of symmetry*, i.e. electron density, shape, force fields, etc., are quite unchanged by reflection in the plane. However, the wave function of the electron, ψ, may either be completely unchanged (symmetrical) or changed in sign only (antisymmetrical) with respect to such a reflection (in either case, of course, ψ^2 is unchanged). The former states are distinguished by a superscript $+$ and the latter by $-$. For several reasons this division is made for Σ states only and nearly all such states are in fact $+$. Certainly all the states of hydrogen are symmetric.

Thus the ground state of molecular hydrogen can be written:

$$\text{Ground state:} \quad (1s\sigma_g)^2 \; ^1\Sigma_g^+$$

A large number of excited singlet states also exist; let us consider some of the lower ones for which one electron only has been raised from the ground state into some higher molecular orbital, i.e. singly excited states. We can ignore promotion into any of the starred states of Fig. 6.14(b), since this would lead to the formation of an unstable molecule and immediate dissociation (cf. Fig. 6.11, where the placing of an electron in each of the σ_g and σ_u^* orbitals produces dissociation into two H atoms). Thus we may consider the three possible excited states $(1s\sigma_g 2s\sigma_g)$, $(1s\sigma_g 2p\sigma_g)$, and $(1s\sigma_g 2p\pi_u)$.

Taking $(1s\sigma_g 2s\sigma_g)$ first: here both electrons are σ electrons; hence $\Lambda = \lambda_1 + \lambda_2 = 0$ and, since we are considering only singlet states, $S = 0$ also. Further, since both constituent orbitals are *even* and *symmetrical*, the overall state will be the same, and we have $(1s\sigma_g 2s\sigma_g) \; ^1\Sigma_g^+$.

Now $(1s\sigma_g 2p\sigma_g)$: here we again have a $^1\Sigma$ state since both electrons are σ, but the overall state is now *odd* (u); this may be rationalized if we think of one electron as rising from a hydrogen atom in the *even* $1s$ state and the other from an *odd* $2p$ state, the combination of an odd and an even state leading to an overall odd state. Thus $(1s\sigma_g 2p\sigma_g) \; ^1\Sigma_u^+$.

Finally the $(1s\sigma_g 2p\pi_u)$: now $\Lambda = \lambda_1 + \lambda_2 = 1$, since one electron is in a π state and, again since one electron originates from a $2p$ orbital, the combined state is u: $^1\Pi_u$.

The energies of these three states increase in the order of the constituent molecular orbitals, as shown in Fig. 6.14(b), i.e.:

$$^1\Sigma_u^+ < \, ^1\Pi_u < \, ^1\Sigma_g^+$$

Similar states are obtained by excitation to the $3s$ and $3p$ states, to the $4s$ and $4p$ states, etc. Also for $n = 3, 4, \ldots$ there exists the possibility of excitation to the nd orbital. It may be shown by methods similar to those above that interaction between $1s$ and nd electrons can lead to the three configurations and state symbols in increasing energy:

$$(1s\sigma \, nd\sigma) \; ^1\Sigma_g^+ < (1s\sigma \, nd\pi) \; ^1\Pi_g < (1s\sigma \, nd\delta) \; ^1\Delta_g$$

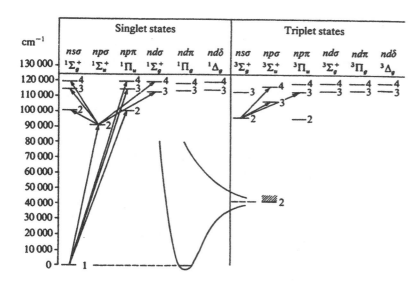

Figure 6.15 The singlet and triplet energy levels of the hydrogen molecule. One electron only is assumed to undergo transitions, the other remaining in the $1s\sigma$ state.

Some of these energy levels are shown at the left of Fig. 6.15. Transitions between them can occur according to the *selection rules*:

1.
$$\Delta\Lambda = 0, \pm 1 \qquad (6.32)$$

Thus transitions $\Sigma \leftrightarrow \Sigma$, $\Sigma \leftrightarrow \Pi$, $\Pi \leftrightarrow \Pi$, etc., are allowed, but $\Sigma \leftrightarrow \Delta$, for example, is not.

2.
$$\Delta S = 0 \qquad (6.33)$$

For the present we are concerned only with singlet states so this rule does not arise.

3.
$$\Delta\Omega = 0, \pm 1 \qquad (6.34)$$

This follows directly from 1 and 2 above.

4. There are also restrictions on symmetry changes. Σ^+ states can undergo transitions only into other Σ^+ states (or, of course, into Π states) while Σ^- go only into Σ^- (or Π). Symbolically:
$$\Sigma^+ \leftrightarrow \Sigma^+ \quad \Sigma^- \leftrightarrow \Sigma^- \quad \Sigma^+ \nleftrightarrow \Sigma^- \qquad (6.35)$$

And finally:
$$g \leftrightarrow u \quad g \nleftrightarrow g \quad u \nleftrightarrow u \qquad (6.36)$$

We show a few allowed transitions from the ground state and the lowest excited state in Fig. 6.15.

Let us now consider some of the triplet states of molecular hydrogen, i.e. those states in which the electron spins are parallel and hence $S = 1$. Plainly both electrons cannot now occupy the same orbital so the state of lowest energy will be either $(1s\sigma_g 2s\sigma_g)$, $(1s\sigma_g 2p\sigma_g)$, or $(1s\sigma_g 2p\pi_u)$. The first two are evidently $^3\Sigma$ states, the third is $^3\Pi$, and, following the rules outlined above, we can write down their state symbols and order of energies as:

$$(1s\sigma_g 2p\sigma_g)\ {}^3\Sigma_u^+ < (1s\sigma_g 2p\pi_u)\ {}^3\Pi_u < (1s\sigma_g 2s\sigma_g)\ {}^3\Sigma_g^+$$

These energy levels are shown on the right of Fig. 6.15, together with those formed by the introduction of $3d$ and $4d$ orbitals. (The very small splitting of the levels into states with different $\Omega = \Lambda + S$ is ignored in the figure.) A few of the allowed transitions are shown from the $(1s\sigma 2s\sigma)$ state, but it should be particularly noted that, because of the selection rule $\Delta S = 0$ given in Eq. (6.33), transitions are not allowed between singlet and triplet states, i.e. between the two halves of Fig. 6.15.

Transitions are not shown from the lowest $^3\Sigma_u^+$ state on Fig. 6.15, i.e. the lowest triplet state. This is not because transitions are forbidden, but because the state is the continuous one shown in the upper half of Fig. 6.11, the $1s\sigma_u$. Thus in this state the molecule immediately dissociates into atoms before further transitions can occur. The energy level shown for this state in Fig. 6.15 is the lower limit, i.e. the dissociation limit, and in fact the state extends continuously from this limit up to the top of the diagram. Part of Fig. 6.11 is reproduced in Fig. 6.15 to underline the relationship between them.

Thus although the hydrogen spectrum will be complicated by the presence of vibrational and rotational structures on each of the transitions sketched in the figure, basically the overall pattern consists of sets of Rydberg-like line series from which the positions of the energy levels can be found.

6.3 ELECTRONIC SPECTRA OF POLYATOMIC MOLECULES

We have seen in Sec. 3.7 that the vibrational frequencies of a particular atomic grouping within a molecule, for example CH_3, $C{=}O$, $C{=}C$, etc., are usually fairly insensitive to the nature of the rest of the molecule. Other bond properties, such as length or dissociation energy, are also largely independent of the surrounding atoms in a molecule. Since all these properties depend, in the final analysis, on the electronic structure of the bond, it is plain that we may, at least as an approximation, discuss the structure, and hence the spectrum, of each bond in isolation. Bonds for which this approximation is adequate are usually said to have 'localized' molecular orbitals, i.e. orbitals embracing a pair of nuclei only; other molecules, for which this approximation is invalid, have non-localized orbitals and are often called 'conjugated'. We shall meet some examples of this latter class shortly.

When each bond may be considered in isolation, it is evident that the complete electronic spectrum of a molecule is the sum of the spectra from each bond. The result will plainly be very complex, but a great deal of information about the molecule is contained within it. Thus if some band series can be recognized for a particular bond we immediately know the vibrational frequency of that bond and probably a good estimate of its dissociation energy also. If the rotational structure is resolved, then we have the moment of inertia (from the line spacing) and hence information about the shape and size of the molecule.

Such detailed information is usually obtainable only for molecules studied in the gas phase: in pure liquids, or in solution, molecular rotation is hindered and no rotational structure will be observed. The blurring of the rotational structure often masks the vibrational line series also, and the electronic spectrum of a liquid is usually rather broad and characterless. However, as we shall discuss shortly, it may still be highly characteristic of a particular molecular grouping both in its frequency and its intensity.

Confining our attention, for the moment, to gas-phase spectra, we hve already remarked that one of the more important advantages of electronic spectroscopy is that the vibrations, rotations, dissociation energies, and structures of molecules may be investigated in their *excited states*, even though a particular molecule may exist in such a state for not much longer than the time it takes to complete a few rotations. We have not the space to discuss this topic in detail, but one aspect is especially interesting—the fact that electronic excitation often leads to a change in shape of the molecule. That this happens can be seen by studying rotational fine structure in the spectra; here we briefly discuss the theoretical basis for its occurrence.

6.3.1 Change of Shape on Excitation

In Fig. 6.16 we show the orbital picture of a hydride H_2A where A is any polyvalent atom, both in its bent configuration (*a*), with a bond angle of 90°, and in a linear form (*b*), bond angle 180°. We have seen in Sec. 5.1 that the *p* orbitals of an atom are at right angles to each other, so we can readily imagine the rectangular molecule to be formed by interaction of two of these *p* orbitals with hydrogen $1s$ orbitals, leaving the third *p* orbital unaffected; the latter is called a 'non-bonding' orbital since it plays no part in bonding A and H together, and in general it will have a higher energy than bonding orbitals. In Fig. 6.16(*a*) we label the non-bonding orbital (which is out of the plane of the paper) N_1, the two bonding molecular orbitals formed from the *p* and $1s$ atomic orbitals as a_1 and a_2, and the unused (and hence non-bonding) atomic orbital *s* of A simply as *s*.

In the linear molecule a new principle must be introduced—that of orbital hybridization. For this, atom A is supposed to mix its *s* orbital with one of its *p* orbitals and, from these two orbitals, to form two new orbitals—hybrids—which, it may be shown theoretically, point at 180° to each other. These *sp* hybrids form rather stronger bonds to other atoms than separate *s*

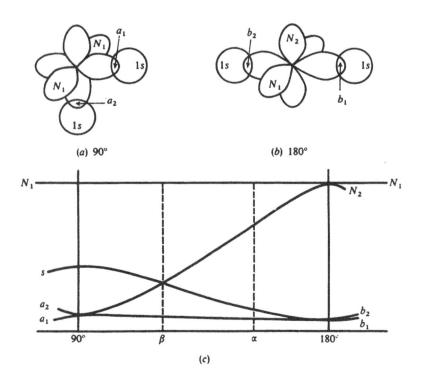

(*a*) 90°   (*b*) 180°

(*c*)

Figure 6.16 Orbital pictures for an AH_2 molecule where the AH bonds are (*a*) at 90° and (*b*) at 180°. In (*c*) is shown qualitatively the change in energy of the various orbitals as the bond angle changes from 90 to 180°. (*Adapted, with kind permission of the author, from A. D. Walsh, J. Chem. Soc., 1953, p. 2262.*)

and p orbitals, so it is energetically favourable, in certain cases, for the atom to 'prepare' hybrid orbitals at the moment of bond formation. In this configuration (Fig. 6.16(b)), there are now two non-bonding p orbitals, labelled N_1 and N_2, and two bonding orbitals formed by overlap between sp hybrids and hydrogen $1s$, called b_1 and b_2.

Remembering that a non-bonding orbital is higher in energy than a bonding orbital and that an sp-bonding orbital is stronger (hence *lower* in energy because the molecule is more stable) than a p-bonding orbital, we can plot the qualitative energy changes for a smooth transition from 90 to 180° bonding. This we do in Fig. 6.16(c), which is constructed as follows:

1. The non-bonding orbital N_1 remains unchanged throughout; hence its energy is constant.
2. The bonding orbital a_1 passes over into the stronger orbital b_1; hence its energy decreases.
3. The bonding orbital a_2 becomes the non-bonding N_2, thus increasing in energy; N_1 and N_2 are identical in energy at 180°.
4. The bonding orbitals b_1 and b_2 are formed by absorption of the non-bonding s into a_1.
5. If we increase the bond angle beyond 180° (or decrease it below 90°) the reverse changes begin to take place, so 180 and 90° represent maxima and minima on the energy curves as shown.

Now let us see the relevance of this to molecular shapes. Consider first the molecule BeH_2, beryllium hydride. Beryllium, we have seen in Sec. 5.3.1, has the electronic ground state configuration $1s^2 2s^2$, i.e. it has two outer electrons with which to form bonds, the two $1s$ electrons being too firmly held by the nucleus to take part in bonding. Each hydrogen atom contributes a further electron, so the BeH_2 molecule must dispose of four electrons into molecular orbitals, with, according to Pauli, a maximum of two electrons per orbital. The most stable state, as can be seen from Fig. 6.16(c), will be for two electrons to go into b_1 and two into b_2, thus producing a *linear* molecule.

When the molecule is excited electronically, the next available orbital to contain the excited electron is N_2 (or N_1), but with a configuration $b_1^2 b_2^1 N_2^1$ it is evident from the figure that the most stable state will be at a bond angle, α, somewhere between 90 and 180°—the increase in the energies of b_1 and b_2 being more than compensated for by the decrease in N_2 until equilibrium occurs at an angle α. Thus we see that the excited state is *bent*. If the electron is so excited as to be ionized completely, leaving the ion BeH_2^+, the three remaining electrons will all be in the b_1 and b_2 orbitals and hence the most stable configuration will again be *linear*.

Now consider the case of water, H_2O. The oxygen atom has an outer electron configuration $2s^2 2p^4$, and so has six electrons to dispose into molecular orbitals. As before each hydrogen contributes one, so water is formed by placing a total of four pairs of electrons into four molecular orbitals. The lowest energy state at which this can be done is shown by the angle β in the figure, which is some angle between 90 and 180° (and is observed experimentally to be about 104°). Thus water is bent in the ground state, with a configuration which may be written $a_1^2 a_2^2 s^2 N_1^2$, since the angle is not far removed from 90°. During excitation one of the N_1 electrons will undergo transitions since these, being of highest energy, are most easily removed. However, since the energy of N_1 does not change with angle, the angle of the remaining $a_1^2 a_2^2 s^2 N_1^1$ state will not change during the transition.

These arguments may be readily extended to other triatomic molecules or to larger poly-atomic molecules, although the energy diagram corresponding to Fig. 6.16(c) is more complicated since more orbitals are involved. The results show, however, and experiment confirms, that linear molecules such as CO_2 and $HC{\equiv}CH$ become bent on excitation, the latter taking up a planar zig-zag conformation.

6.3.2 Chemical Analysis by Electronic Spectroscopy

Although rotational and sometimes vibrational fine structure do not appear in the liquid or solid state, both the position and intensity of the rather broad absorption due to an electronic transition is very characteristic of the molecular group involved. In this branch of spectroscopy the position of an absorption is almost invariably given as the *wavelength* at the point of maximum absorption, $\lambda_{max.}$, quoted either in ångström units (1 Å $= 10^{-10}$ m) or in nanometres (1 nm $= 10^{-9}$ m), the latter being more usual. It should be particularly noted that a *large* energy change, corresponding to a high frequency or wavenumber, is represented by a *small* wavelength. For practical reasons the electronic spectrum is divided into three regions: (1) the visible region, between 400 and 750 nm (4000–7500 Å or 25 000–13 300 cm^{-1}), (2) the near-ultra-violet region, between 200 and 400 nm (2000–4000 Å or 50 000–25 000 cm^{-1}), and (3) the far (or vacuum) ultra-violet, below 200 nm (below 2000 Å or above 50 000 cm^{-1}). The latter is so called because absorption by atmospheric oxygen is considerable in this region and spectra can only be obtained if the whole spectrometer is carefully evacuated. Thus commercial instruments extend only down to about 185 nm and absorptions below this range are little used for routine chemical purposes.

The *intensity* of an electronic absorption is given by the simple equation:

$$\varepsilon = \frac{1}{cl} \log_{10} \frac{I_0}{I} \quad l\,\mathrm{mol}^{-1}\,\mathrm{cm}^{-1} \tag{6.37}$$

where c and l are the concentration and path length of the sample (in mol l^{-1} and in cm^{-1}, respectively), I_0 is the intensity of light of wavelength $\lambda_{max.}$ falling on the sample, and I is the intensity transmitted by the sample. ε is the molar absorption coefficient and ranges from some 5×10^5 for the strongest bands to 1 or less for very weak absorptions.

Electrons in the vast majority of molecules fall into one of the three classes: σ electrons, π electrons, and non-bonding electrons (called n electrons). The first two classes were discussed in Sec. 6.2.2 and the third, which plays no part in the bonding of atoms into molecules, was mentioned briefly in Sec. 6.3.1. In chemical terms a single bond between atoms, such as C—C, C—H, O—H, etc., contains only σ electrons, a multiple bond, C=C, C≡C, C=N, etc., contains π electrons in addition, while atoms to the right of carbon in the periodic table, notably nitrogen, oxygen, and the halogens, possess n electrons. In general the σ electrons are most firmly bound to the nuclei and hence require a great deal of energy to undergo transitions, while the π and n electrons require less energy, the n electrons usually (but not invariably) requiring less than the π. Thus, in an obvious notation, $\sigma \rightarrow \sigma^*$ transitions fall into the vacuum ultra-violet, $\pi \rightarrow \pi^*$ and $n \rightarrow \sigma^*$ appear near the borderline of the near- and far-ultra-violet, and $n \rightarrow \pi^*$ come well into the near-ultra-violet and visible regions. These generalizations are indicated schematically in Fig. 6.17, which also shows the relationship between the nanometer and wavenumber scales.

Saturated hydrocarbon molecules, then, which can only undergo $\sigma \rightarrow \sigma^*$ transitions, do not give rise to spectra with any analytic interest since they fall outside the generally available range; examples are the $\sigma \rightarrow \sigma^*$ transitions of methane CH_4 and ethane C_2H_6 which are at 122 and 135 nm, respectively.

The insertion of a group containing n electrons, e.g. the NH_2 group, allows the possibility of $n \rightarrow \sigma^*$ transitions in addition and also tends to increase the wavelength of the $\sigma \rightarrow \sigma^*$ absorption; for example CH_3NH_2: $\sigma \rightarrow \sigma^*$ 170 nm, $n \rightarrow \sigma^*$ 213 nm. It is unsaturated molecules, i.e. molecules containing multiple bonds, which give rise to the most varied and interesting spectra, however. We cannot here discuss the large mass of data in any detail but must be content to indicate a few of the more important generalizations. More detail is to be found in the book by Williams and Fleming, listed in the bibliography at the end of this chapter.

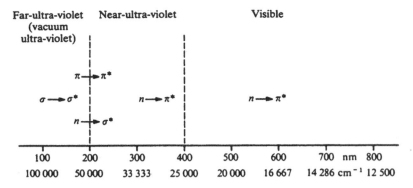

Figure 6.17 The regions of the electronic spectrum and the type of transition which occurs in each.

Consider, first, *isolated* multiple bonds within a molecule; the most important factor determining the position of the absorption maxima is, of course, the nature of the atoms which are multiply-bonded. From the following table we see that the $\pi \to \pi^*$ transitions are relatively insensitive to those atoms while the $n \to \pi^*$ transitions vary widely:

	$\pi \to \pi^*$ (strong) (nm)	$n \to \pi^*$ (weak) (nm)
$>$C$=$C$<$	170	—
—C$\equiv$C—	170	—
$>$C$=$O	166	280
$>$C$=$N$\diagdown$	190	300
$\diagup$N$=$N$\diagdown$	?	350
$>$C$=$S	?	500

This behaviour is very reasonable since the n electrons play no part in the bonding and control of them is retained by the atom (O, N, or S) contributing them. The above data are approximate only since different substituents on the A$=$B group produce slight variations in the wavelength of the $n \to \pi^*$ transition. Thus, considering ketones alone, $\lambda_{max.}$ varies from 272 nm for CH_3COCH_3 to 290 nm for cyclohexanone, and even higher if halogen substituents are included. From the mass of empirical data already assembled, a great deal of information about the substituents to a particular group is obtainable from the electronic spectrum. Figure 6.18 shows the ultra-violet/visible spectrum of propanone. Then $n \to \pi^*$ transition at 272 nm is clearly seen as a broad feature.

More pronounced changes occur when two or more multiple bonds are *conjugated* in the molecule, i.e. when structures having alternate single and multiple bonds arise, for example —C$=$C—C$=$C— or —C$=$C—C$=$O. In this case the $\pi \to \pi^*$ and $n \to \pi^*$ transitions both increase considerably in wavelength and intensity, the increase being greater the more conjugate linkages there are. As a simple example we have the following approximate data for $\pi \to \pi^*$ transitions in carbon–carbon bonds:

	$\lambda_{max.}$ (nm)	ε
—C$=$C—	170	16 000
—C$=$C—C$=$C—	220	21 000
—C$=$C—C$=$C—C$=$C—	260	35 000

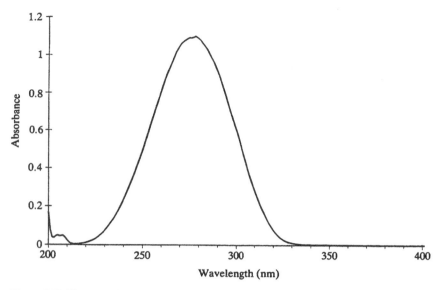

Figure 6.18 The ultra-violet spectrum of propanone. (*Thanks are due to Mr P. Gibbs of Queen's University, Belfast, for this spectrum.*)

while for oxygen-containing molecules we have both $\pi \rightarrow \pi^*$ and $n \rightarrow \pi^*$ transitions:

	$\pi \rightarrow \pi^*$ (strong) (nm)	$n \rightarrow \pi^*$ (weak) (nm)
—C=O	166	280
—C=C—C=O	240	320
—C=C—C=C—C=O	270	350
O=⬡=O	245	435

Thus we see that conjugation immediately brings the very intense $\pi \rightarrow \pi^*$ transition into the easily available region of ultra-violet spectrometers. For this reason these techniques are particularly well adapted to the study of conjugated and aromatic systems.

In the last example given above we see that the $n \rightarrow \pi^*$ absorption of *p*-benzoquinone, at 435 nm, has shifted into the blue region of the visible spectrum. When the substance is seen in the ordinary way the complementary colour—yellow—is observed. Colour in large organic molecules is invariably due to the existence of considerable conjugation raising the transition wavelength into the visible region—a fact on which the chemistry of dyestuffs is based.

Figure 6.19 shows part of the ultra-violet/visible spectrum of benzene, the series of bands around 250 nm being due to a weak $\pi \rightarrow \pi^*$ transition. This transition slightly weakens the bond and gives rise to a vibrational structure, the origin of which is shown in Fig. 6.3(*c*).

Substituents on conjugated systems also perturb the ultra-violet transitions in a systematic way; a great deal of empirical data has led to the formulation of rules to predict the effects. These are known as Woodward's rules, but they have undergone considerable modification and extension since they were first formulated by Woodward in 1941. As a simple example, consider butadiene, CH_2=CH . CH=CH_2, which has a strong absorption at 217 nm due to the $\pi \rightarrow \pi^*$

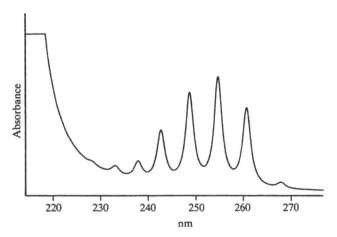

Figure 6.19 Part of the ultra-violet spectrum of benzene showing vibrational fine structure.

transition. This molecule may be considered to be the 'parent' of a whole series of molecules containing the trans conjugated ∿∿ group, whether in hydrocarbon chains or in ring systems. Any substituents or modifications to this parent fragment have each been assigned a positive or negative value which must be added to the basic absorption at 217 nm in order to arrive at the expected absorption frequency of the substituted molecule. Thus the increment for a chlorine atom is 5 nm and for an —OCH$_3$ group 6 nm; if both of these substituents occur together, the molecule would absorb at $217 + 5 + 6 = 228$ nm. Excellent accounts of these rules, and tabulations of the values for various substituents, etc., are to be found in the book by Williams and Fleming, mentioned in the bibliography.

6.3.3 The Re-emission of Energy by an Excited Molecule

After a molecule has undergone an electronc transition into an excited state there are several processes by which its excess energy may be lost; we discuss some of these briefly below.

1. *Dissociation.* The excited molecule breaks into two fragments. This was discussed in some detail for the particular case of a diatomic molecule dissociating into atoms in Sec. 6.1.4. No spectroscopic phenomena, beyond the initial absorption spectrum, are observed unless the fragments radiate energy by one of the processes mentioned below.
2. *Re-emission.* If the absorption process takes place as shown schematically in Fig. 6.20(*a*), then re-emission is just the reverse of this, as in Fig. 6.20(*b*). The radiation emitted, which may be collected and displayed as an emission spectrum, is identical in frequency with that absorbed.
3. *Fluorescence.* If, as in Fig. 6.20(*a*), the molecule is in a high *vibrational* state after electronic excitation, then excess vibrational energy may be lost by intermolecular collisions; this is illustrated in Fig. 6.20(*c*). The vibrational energy is converted to kinetic energy and appears as heat in the sample; such transfer between energy levels is referred to as 'radiationless'. When the excited molecule has reached a lower vibrational state (for example $v' = 0$), it may then emit radiation and revert to the ground state; the radiation emitted, called the *fluorescence spectrum*, is normally of lower frequency than that of the initial absorption, but under certain conditions it may be of higher frequency. The time between initial absorption and return to the ground state is very small, of the order of 10^{-8} s.

(a) Absorption   (b) Re-emission   (c) Fluorescence   (d) Phosphorescence

Figure 6.20 Showing the various ways in which an electronically excited molecule can lose energy.

4. *Phosphorescence*. This can occur when two excited states of different total spin have comparable energies. Thus in Fig. 6.20(d), we imagine the ground state and one of the excited states to be singlets (that is $S = 0$), while the neighbouring excited state is a triplet ($S = 1$). Although the rule $\Delta S = 0$ forbids *spectroscopic* transitions between singlet and triplet states, there is no prohibition if the transfer between the excited states occurs *kinetically*, i.e. through radiationless transitions induced by collisions. Such transfer, known as inter-system crossing, can only occur close to the cross-over point of the two potential curves (cf. Sec. 6.1.7), and once the molecule has arrived in the triplet state and undergone some loss of vibrational energy in that state, it cannot return to the excited singlet state. It will, therefore, eventually reach the $v' = 0$ level of the *triplet* state. Now although a transition from here to the ground state is spectroscopically forbidden, it *may* take place but much more slowly than an allowed electronic transition. Thus it is that a phosphorescent material will continue to emit radiation seconds, minutes, or even hours after the initial absorption. The phosphorescence spectrum, as a rule, consists of frequencies lower than that absorbed.

Considerable confusion often occurs between the Raman effect, discussed in Chapter 4, and the phenomena of fluorescence or phosphorescence. The main points of difference are as follows:

(a) In fluorescence and phosphorescence, radiation must be absorbed by the molecule and an excited electronic state formed; in Raman spectroscopy energy is merely transferred from radiation to molecule (in the form of an electrical perturbation of the ground state electron distribution), or vice versa, but no excited electronic state is formed.

(b) The exciting radiation for fluorescence or phosphorescence must be just that equivalent to the energy difference between electronic states. The exciting radiation for classical Raman spectroscopy can be of any frequency *except* that which would induce an electronic transition; in the latter case absorption would occur, rather than scattering. Resonance Raman spectroscopy arises from the intermediate case.

5. *Stimulated emission*. This increasingly important mechanism for removal of excess energy can lead to the production of laser radiation, as discussed in Sec. 1.10. The helium–neon laser, which emits light in the visible region of the spectrum, was considered in Sec. 5.4.6.

6.4 TECHNIQUES AND INSTRUMENTATION

The simple techniques of electronic spectroscopy are familiar to anyone studying physics—a glass prism, some sort of telescope, a bunsen burner, and a pinch of common salt are sufficient apparatus for observing part of the emission spectrum of sodium. In fact, a great deal of rapid and precise analytical work, both qualitative and quantitative, is carried out using flame spectrophotometers not very much more sophisticated in construction than this, except that a photomultiplier or photographic plate is used instead of the rather inaccurate human eye. However, for high-resolution work or for absorption studies, the practical requirements are more stringent.

The choice of a suitable source was formerly one of the main difficulties. The prime requirements of a source are that it should be *continuous* over the region of interest (i.e. there must be no wavelengths at which it does not emit) and it should be as *even* as possible (i.e. there must be no intense emission lines). In the visible region and just into the near-ultra-violet—say between 350 and 800 nm—an ordinary tungsten filament lamp is quite suitable. The filament acts as a black body radiator and has sufficient intensity for measurements in this region. Below this a hydrogen discharge lamp proves adequate, down to about 190 nm, while below this again discharge lamps containing rare gases, such as xenon, must be used. Thus we see that, in contrast to the other forms of spectroscopy discussed in previous chapters, no one source is suitable throughout the region.

Transparent materials for windows and sample cells present no problem, at least in the visible and near-ultra-violet regions, since good-quality glass or quartz transmit down to 200 nm or better. Below this region alkali fluorides, such as lithium fluoride or calcium fluoride, must be used; these are transparent down to about 100 nm. Prisms, if used, can be made of the same materials. It is common nowadays to use reflection gratings in ultra-violet/visible spectrometers, rather than prisms, since the former gives better dispersion and so allows more precise wavelength selection and measurement.

The detector for visible and ultra-violet studies is either a photographic plate or a photomultiplier tube. The chief disadvantage of the photographic method is that the resolving power is limited by the graininess of the image; on the other hand there is no other detector which can record the complete spectrum simultaneously in a small fraction of a second. When studying short-lived species, such as free radicals, it would be quite impossible to scan the complete spectrum using a photomultiplier. Also, at the other end of the time-scale the photographic plate is an efficient integrator of very weak signals—exposure times can be extended to many hours or even days to record a weak emission or absorption. For most routine purposes, however, where the spectrum of a stable material is to be recorded in a time of several minutes, a photomultiplier detector coupled to an amplifier and computer is the most flexible and useful combination. In the long-wavelength end of the visible region, photoconductive detectors can also be used.

6.5 MOLECULAR PHOTOELECTRON SPECTROSCOPY

The principles of photoelectron spectroscopy (PES) and its application to atoms were described in Sec. 5.5. In this section we will discuss applications to *molecular* species.

In PES an electron is ejected from an atom or molecule following irradiation by ultra-violet or X-ray radiation, leaving a singly charged ion. Ejection of the electron can take place from the valence levels (which have energies corresponding to ultra-violet radiation) or from the core levels (corresponding to X-ray excitation). For atomic species, the features observed in both XPES and UPES spectra are characteristic of the elements, providing a 'fingerprint' of a given element. For molecular species this is still true of core-level (XPES) spectra, but UPES spectra arise from the outer valence-related molecular orbitals and are more characteristic of the molecule as a whole.

6.5.1 Ultra-violet Photoelectron Spectroscopy (UPES)

Typical values for the binding energies of electrons in the valence level of atoms fall in the range 5–30 eV. It is, of course, possible to eject such electrons with X-ray radiation, but considerably higher resolution is obtained with an ultra-violet source. Recognition and assignment of observed binding energies can usually be made by comparison with chemically similar species, but the fine structure of bands, which can often be observed at high resolution and which results from transitions to more than one vibrational level in the ion, gives further information.

An ionized molecule is short-lived but normally relatively stable and, in the absence of electron capture, it will exist for long enough to undergo vibrations and rotations like any other molecule. Typically, transitions originate from the $v = 0$ state in the neutral molecule to one of several vibrational states of the ion, resulting in a series of approximately equally spaced bands in the spectrum. The transitions are governed by the Franck–Condon principle (see Sec. 6.1.3 and, in particular, Fig. 6.3), and the relative intensities of the lines depend on the positions of the potential wells for the neutral and ionized species, giving the highest intensities where the bond length in the ion is closest to that in the neutral species. The bond length, and other properties of the ion, in turn depend on what sort of electron is ionized from the neutral molecule, as follows:

1. If a *non-bonding* electron is removed, there is virtually no change in bond strength or length and, as shown in Fig. 6.3(*a*), the intensity of the transition is concentrated almost entirely into the 0,0 line.
2. If an *antibonding* electron is ionized, the bond becomes stronger and hence shorter; this, as shown in Fig. 6.3(*b*), results in transitions to several vibrational states in the ion, and consequently a spectrum showing vibrational fine structure. In this case, too, the potential well for the ion is generally rather deeper and narrower than that of the molecule, and the vibrational frequency increases; this is a consequence of the increase in bond strength or force constant.
3. The removal of a *bonding* electron gives a weaker, longer bond (Fig. 6.4(*c*)), with a spectrum otherwise similar to that of 2 above. Here, however, the decrease in bond strength results in a wider and shallower potential well, with a lower vibrational frequency than that of the neutral molecule.

As can be seen in Fig. 6.3, the spacing of the vibrational fine-structure lines reflects the separations between the vibrational level of the *ion* rather than those of the neutral molecule. Thus observation of the fine structure immediately leads to an approximate value for the vibrational frequency of the ion concerned.

Figure 6.21 shows the UPES spectrum of carbon monoxide. This molecule has a total of ten valence electrons—four from the carbon and six from the oxygen—the outer eight of which occupy the molecular orbitals $2\sigma^2 2\pi^4 2n^2$, where n^2 represents a pair of electrons in a non-bonding orbital. It should be noted that this structure is amply confirmed by the photoelectron spectrum, as we show below.

Three main features appear in the spectrum, at 14.0, 16.5, and 19.7 eV. The 14 eV band has virtually no fine structure—only two peaks appear, one of them very weak—and it clearly arises from removal of an electron from the non-bonding orbital. The band at 16.5 eV with its

Figure 6.21 The photoelectron spectrum of carbon monoxide.

associated fine structure arises from the $2p\pi$ bonding orbital, which is higher in energy (i.e. has a weaker binding energy) than the $2s\sigma$; the latter gives rise to the 19.7 eV peak. Confirmation of this assignment can be made by measuring the vibrational spacing. This is observed to be $1549\,\mathrm{cm}^{-1}$ for the 16.5 eV band and $1706\,\mathrm{cm}^{-1}$ for the 19.7 eV band; both of these values are significantly lower than that of the fundamental frequency of the free molecule ($2157\,\mathrm{cm}^{-1}$), indicating removal of an electron from a bonding orbital. The spacing of the two lines for the 14.0 eV band is about $2200\,\mathrm{cm}^{-1}$, which is reasonably close to the neutral molecule and confirms a non-bonding case.

Two further effects should be mentioned. Firstly, spin–orbit coupling may sometimes result in the additional splitting of bands, just as for atoms. Secondly, if an electron is removed from a degenerate level, the resulting ion may initially also be in a degenerate state; subsequent distortion of the ion may lead to a removal of the degeneracy and hence a splitting of the energy levels; this is known as the Renner–Teller effect for linear molecules and as the Jahn–Teller effect for non-linear ones. For example, if an electron is removed from one of the four equivalent hybridized orbitals of methane, the resulting CH_4^+ ion is not tetrahedral. The remaining three orbitals are no longer degenerate and so each contributes separately to the spectrum, producing three maxima. In this case the splitting is small, and often observed merely as a broadening, rather than a splitting, of the photoelectron peak.

6.5.2 X-Ray Photoelectron Spectroscopy (XPES)†

X-ray photoelectron spectroscopy probes the binding energies of core electrons in an atom. Although such electrons usually play little part in chemical bonding, different chemical environments can induce small changes in their binding energies; this is because the formation of a bond changes the distribution of electrons in the system and hence, by modifying nuclear shielding, produces changes in the effective nuclear charge of the bound atoms.

Consider, as an example, the case of carbon in the C — F bond. The highly electronegative fluorine attracts electrons and so reduces the electron density around the carbon nucleus; the carbon is thus less shielded than in the free state, and its effective nuclear charge is increased.

† In the earlier literature XPES was called electron spectroscopy for chemical analysis and abbreviated to ESCA.

Figure 6.22 The photoelectron spectrum of N_2O.

This means that core carbon electrons experience an increased nuclear force and so have a higher binding energy. This phenomenon is known as the *chemical shift*.

Another example is N_2O; we show the nitrogen $1s$ region of the XPES spectrum of the molecule in Fig. 6.22. The molecule is linear with the structure $N-N-O$ and so the central nitrogen experiences a greater reduction of electron density, and hence more strongly bound core electrons, from the electronegative oxygen than does the end one. Clearly, then, the central atom is assigned to the 412.6 eV peak and the end atom to the 408.7 eV peak, as shown.

In principle it would seem to be possible to use the chemical shift to determine the effective charge on an individual atom in the molecule; in practice this process is sometimes complicated by a number of experimental factors which will not be dealt with here.

A further point worth mentioning is that, as for UPES measurements, XPES spectra may show the effect of spin–orbit coupling. In the case of XPES, however, the splittings are usually so small that they cannot be resolved and merely appear as slight broadening of the bands.

BIBLIOGRAPHY

Ebsworth, E. A. V., D. W. H. Rankin, and S. Cradock: *Structural Methods in Inorganic Chemistry*, 2nd ed., Blackwell, 1991.

Herzberg, G.: *Molecular Spectra and Molecular Structure*: vol. 1, *Spectra of Diatomic Molecules*, 2nd ed., Van Nostrand, 1950.

Herzberg, G.: *Molecular Spectra and Molecular Structure*: vol. 3, *Electronic Spectra and Electronic Structure of Polyatomic Molecules*, Van Nostrand, 1967.

Turner, D. W., C. Baker, A. D. Baker, and C. R. Brundle: *Molecular Photoelectron Spectroscopy*, Wiley–Interscience, 1970.

Williams, D. H., and I. Fleming: *Spectroscopic Methods in Organic Chemistry*, 4th ed., McGraw-Hill, 1989.

PROBLEMS

(Useful constant: $1\,cm^{-1} \equiv 11.958\,J\,mol^{-1}$.)

6.1 Using the data of Fig. 6.5 (some of which is tabulated below), estimate the dissociation energy of the I_2 molecule:

v:	0	5	10	30	50	70	75	80
$\Delta\varepsilon(cm^{-1})$:	213.3	207.2	200.7	172.1	134.7	82.3	67.0	52.3

6.2 The absorption spectrum of O_2 shows vibrational structure which becomes a continuum at $56\,876\,cm^{-1}$; the upper electronic state dissociates into one ground state atom and one excited atom (the excitation energy of which, measured from the atomic spectrum, is $15\,875\,cm^{-1}$). Estimate the dissociation energy of ground state O_2 in $kJ\,mol^{-1}$.

6.3 The values of $\bar\omega_e$ and x_e in the ground state ($^3\Pi_u$) and a particular excited state ($^3\Pi_g$) of C_2 are:

	$\bar\omega_e$ (cm^{-1})	x_e
Ground state	1641.4	7.11×10^{-3}
Excited state	1788.2	9.19×10^{-3}

Use Eq. (6.10) to find the number of vibrational energy levels below the dissociation limit and hence the dissociation energy of C_2 in both states.

6.4 The spectrum arising from transitions between the two states of C_2 in Prob. 6.3 shows the v_{00} line at $19\,378\,cm^{-1}$ and a convergence limit at $39\,231\,cm^{-1}$. The dissociation is into one ground state atom and one excited atom, the excitation energy of the latter being $10\,308\,cm^{-1}$; calculate the exact dissociation energies of the two states and compare your answers with those of Prob. 6.3. Explain any discrepancy.

6.5 The band origin of a transition in C_2 is observed at $19\,378\,cm^{-1}$, while the rotational fine structure indicates that the rotational constants in excited and ground states are, respectively, $B' = 1.7527\,cm^{-1}$ and $B'' = 1.6326\,cm^{-1}$. Estimate the position of the band head. Which state has the larger internuclear distance?

6.6 Describe qualitatively what you would expect to see in the carbon region of the XPES spectrum of ethyl trifluoroacetate, giving reasons. The structure of the molecule is:

$$
\begin{array}{ccccccc}
 & F & & O & & H & H \\
 & | & & \| & & | & | \\
F & - C & - C & - O & - C & - C & - H \\
 & | & & & & | & | \\
 & F & & & & H & H
\end{array}
$$

6.7 A diatomic molecule, AB, has a UPES spectrum which contains two main features. The first shows only one sharp line, while the other is a cluster of peaks separated by $2300\,cm^{-1}$. Interpret the spectrum in terms of the orbitals from which the features arise, given that the vibrational frequency of the neutral molecule in the ground state is about $2100\,cm^{-1}$.

SPIN RESONANCE SPECTROSCOPY

We have seen in earlier chapters that all electrons and some nuclei possess a property conveniently called 'spin'. Electronic spin was introduced in Chapter 5 to account for the way in which electrons group themselves about a nucleus to form atoms and we found that the spin also accounted for some fine structure, such as the doublet nature of the sodium D line, in atomic spectra. Equally, in Sec. 5.6 it was necessary to invoke a nuclear spin to account for very tiny effects, called hyperfine structure, observed in the spectra of some atoms.

In this chapter we shall consider these spins in rather more detail and discuss the sort of spectra they can give rise to directly rather than their influence on other types of spectra. After an introduction discussing the interaction of spin with an external magnetic field we shall consider in some detail the spectra of particles with a spin of $\frac{1}{2}$ (i.e. electrons and some nuclei such as hydrogen, fluorine, or phosphorus), then a brief discussion of some other nuclei whose spins are greater than $\frac{1}{2}$, and finally a few words on the techniques involved in producing electronic and nuclear spin spectra.

7.1 SPIN AND AN APPLIED FIELD

7.1.1 The Nature of Spinning Particles

We have seen that all electrons have a spin of $\frac{1}{2}$, that is they have an angular momentum of $\sqrt{[\frac{1}{2}(\frac{1}{2}+1)]}(h/2\pi) = \sqrt{3/2}$ units. Many nuclei also possess spin although the angular momentum concerned varies from nucleus to nucleus.

The simplest nucleus is that of the hydrogen atom, which consists of one particle only, the *proton*. The protonic mass (1.67×10^{-27} kg) and charge ($+1.60 \times 10^{-19}$ C) are taken as the units of atomic mass and charge, respectively; the charge is, of course, equal in magnitude but opposite in sign to the electronic charge. The proton also has a spin of $\frac{1}{2}$.

Another particle which is a constituent of all nuclei (apart from the hydrogen nucleus) is the *neutron*; this has unit mass (i.e. a mass equal to that of the proton), no charge, and, again, a spin of $\frac{1}{2}$.

Thus if a particular nucleus is composed of p protons and n neutrons its total mass is $p + n$ (ignoring the small mass defects associated with nuclear binding energy), its total charge is $+p$, and its total spin will be a vector combination of $p + n$ spins each of magnitude $\frac{1}{2}$. The atomic mass is usually specified for each nucleus by writing it as a prefix to the nuclear symbol; for example ^{12}C indicates the nucleus of carbon having a mass of 12. Since the atomic charge is six for this nucleus we know immediately that the nucleus must contain six protons and six neutrons to make up a mass of 12. The nucleus ^{13}C (an *isotope* of carbon) has six protons and seven neutrons.

Each nuclear isotope, being composed of a different number of protons and neutrons, will have its own total spin value. Unfortunately, the laws governing the vector addition of nuclear spins are not yet known so the spin of a particular nucleus cannot be predicted in general. However, observed spins can be rationalized and some empirical rules have been formulated.

Thus the spin of the hydrogen nucleus (^{1}H) is $\frac{1}{2}$ since it consists of one proton only; deuterium, an isotope of hydrogen containing one proton and one neutron (that is ^{2}H) might have a spin of 1 or 0 depending on whether the proton and neutron spins are parallel or opposed: it is observed to be 1. The helium nucleus, containing two protons and two neutrons (^{4}He), has zero spin, and from these and other observations stem the following rules:

1. Nuclei with both p and n even (hence charge *and* mass even) have zero spin (for example ^{4}He, ^{12}C, ^{16}O, etc.).
2. Nuclei with both p and n odd (hence charge *odd* but mass $= p + n$, *even*) have integral spin (for example ^{2}H, ^{14}N (spin $=$ 1), ^{10}B (spin $=$ 3), etc.).
3. Nuclei with odd mass have half-integral spins (for example ^{1}H, ^{15}N (spin $= \frac{1}{2}$), ^{17}O (spin $= \frac{5}{2}$), etc.).

The spin of a nucleus is usually given the symbol I, called the *spin quantum number*. Quantum mechanics shows that the angular momentum of a nucleus is given by the expression:

$$\text{Angular momentum } \mathbf{I} = \sqrt{I(I+1)}(h/2\pi) = \sqrt{I(I+1)} \text{ units} \tag{7.1}$$

where I takes, for each nucleus, *one* of the values, $0, \frac{1}{2}, 1, \frac{3}{2}, \ldots$. We can conveniently include the spin quantum number and angular momentum of an *electron* in Eq. (7.1) if we agree to label its spin quantum number I (instead of s as in Chapter 5), and remember that I can be $\frac{1}{2}$ only for an electron. Thus Eq. (7.1) represents the angular momentum of nuclei and of electrons once the appropriate value of I is inserted.

We may note here that many texts use a simpler form of Eq. (7.1), viz.:

$$\mathbf{I} = I\,\frac{h}{2\pi} = I \text{ units}$$

This equation is not, however, strictly correct from a quantum mechanical point of view and we shall use the more rigorous equation (7.1) throughout this chapter.

By now the reader will be familiar with the idea that the angular momentum vector $\mathbf{I}$ cannot point in any arbitrary direction, but can point only so that its components along a particular reference direction are either all integral (if I is integral) or all half-integral (if I is half-integral). Thus we can have components along a particular direction z, of:

$$I_z = I, I-1, \ldots, 0, \ldots, -(I-1), -I \quad \text{(for } I \text{ integral)}$$

or $\qquad I_z = I, I-1, \ldots, \tfrac{1}{2}, -\tfrac{1}{2}, \ldots, -I \qquad \text{(for } I \text{ half-integral)} \qquad (7.2)$

giving $2I + 1$ components in each case. These components are normally degenerate—i.e. they all have the same energy—but the degeneracy may be lifted and $2I + 1$ different energy levels result if an external magnetic field is applied to define the reference direction. We shall now consider the effect of such a field.

7.1.2 Interaction between Spin and a Magnetic Field

In general, a charged particle spinning about an axis constitutes a circular electric current which in turn produces a magnetic dipole. In other words the spinning particle behaves as a tiny bar magnet placed along the spin axis. The size of the dipole (i.e. the strength of the magnet) for a point charge can be shown to be:

$$\mu = \frac{q}{2m} \mathbf{I} = \frac{q\sqrt{I(I+1)}}{2m} \frac{h}{2\pi} = \frac{qh}{4\pi m} \sqrt{I(I+1)} \quad \text{A m}^2$$

where q and m are the charge and mass of the particle. The magnetic moment is here expressed in the appropriate fundamental SI units, ampere square metre (A m^2); it is useful for later arguments, however, to express the magnetic moment in terms of the magnetic flux density (colloquially 'magnetic field strength'), the SI unit of which is the tesla (symbol T, units $\text{kg s}^{-2} \text{A}^{-1}$), where $1\,\text{T} \equiv 10\,000$ gauss. The conversion is:

$$\text{A m}^2 \equiv (\text{kg s}^{-2} \text{T}^{-1})\,\text{m}^2 \equiv \text{J T}^{-1} \text{ (joules per tesla)}$$

So we may write:

$$\mu = \frac{qh}{4\pi m} \sqrt{I(I+1)} \quad \text{J T}^{-1} \qquad (7.3)$$

and it is this form we shall use in our subsequent discussion. (It should be pointed out that, when the magnetic moment is expressed in c.g.s. units, as in older textbooks, the right-hand side of Eq. (7.3) will be divided by c, the velocity of light.)

When we remove the fiction that electrons and nuclei are *point* charges, Eq. (7.3) becomes modified by the inclusion of a numerical factor G:

$$\mu = \frac{Gqh}{4\pi m} \sqrt{I(I+1)} \quad \text{J T}^{-1} \qquad (7.4)$$

For electrons, we have seen (cf. Sec. 5.6) that G is given the symbol g and called the Landé splitting factor; its value depends on the quantum state of the electron and may be calculated from the L, S, and J quantum numbers (cf. Eq. (5.28)). Nuclear G factors, on the other hand, cannot be calculated in advance and are obtainable only experimentally.

For electrons, Eq. (7.4) is usually written:

$$\mu = -g\beta\sqrt{I(I+1)} \quad \text{J T}^{-1} \qquad (7.5)$$

where we have expressed the set of constants $qh/4\pi m$ as a (positive) constant β, called the *Bohr magneton*; replacing the electronic charge (1.60×10^{-19} C) and mass (9.11×10^{-31} kg) in this expression, we can calculate $\beta = 9.273 \times 10^{-24}\,\text{J T}^{-1}$.

Nuclear dipoles, on the other hand, are conveniently expressed in terms of a *nuclear magneton* β_N, which is defined in terms of the mass and charge of the *proton*:

$$\beta_N = \frac{eh}{4m_p\pi} = 5.050 \times 10^{-27} \text{ J T}^{-1}$$

Thus for a nucleus of mass M and charge pe (where p is the number of protons) we would write:

$$\mu = \frac{Gpe}{2M}\sqrt{I(I+1)}\,\frac{h}{2\pi} = \frac{Gm_pp}{M}\,\beta_N\sqrt{I(I+1)}$$
$$= g\beta_N\sqrt{I(I+1)} \quad \text{J T}^{-1} \tag{7.6}$$

where we have collected the parameters Gm_pp/M, in which m_p is the protonic mass, into a factor g which is characteristic of each nucleus. This factor has values up to about six and is positive for nearly all known nuclei (see Table 7.1 later).

Thus the analogous equations (7.5) and (7.6) define the equivalent spin dipole for any spinning particle. The dipole will plainly have components along a reference direction governed by the I_z values:

$$\left.\begin{array}{ll}\mu_z = -g\beta I_z & \text{(for electrons)} \\ \mu_z = g\beta_N I_z & \text{(for nuclei)}\end{array}\right\} \tag{7.7}$$

where the I_z are given by Eq. (7.2) for a particular particle, and the dipole will interact to different extents, depending on its magnitude, with a magnetic field. The situation for a *nucleus* with $I = 1$ is shown in Fig. 7.1. The angular momentum of the particle (Eq. (7.1)) is:

$$\mathbf{I} = \sqrt{1 \times 2} = \sqrt{2} \text{ units}$$

and if we consider a semicircle with this radius, it is plain that the vector arrow corresponding to $\mathbf{I}$ can point so as to have z components of $+1$, 0, or -1 (the z direction is counted positive towards the top of the paper). This is shown in Fig. 7.1(a).

Equation (7.7) shows that μ_z and I_z have the same sign (i.e. point in the same direction) for the many nuclei which have positive g values. The three μ_z values for this system are represented by arrows in Fig. 7.1(b) where, conventionally, the lines of force inside the magnet are drawn

| (a) | (b) | (c) |

Figure 7.1 Showing (a) the three orientations of the spin of a nucleus with spin quantum number $I = 1$; (b) the resulting magnetic dipole, μ_z, oriented in an applied magnetic field B_z; and (c) the three energy levels allowed to the nucleus.

with an arrow pointing to the N pole. If we imagine the applied magnetic field to be produced by a horsehoe magnet and apply the same convention, it is clear that the lines of force *external* to the magnet will be shown pointing from the N to the S pole and, if we require these to be in the positive z direction, we arrive at the configuration given in Fig. 7.1(b). Thus we see that the state $I_z = -1$ represent a nuclear dipole *opposed* to the magnetic field (i.e. of high energy) while $I_z = +1$ is in the same direction as the applied field and is, therefore, of low energy. The state $I_z = 0$ has no net dipole along the field direction and is therefore unchanged in energy whether the field is applied or not. This is shown in Fig. 7.1(c).

Of course, if the nuclear g factor is negative, μ_z has a sign opposite from I_z, and the order of labelling the energy levels of Fig. 7.1 will be reversed. Similarly, of the two energy levels allowed to an electron in a magnetic field, the lower will be associated with $I_z = -\frac{1}{2}$, the upper with $I_z = +\frac{1}{2}$.

The extent of interaction between a magnetic dipole and a field of strength B_z applied along the z axis is equal to the product of the two:

$$\text{Interaction} = \mu_z B_z$$

Thus the separation between neighbouring energy levels (where I_z differs by unity) is:

$$\Delta E = |E_{I_z} - E_{(I_z - 1)}| = |g\beta_N I_z B_z - g\beta_N (I_z - 1)B_z|$$
$$= |g\beta_N B_z| \quad \text{J} \quad \text{(when } B_z \text{ is expressed in tesla)} \tag{7.8}$$

Thus in hertz:

$$\frac{\Delta E}{h} = \left|\frac{g\beta_N B_z}{h}\right| \quad \text{Hz} \tag{7.9}$$

(where the modulus sign, $|\dots|$, indicates that *positive* differences only should be considered). Here, then, is the basis for a spectroscopic technique: a transition of electron or nuclear spins between energy levels (loosely referred to as 'a change of spin') may be associated with the emission or absorption of energy in the form of radiation at the frequency of Eq. (7.9). Further, since the frequency is proportional to the applied field we can arrange, in principle, to study spin spectra in any region of the electromagnetic spectrum, merely by choosing an appropriate field. However, for practical reasons, the fields used are normally of the order of 1–5 tesla for nuclei and 0.3 tesla for electrons. Let us calculate the approximate frequency to be expected under these circumstances.

For nuclei: We have already $\beta_N = 5.05 \times 10^{-27}$ J T^{-1}, and if we choose the rather specific value of $B_z = 2.3487$ T and the g factor of hydrogen, $g = 5.585$, we calculate:

$$\frac{\Delta E}{h} = \frac{5.585 \times 5.05 \times 10^{-27} \times 2.3487}{6.63 \times 10^{-34}} = 100 \times 10^6 \text{ Hz} \tag{7.10}$$

and we see that the appropriate frequency for protons, 100 MHz, falls in the short-wave radio-frequency region; in fact a great many nuclear magnetic resonance spectrometers operate at just this frequency, which is why we chose such a precise value of B_z for the calculation. All other nuclei (except tritium) have smaller g factors, and their spectra fall between 1 and 100 MHz for the same applied field. Table 7.1 collects some data for a few of the more important nuclei.

For electrons: Here $\beta = 9.273 \times 10^{-24}$ J T^{-1}, and let us assume $g = 2$ and $B_z = 0.34$ T. Then:

$$\frac{\Delta E}{h} = \frac{2 \times 9.273 \times 10^{-24} \times 0.34}{6.63 \times 10^{-34}} \approx 9500 \times 10^6 \text{ Hz} \tag{7.11}$$

Thus electron spin spectra fall at a considerably higher frequency, which is on the long wave-length edge of the microwave region. Because of this difference, techniques of nuclear and

Table 7.1 Properties of some nuclei with non-zero spin

Nucleus	Spin	Resonance frequency (MHz) in field of 2.3487 T	g value
^{1}H	$\frac{1}{2}$	100.00	5.585
^{10}B	3	10.75	0.6002
^{11}B	$\frac{3}{2}$	32.08	1.792
^{13}C	$\frac{1}{2}$	25.14	1.404
^{14}N	1	7.22	0.4036
^{15}N	$\frac{1}{2}$	10.13	−0.5660
^{17}O	$\frac{5}{2}$	13.56	−0.7572
^{19}F	$\frac{1}{2}$	94.07	5.255
^{29}Si	$\frac{1}{2}$	19.87	−1.110
^{31}P	$\frac{1}{2}$	40.48	2.261
^{35}Cl	$\frac{3}{2}$	9.80	0.5472
^{37}Cl	$\frac{3}{2}$	8.16	0.4555
^{107}Ag	$\frac{1}{2}$	4.05	−0.2260
^{119}Sn	$\frac{1}{2}$	37.27	−2.082
^{127}I	$\frac{5}{2}$	20.00	1.118
^{199}Hg	$\frac{1}{2}$	17.83	0.996

electronic spin spectroscopy differ considerably, as we shall see later, although in principle they are concerned with very similar phenomena.

Perhaps we should note that practitioners of nuclear magnetic resonance (n.m.r.) spectroscopy normally refer to the applied magnetic field as B_0 rather than the B_z which we have used above. For the remainder of Sec. 7.1, however, it is convenient to retain B_z (and later to introduce B_x and B_y) in order to emphasize the interaction of the spin vectors I_z with these fields. We will use B_0 for B_z (and B_1 for B_x) from Sec. 7.2 onwards.

7.1.3 Population of Energy Levels

When first confronted with nuclear and electron spin spectroscopy the student (who has experimented earlier with bar magnetics in the earth's field) usually asks: why don't the nuclear (or electronic) magnetic moments immediately line themselves up in an applied field so that they all occupy the lowest energy state?

There are several facets to this question and its answer. Firstly, if we take 'immediately' to refer to a period of some seconds, then spin magnetic moments *do* immediately orientate themselves in a magnetic field, although they do *not* all occupy the lowest available energy state. This is a simple consequence of thermal motion and the Boltzmann distribution. We have seen that spin energy levels are split in an applied field, and their energy separation (Eq. (7.8)) is ΔE joules. Let us confine our attention to particles with spin $\frac{1}{2}$ (and hence just two energy levels) for simplicity—our remarks, however, are easily extended to cover the general case. Classical theory states that at a temperature T K the ratio of the populations of such levels will be given by:

$$\frac{N_{\text{upper}}}{N_{\text{lower}}} = \exp\left(-\frac{\Delta E}{kT}\right) \tag{7.12}$$

where k is the Boltzmann constant. Thus at all temperatures above absolute zero the upper level will always be populated to some extent, although for large ΔE the population may be insignificant. In the case of nuclear and electron spins, however, ΔE is extremely small:

$$\Delta E \text{ nuclei} \quad \approx 7 \times 10^{-26}\,\text{J} \quad \text{in a 2.3487 T field}$$

$$\Delta E \text{ electrons} \approx 6 \times 10^{-24}\,\text{J} \quad \text{in a 0.34 T field}$$

and since $k = 1.38 \times 10^{-23}\,\text{J K}^{-1}$, we have at room temperature ($T = 300\,\text{K}$):

$$\frac{N_{\text{upper}}}{N_{\text{lower}}} \approx \exp\left(-\frac{7 \times 10^{-26}}{4.2 \times 10^{-21}}\right) \approx \exp(-1 \times 10^{-5})$$

$$\approx 1 - (1 \times 10^{-5}) \quad \text{for nuclei}$$

and

$$\frac{N_{\text{upper}}}{N_{\text{lower}}} \approx \exp\left(-\frac{6 \times 10^{-24}}{4.2 \times 10^{-21}}\right) \approx \exp(-1 \times 10^{-3})$$

$$\approx 1 - (1 \times 10^{-3}) \quad \text{for electrons}$$

In both cases the ratio is very nearly equal to unity and we see that the spins are almost equally distributed between the two (or, in general, $2I + 1$) energy levels.

We need to discuss now the nature of the interaction between radiation and the particle spins which can give rise to transitions between these levels.

7.1.4 The Larmor Precession

We have seen (Eq. (7.6)) that the dipole moment of a spinning nucleus is

$$\mu = g\beta_N \sqrt{I(I+1)} \quad \text{J T}^{-1}$$

and that, according to quantal laws, the vector represented by μ can be oriented only so that its components are integral or half-integral in a reference direction. The corollary to this is that, since $\sqrt{I(I+1)}$ cannot be integral or half-integral if I is integral or half-integral, the vector arrow can *never* be exactly in the field direction. We see one example of this in Fig. 7.1(*a*) and another for a nucleus with spin $\frac{1}{2}$ in Fig. 7.2(*a*). For such a particle:

$$\mu = g\beta_N \sqrt{3/2} \quad \text{and} \quad \mu_z = \pm\tfrac{1}{2}g\beta_N \qquad \text{only}$$

Thus, whichever energy state a spinning nucleus or electron is in, it will always lie more or less across the field and will therefore be under the influence of a couple tending to turn it into the field direction.

Now the behaviour of a spinning nucleus or electron can be considered analogous to that of a gyroscope running in friction-free bearings. Experiments convince us that the application of a couple to a gyroscope does not cause its axis to tilt but merely induces a *precession* of the axis about the direction of the couple. Essentially the same occurs with a spinning particle and the precession, known as the Larmor precession, is sketched in Fig. 7.2(*b*). The precessional frequency (or Larmor frequency) is given by:

$$\omega = \frac{\text{magnetic moment}}{\text{angular momentum}} \times B_z \quad \text{rad s}^{-1}$$

$$= \frac{\mu B_z}{2\pi I} \quad \text{Hz}$$

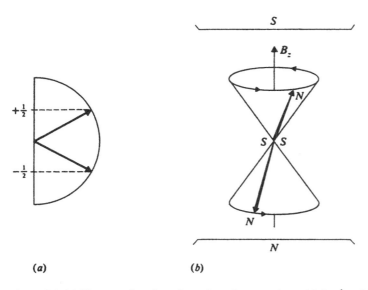

(a)   (b)

Figure 7.2 (a) The two spin orientations allowed to a nucleus with $I = \frac{1}{2}$ and (b) the Larmor precession of such a nucleus.

Replacing μ and $\mathbf{I}$ by their expressions in Eqs (7.6) and (7.1):

$$\omega = \frac{g\beta_N \sqrt{I(I+1)}}{\sqrt{I(I+1)}\,(h/2\pi)} \frac{B_z}{2\pi} = \frac{g\beta_N B_z}{h} \quad \text{Hz}$$

and, comparing with Eq. (7.9), we see that the Larmor precessional frequency is just the frequency separation between energy levels.

This, then, is a mechanism by which particle spins can interact with a beam of electromagnetic radiation. If the beam has the same frequency as that of the precessing particle, it can interact coherently with the particle and energy can be exchanged; if it is of any other frequency, there will be no interaction. The phenomenon, then, is one of *resonance*. For nuclei it is referred to as *nuclear magnetic resonance* (n.m.r.), while for electrons it is called *electron spin resonance* (e.s.r.) or, sometimes, *electron paramagnetic resonance* (e.p.r.).

Experimentally there is a choice of two arrangements. We might either apply a fixed magnetic field to a set of identical nuclei so that their Larmor frequencies are all, say, 100 MHz; if the frequency of the radiation beam is then swept over a range including 100 MHz, resonance absorption will occur at precisely that frequency. On the other hand, we could bathe the nuclei in radiation at a fixed frequency of 100 MHz and sweep the applied field over a range until absorption occurs.

The probability of transitions occurring from one spin state to another is directly proportional to the population of the state from which the transition takes place. We have seen in the previous section that these populations are very nearly equal, and so during resonance, upward and downward transitions are induced to almost the same extent. However, while the lower state is more populated than the upper (e.g. at equilibrium in the absence of radiation), upward transitions predominate slightly, and a net (but very small) absorption of energy occurs from the radiation beam. When the populations become equal, upward and downward transitions are equally likely, no further absorption can take place, and the system is said to be *saturated*. The equilibrium populations can be re-established if the system loses its absorbed energy; this it cannot do spontaneously, but only as a result of interaction with radiation or with fluctuations in surrounding magnetic fields of the appropriate frequency. The emitted energy can be collected and displayed as an emission spectrum which, since the emission is induced and not spontaneous, is called the *nuclear induction spectrum*.

7.1.5 Relaxation Times

Let us return to the question posed at the beginning of Sec. 7.1.3, and in particular consider the word 'immediately'. If an external field were suddenly applied to a set of nuclei in bulk material, and if such nuclei can be considered as completely frictionless gyroscopes, then they could not change their orientation in order to produce the correct statistical population of the energy levels unless radiation of the appropriate frequency were present. Without such radiation there would be no mechanism by which the excess energy of the nuclei could be removed from the system (and one would speak of it as having a high 'spin temperature'). However, it is a fact that such nuclei do orient themselves to give the appropriate populations of states for a given temperature without the presence of radiation. The mechanism by which excess spin energy is shared either with the surroundings or with other nuclei is referred to generally as a *relaxation process*; the time taken for a fraction $1/e = 0.37$ of the excess energy to be dissipated is called the *relaxation time*.

Two different relaxation processes can occur for nuclei. In the first, the excess spin energy equilibrates with the surroundings (the *lattice*) by spin-lattice relaxation having a *spin-lattice relaxation time* (or *longitudinal relaxation time*) T_1. Such relaxation comes about by lattice motions (e.g. atomic vibrations in a solid lattice, or molecular tumbling in liquids and gases) having approximately the right frequency to interact coherently with nuclear spins. T_1 varies greatly, being some 10^{-2}–10^4 s for solids and 10^{-4}–10 s for liquids, the overall shorter times for liquids being due to the greater freedom of molecular movement leading to larger fluctuations of magnetic field in the vicinity of the nuclei. We discuss the measurement of T_1 in Sec. 7.1.8.

Secondly, there is a sharing of excess spin energy directly between nuclei via *spin-spin* (or *transverse*) *relaxation*, the symbol for the time of which is T_2. For solids T_2 is usually very short, of the order of 10^{-4} s, while for liquids $T_2 \approx T_1$. The measurement of T_2 is described in Sec. 7.1.7.

T_1 and T_2 have a marked effect on the widths of n.m.r. spectral lines since they reflect the lifetime of a particular spin state. Thus a long relaxation time (both T_1 and T_2 large) means that an excited nuclear spin reverts rather slowly to a lower state (i.e. has a long lifetime), and we know from Heisenberg's uncertainty principle (Eq. (1.10)) that this results in only a small uncertainty in the excited state energy level. Thus taking a typical value of one second for the relaxation time of a nucleus in a liquid, we would calculate:

$$\delta E \approx h/2\pi\delta t \approx 10^{-34} \text{ J}$$

(which uncertainty is to be compared with the spacing between adjacent energy levels for nuclei in a 2.3487 T field, shown in Sec. 7.1.3 to be about 10^{-26} J). The corresponding uncertainty in the radiation frequency associated with the transition would be:

$$\delta v = \delta E/h \approx 0.1 \text{ Hz}$$

In general n.m.r. spectrometers are not capable of resolving lines closer than about 0.5 Hz apart, so that a line width of only 0.1 Hz represents a narrow spectral line.

When either T_1 or T_2 is short the uncertainty is correspondingly larger; for example taking $\delta t = 10^{-4}$ s, as for T_2 in a typical solid, the calculation yields $\delta v \approx 1000$ Hz, which is clearly a very broad line compared with the resolving power. Thus n.m.r. experiments are divided into two main classes: broad-line, usually comprising solid samples, and high-resolution, usually of liquids or gases. There are techniques, however, which yield high-resolution spectra for some solids (see Chapter 8, Sec. 8.3, for brief details), and some liquids may give spectra containing broad lines if both T_1 and T_2 are small; such liquids are generally very viscous or contain paramagnetic ions which increase the efficiency of relaxation processes.

Line-width measurement is not a wholly satisfactory method of obtaining relaxation times, for several reasons: line widths themselves cannot always be accurately measured, particularly in a spectrum with overlapping lines; other factors, like magnetic field inhomogeneity or chemical exchange processes, also affect the line width; and a line width can only give an estimate of the efficiency of the whole relaxation process—it does not distinguish between spin-spin and spin-lattice effects. These drawbacks can largely be overcome by using Fourier transform techniques which, in n.m.r., are invariably carried out by applying one or more pulses of exciting radiation to the sample and observing the signal emitted by excited nuclei. The following sections will be devoted to discussing how these measurements are made; we shall discuss some of the applications of relaxation times in Sec. 7.3.2.

7.1.6 Fourier Transform Spectroscopy in N.M.R.

Consider again Fig. 7.2(b), which shows the precession of a nuclear spin in an applied field B_z, where the spin may be either with or against the field. In what follows we shall implicitly consider hydrogen nuclei in a field of 2.5 T, precessing at a frequency of 100 MHz,† that is 10^{-8} times per second. Remember that there is really a very large number of nuclear spins in the sample and, on average, they will be spread evenly around the precessional 'cone' as in Fig. 7.3(a). Because of this, the total resultant magnetic effect of these spins will be aligned along the cone axis and, since there are slightly more 'up' spins than 'down', we represent the net *bulk*

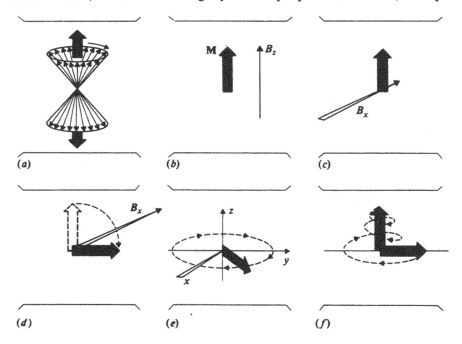

Figure 7.3 (a) The precessional 'cone' of a group of spinning nuclei; (b) the resulting bulk magnetic moment vector, **M**; (c) the application of a 90° pulse along the x axis; (d) the precession of **M** through 90° about the x axis; (e) the additional precession of **M** about the main field axis; and (f) the overall spiral motion of **M**.

† As we saw earlier, hydrogen nuclei precess at 100 MHz in a 2.3487 T field, but since we need only approximate calculations in this section, we will use the simpler figure of 2.5 T

magnetic moment vector of the spins by an upward arrow, as in Fig. 7.3(*b*). It is sometimes helpful to think of the behaviour of the collection of nuclear spins in terms of this bulk magnetic moment, which we call **M**.

We are now ready to consider the Fourier transform experiment in n.m.r. Imagine that we apply a magnetic field of strength B_x along the x axis (i.e. vertical to the plane of the paper), as in Fig. 7.3(*c*); experimentally this would be achieved by passing a radiofrequency current (of 100 MHz to maximize the resonance interaction between the field and nuclei) through a coil whose axis is aligned along the x direction. Such a field will try to turn the bulk magnetic moment, **M**, into the x direction. However, **M**, being composed of spinning magnets, behaves like a gyroscope and cannot be turned in this way; instead it precesses about the field direction, i.e. about the x axis, as shown in Fig. 7.3(*d*). The *rate* at which it precesses will be directly proportional to the magnitude of B_x. For simplicity, let us imagine B_x to be 2.5×10^{-4} T, i.e. just 10^{-4} of the main field B_z, so **M** will precess at 10^{-4} times the precessional frequency induced by the main field, i.e. at $10^8 \times 10^{-4} = 10^4$ Hz; thus we see that **M** will take just 10^{-4} seconds (100 microseconds) to complete a revolution. If we apply B_x for a very short period of time, **M** will not be able to complete a whole revolution; for example, we could apply B_x for 50 micro-seconds, in which case **M** will complete only half a revolution and then stop precessing—this is referred to as a 180° pulse of radiation. Similarly a 25 microsecond pulse at an intensity of 2.5×10^{-4} T is a 90° pulse, the effect of which is shown in Fig. 7.3(*d*). Remember that the actual length, in microseconds, of a 90° pulse will depend on the strength of the pulse field. Increasing B_x by a factor of 10 will increase the rate of **M**'s precession by 10, and so a 90° pulse at 2.5×10^{-3} T would be 2.5 microseconds long.

In an n.m.r. spectrometer the transmitter and receiver coils are carefully aligned with their axes at right angles to each other so that there is no direct coupling between them. Conventionally we take the transmitter to be aligned along the x axis and the receiver along the y axis (horizontally in the paper), with the main field in the z axis. Now while the bulk magnetic spin vector is along the z axis there is no fluctuation of magnetic field along y so the receiver coil sees no signal. When the 90° pulse is applied, however, **M** is 'flipped' into the xy plane, and three things immediately start to happen. Firstly, **M** experiences a force from the main applied field trying to turn it back into the z direction; as a gyroscope, however, its only response to this force is precession about the force axis, so **M** will begin to precess in the xy plane, as in Fig. 7.3(*e*), at 100 MHz. Secondly, and because of this precession, there *is* now a fluctuating magnetic effect along the x axis—the precessing **M**— so the detector receives a signal, also at 100 MHz. Finally, **M** is in an unstable state—it would 'prefer' to be aligned with the main field axis—and so normal relaxation processes occur to dissipate its excess energy, and **M** gradually (i.e. in a period depending on the relaxation time) returns to the vertical position. The overall motion of **M** after the 90° pulse can be visualized as a spiral, shown in Fig. 7.3(*f*), during which the detector will receive a decreasing signal until **M** is vertical once more. (Strictly, with a precessional frequency of 10^8 Hz and a relaxation time of, perhaps, some 10^{-2} seconds, **M** will precess many thousands or even millions of times during the decay; Fig. 7.3(*f*) should be treated as purely diagrammatic.) This signal can be Fourier transformed (cf. Sec. 1.8) and displayed as the frequency spectrum of the nuclei. Note that the decreasing signal at the detector arises here because the pulse along x excites nuclei to the upper state from which they revert to the lower. Once equilibrium is re-established (vertical **M**) no more signal is emitted unless a further energizing pulse is applied. In fact, of course, all the sample nuclei do *not* emit at precisely 100 MHz—a package of frequencies is emitted (again cf. Chapter 1) which will also, by interference, contribute towards the decay of the signal.

The important point to note is that FT spectroscopy in n.m.r. requires the application of a 90° pulse, observation of the decaying emission signal (often called the *free induction decay*, or FID), and then transformation of the signal for display. Although relaxation processes decide the rate of signal decay, this single experiment is no more suitable than conventional 'frequency domain' methods for measuring relaxation times, and suffers from the deficiencies mentioned at the end of the preceding section. In order to measure relaxation times we need to consider *multiple-pulse* techniques.

7.1.7 Multiple-Pulse FT: Spin-Spin Relaxation

Let us consider in rather more detail the effect of the 90° pulse described above. Essentially for **M** to be tilted into the horizontal (*xy*) plane, individual nuclear magnets must have been disturbed from their even distribution round the precessional cone, and have bunched to one side, as shown in Fig. 7.4(*a*). Their precessional motion still exists, of course, so one should think of the bunch precessing as a whole, thus causing **M** to precess in the *xy* plane, as indicated earlier. Now for **M** to relax back to equilibrium implies that the energy of the spins must be shared evenly again, and for this to happen we require an exchange of energy *between* the spins. In other words the relaxation process which allows horizontal **M** to become vertical **M** again is *spin-spin relaxation*. Since **M** decays *across* the main field direction, this process is also referred to as *transverse* relaxation.

In order to measure the spin-spin relaxation time (T_2) we need to disentangle this relaxation process from other processes which can cause the spins to precess at different frequencies (mainly field inhomogeneities) and so can also unscramble the bunching. To understand the method, let us imagine looking along the main field direction from the top (i.e. along the axis of the precessional cone). We show the total magnetic vector, seen in this way, in Fig. 7.4(*b*). Now **M** actually rotates at 100 MHz due to its precession; but to simplify the pictures which follow, imagine that we have 'killed' this rotation by, for example, observing the system with stroboscopic light, flashing at 100 MHz. Thus we shall essentially see **M** once in each rotation at the same place. Field inhomogeneities will mean that some nuclei precess slightly faster, some slightly slower, than average. If we observe the system for a time τ the slower nuclei will appear

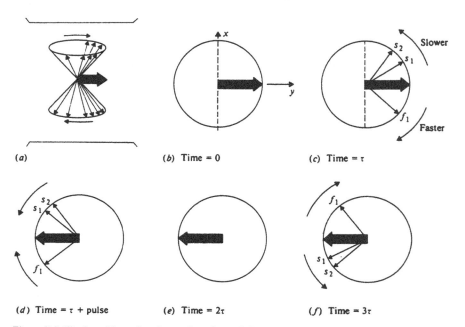

(*a*) (*b*) Time = 0 (*c*) Time = τ

(*d*) Time = τ + pulse (*e*) Time = 2τ (*f*) Time = 3τ

Figure 7.4 The bunching of nuclear spins after a 90° pulse and the effect of a subsequent 180° pulse.

to lag behind the majority, the faster to drift ahead, as shown in Fig. 7.4(c), where f_1 is a faster nucleus, s_1 and s_2 are slower.

At this point another pulse is applied along the x axis via the transmitter coil, but this time a 180° pulse. Its effect is to rotate all spins by 180° about the x axis (*not* about the centre of the circle), and the net result will be as shown in Fig. 7.4(d): noting carefully where the slower and faster nuclei arrive, we see that, since we have done nothing to change their precessional velocities, the faster nuclei will now catch up with the majority while the slower will drift back towards them. Since it took a time τ for the bunch to fan out to Fig. 7.4(c), it will take a further time τ (total time 2τ) for the fanning out to be reversed. The effect of all this on the detector is as follows: immediately after the original 90° pulse the detector receives a maximum signal, which decays due to the fanning out of nuclear spins from field inhomogeneities. The 180° pulse is applied at τ, the fanning out reverses, and the signal builds up again until at 2τ, when the spins are bunched again, the signal reaches a maximum once more. The process continues—at 3τ the fanning out will look like Fig. 7.4(f), and a further 180° pulse will then cause rebunching of the spins at 4τ. This is all collected into Fig. 7.5, where (a) shows the pulse timing and (b) the detector output. Not surprisingly the technique is often referred to as 'spin-echo', or as the 90°–τ–180° sequence.

Now although the 180° pulses effectively cancel out those *external* effects, such as field inhomogeneity, which produce different precessional frequencies, they cannot cancel the internal decay of **M** from spin-spin interactions which also equalize the distribution of spins round the precessional cone. Thus each spin-echo maximum is smaller than the preceding one as **M** shrinks. The exponential decay of these maxima, drawn in on Fig. 7.5(b), can thus be used to measure T_2. Note that we are here dealing with the decay of the 'raw' detector signal—there is no need to Fourier transform the signal for display as a frequency spectrum, since we are only interested in the change in its overall intensity.

In fact the T_2 measurement described above is by no means simple to carry out—very precise timing of the microsecond pulses is essential otherwise the spin system will rapidly become hopelessly scrambled, and the presence of nuclei in the sample with different precessional frequencies (see the discussion of chemical shifts later) will mask the echo pattern. Thus T_2 measurements are of little practical use, as yet, in structure determination. However, the principle of their measurement is simple to understand, and should be helpful when we now consider spin-lattice relaxation, giving T_1 values which are becoming of increasing practical importance.

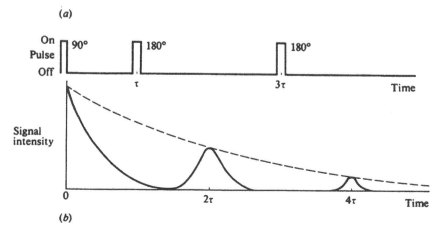

Figure 7.5 (a) The pulse timing of a 90°–τ–180° sequence and (b) the resultant spin echoes.

7.1.8 Multiple-Pulse FT: Spin-Lattice Relaxation

Instead of starting with a 90° pulse, consider applying a 180° pulse initially to the spin system. Figure 7.6(*a*) and (*b*) shows that this will *reverse* the direction of M, and will leave the spin system in an unstable state since more spins point against the field than with it. As excited nuclei give up their energy (i.e. relax) M will become smaller, pass through zero, then grow upwards until it reaches its equilibrium value once more—an exponential change which is illustrated in Fig. 7.6(*c*). Remembering that the relaxation time is the time required for $1/e = 0.37$ of the original excitation energy to be dissipated, we can indicate this time on the curve.

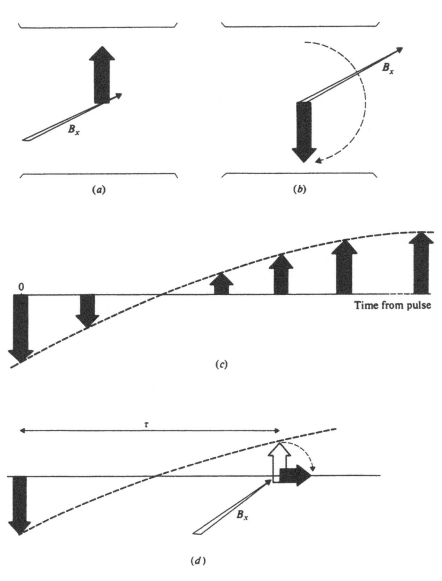

Figure 7.6 A 180°–τ–90° pulse sequence and its effect on M.

Two points are important here. Firstly, the excess spin energy is given up by the nuclei to their surroundings, and so the time involved is the spin-*lattice* relaxation time, and we can see that the alternative name for this, the *longitudinal* relaxation time, is because M relaxes *along* the main field axis. Secondly, all the changes in M sketched in Fig. 7.6(*c*) are confined to the *z* axis, and so will produce no signal in the receiver coil along *y*.

We know that we can induce signal emission, however, by applying a subsequent 90° pulse to the system after some delay time τ, as shown in Fig. 7.6(*d*), thus reaching the situation discussed earlier in Fig. 7.3(*c*) to (*f*). M will spiral up to equilibrium, emitting a decaying signal, but the *initial* intensity of the signal, immediately after the 90° pulse, will be directly proportional to the magnitude of M at the time of that pulse. If we lengthen τ, M will have grown more before being tipped into the *xy* plane, and a more intense signal will result. Shorten τ a little, and the signal will be weaker because M has not grown so much. Indeed, make τ much shorter, so that M is still pointing in the 'wrong' direction when it is tipped into the *xy* plane, and the signal will be *negative*, i.e. it will be an out-of-phase signal which, on Fourier transformation, will give a negative spectrum peak rather than a positive one. In order to measure spin-lattice relaxation, then, we must carry out several 180°–τ–90° experiments with different values of the delay time τ between the pulses, and then plot the intensity of the signal, observed immediately after the 90° pulse, against τ. Typically τ will be of the order of seconds.

Of course the relaxation of M into its final equilibrium position and magnitude continues after we have observed the initial signal. We are not interested in this, however—indeed we saw in the previous section that subsequent relaxation is also affected by field inhomogeneities and spin-spin relaxation. We can best think of the 90° pulse as 'freezing' the spin-lattice relaxation of M at various times after reversal so that we can measure how far relaxation has progressed each time.

Naturally the computer attached to the spectrometer is used to control the whole process of measurement. It times the original 180° pulse, selects increasing τ values, times the 90° pulse, and collects and measures the signal intensity. From these data it can readily calculate and report a value for T_1 directly, and there is again strictly no reason to Fourier transform the signal and produce frequency spectra. It has become conventional, however, to continue with this final step and to display the resulting spectra in a particularly graphic form giving a pseudo three-dimensional picture showing the frequency of lines in the spectrum and the relationship of their intensities to the delay time τ. Typically the computer plots a set of 'stacked' spectra in the form of Fig. 7.7, where each spectrum is offset vertically and to the right by an amount proportional to its delay time. Such a picture shows very pleasingly the initial negative signal from the peak, its diminishment through zero, and subsequent growth to a maximum. On the figure we have joined the peak maxima with a dashed line to show the exponential change in height. Comparison of this figure, viewed sideways, with Fig. 7.6(*c*), shows the very precise correspondence between the magnitude and direction of M and its n.m.r. signal. An approximate value for the spin-lattice relaxation time, T_1, can be found very simply from such a display—in fact T_1 is about $1\frac{1}{2}$ times the time taken for the peak to decay to zero intensity. In this particular (hypothetical) example, the signal becomes zero at about 7 seconds, so the relaxation time is 1.5×7 or about 10 seconds.

In summary, then, T_1 measurements are relatively easy and usually fairly fast. Some half a dozen 180°–τ–90° experiments, each lasting a few seconds or a minute, can establish T_1 with good accuracy. Small errors in pulse timings are not significant, nor are field irregularities. Complicated spectra, containing several resonances, are no more difficult to tackle than simple ones. Small wonder, then, that T_1 measurements are increasingly being used to assist in spectral assignments and structure determination by n.m.r. We shall return briefly to this topic in Sec. 7.3.2.

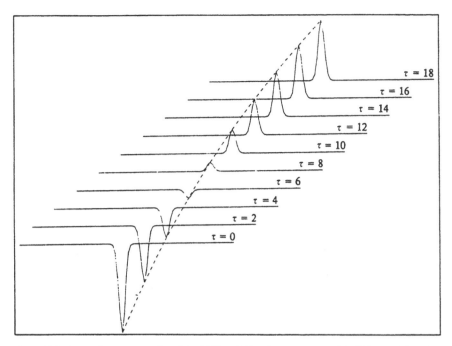

Figure 7.7 A set of spectra produced by 180°–τ–90° pulse sequences with different delay times τ.

7.2 NUCLEAR MAGNETIC RESONANCE SPECTROSCOPY: HYDROGEN NUCLEI

We have seen that a particular chemical nucleus placed in a magnetic field gives rise to a resonance absorption of energy from a beam of radiation, the resonance frequency being characteristic of the nucleus and of the strength of the applied field (some data have already been referred to in Table 7.1). Thus n.m.r. techniques may be used to detect the presence of particular nuclei in a compound and, since for a given nuclear species the strength of the n.m.r. signal is directly proportional to the number of resonating nuclei, to estimate them quantita-. tively. However, an n.m.r. spectrometer is an expensive instrument, and there are many simpler and cheaper methods available to detect the presence or absence of a particular atom in a molecule. Two other characteristics of n.m.r. spectra which have not so far been mentioned make the technique far more powerful and useful; these are the *chemical shift* and the *coupling constant*, which we shall discuss in the following sections.

The vast majority of substances of interest to chemists contain hydrogen atoms and, as this nucleus has one of the strongest resonances, it is not surprising that n.m.r. has found its widest application to these substances. When discussing chemical shifts and nuclear coupling it is convenient to use one type of nucleus as an example, although all spinning nuclei show these phenomena, and in what follows we shall consider the spectra of hydrogen-containing substances only, leaving discussion of other nuclei to Sec. 7.3.

7.2.1 The Chemical Shift

Up to now we have considered the behaviour of an isolated nucleus in an applied field. Such a situation is not, of course, realizable in practice since all nuclei are associated with electrons in atoms and molecules. When placed in a magnetic field the surrounding electron cloud tends to circulate in such a direction as to produce a field *opposing* that applied (so-called diamagnetic circulation), as shown for a single atom in Fig. 7.8. Plainly the total field experienced by the nucleus is:

$$B_{\text{effective}} = B_{\text{applied}} - B_{\text{induced}}$$

and, since the induced field is directly proportional to the applied field

$$B_{\text{induced}} = \sigma B_{\text{applied}}$$

where σ is a constant, we have:

$$B_{\text{effective}} = B_0(1 - \sigma) \tag{7.13}$$

where B_0 is the applied magnetic field. (Note that we here use the n.m.r. spectroscopist's normal notation of B_0 for the applied field, rather than B_z used earlier, since henceforth we shall not be concerned with the details of spin vector interaction with the field.) Thus the nucleus can be said to be *shielded* from the applied field by diamagnetic electronic circulation. The extent of the shielding will be constant for a given atom in isolation, but will vary with the electron density about an atom in a molecule; thus we may generalize Eq. (7.13) by writing:

$$B_i = B_0(1 - \sigma_i) \tag{7.14}$$

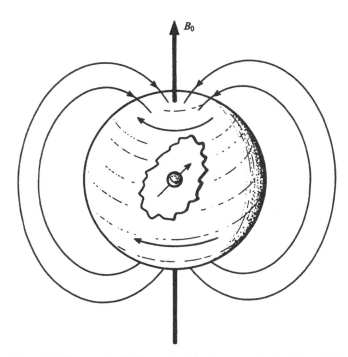

Figure 7.8 Showing the field, produced by diamagnetic circulation of the electron cloud about a nucleus, which opposes the applied field B_0.

where B_i is the field experienced by a particular nucleus i whose shielding constant is σ_i. As an example, since we know that oxygen is a much better electron acceptor than carbon (since oxygen has the greater electronegativity), then the electron density about the hydrogen atom in C—H bonds should be considerably higher than in O—H bonds. We would thus expect $\sigma_{CH} > \sigma_{OH}$ and hence

$$B_{CH} = B_0(1 - \sigma_{CH}) < B_{OH} = B_0(1 - \sigma_{OH})$$

Thus the field experienced by the hydrogen nucleus in O—H bonds is greater than that at the same nucleus in C—H bonds and, for a given applied field, the CH hydrogen nucleus will precess with a smaller Larmor frequency than that of OH. Put conversely, in order to come to resonance with radiation of a particular frequency (for example 100 MHz), a CH hydrogen requires a greater applied field than OH.

The effect of a steadily increasing field on the energy levels of the CH$_3$ and OH hydrogens in CH$_3$OH is shown in Fig. 7.9. The OH hydrogen nucleus, having a smaller shielding constant, experiences a greater field; hence its energy levels are more widely spaced than those of the more shielded CH$_3$ nuclei at any given applied field. If the system is irradiated with a beam of radiation at, say, 100 MHz while the applied field is increased from zero, the OH nucleus will come into resonance first and absorb energy from the beam, the CH$_3$ nuclei absorbing at a higher field. This is shown in the spectrum of methanol, CH$_3$OH, at the foot of the figure. The fact that the ratio of the absorption intensities (strictly the ratio of the areas under the peaks) is 1 : 3 immediately allows us to identify the smaller peak with the single hydrogen nucleus in the OH group, the larger with the CH$_3$ group. Since neither carbon nor oxygen have nuclei with spin, they do not contribute to the spectrum.

Two very important facets of n.m.r. spectroscopy appear in Fig. 7.9: (1) identical nuclei (that is H nuclei) give rise to different absorption *positions* when in different *chemical* surroundings (for this reason the separation between absorption peaks is usually referred to as their *chemical shift*) and (2) the *area* of an absorption peak is proportional to the number of *equivalent* nuclei (i.e. nuclei with the same *chemical shift position*) giving rise to the absorption. We see here the basis of a qualitative and quantitative analytical technique.

It should be noted that electron density is not the only factor determining the value of the shielding constant. Another, frequently very important, contribution to shielding arises from the field-induced circulation of electrons in neighbouring parts of a molecule which gives rise to a small magnetic field acting in opposition to the applied field. Two of the many possibilities are shown in Fig. 7.10. In Fig. 7.10(*a*) the circulation of the cylindrical charge cloud comprising an acetylenic triple bond is shown to *reduce* the effect of the applied field at a hydrogen nucleus on the axis of the circulating charge. The nucleus is thus shielded and will resonate to high applied field. In Fig. 7.10(*b*), on the other hand, we show the circulation of the annular cloud of π-orbital electrons around a benzene ring (only the top annulus is shown, to simplify the diagram, but the identical annulus underneath the ring circulates in the same direction). The induced field is here in the *same direction* as the applied field in the vicinity of the hydrogen nuclei, and these nuclei are consequently deshielded and resonate to low applied field. Obviously both the effects noted here will be somewhat reduced by molecular tumbling, but they do not average to zero.

Thus both shielding and deshielding may arise from induced electron circulation, and the total of such effects at any particular site, together with an electron density contribution, is included in the shielding constant σ.

There are several ways in which the chemical shift difference, Δ, between the OH and CH$_3$ signals of Fig. 7.9 may be expressed. Firstly, since we have imagined the spectrum to be recorded by varying the applied field, we could attach a tesla scale (or, rather, a microtesla (μT) scale, since the separation is very small) to the spectrum, and observe that $\Delta = 3.26 \, \mu$T. Or,

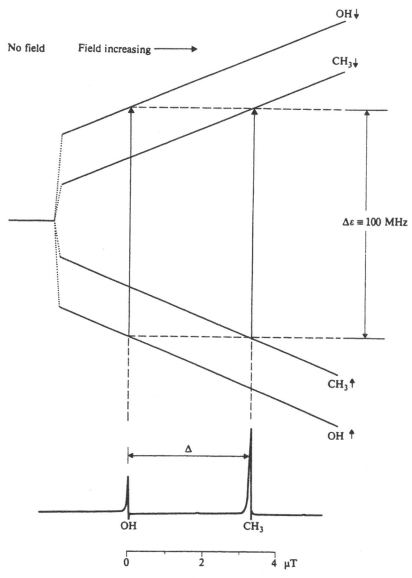

Figure 7.9 The effect of an applied magnetic field on the energy levels of the methyl and hydroxyl nuclei of methanol, CH_3OH. The applied field is increased rapidly initially (dotted portion) until near resonance at 2.3487 T and then the increase is much slower. The n.m.r. spectrum of methanol is shown at the foot of the figure.

remembering that the spectrum could equally well have been obtained by varying the applied frequency at constant field, we could quote the chemical shift difference in hertz as follows:

$$2.3487 \text{ T} \equiv 100 \text{ MHz}$$

Hence:

$$3.26 \text{ } \mu\text{T} \equiv 139 \text{ Hz} = \Delta$$

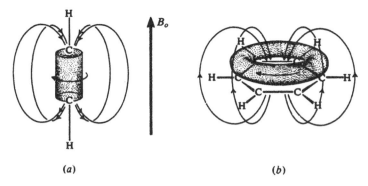

Figure 7.10 The field-induced electronic circulation in (*a*) ethyne and (*b*) benzene, showing shielding and deshielding, respectively, at nearby hydrogen nuclei.

In practice, however, neither of the above chemical shift units is entirely satisfactory, although both appear in the older literature. The difficulty can be seen as follows: rewriting Eq. (7.9) to take account of shielding at nucleus i and combining with Eq. (7.14):

$$\frac{\Delta E_i}{h} = \frac{g\beta_N B_i}{h} = \frac{g\beta_N B_0 (1 - \sigma_i)}{h} \quad \text{Hz}$$

and so, in hertz:

$$\Delta = \frac{g\beta_N B_0}{h} (\sigma_{CH} - \sigma_{OH}) \quad \text{Hz} \tag{7.15}$$

Now the shielding constant, σ_i, is independent of the applied field or frequency, so plainly chemical shift separation measured in hertz or microtesla is directly proportional to the applied field, B_0: it is, at the least, inconvenient to measure a quantity which changes with the operating conditions of the instrument, particularly since different instruments use fields varying between 0.6 and 15 tesla.

The difficulty can be overcome if we quote chemical shifts as a *fraction* of the applied field or frequency. Thus for methanol, $\Delta = 139\,\text{Hz}$ at 100.0 MHz is equivalent to $\Delta = 1.39 \times 10^{-6}$ or 1.39 p.p.m. (parts per million) of the operating frequency; similarly, $\Delta = 3.26\,\mu\text{T}$ at 2.3487 T is also 1.39 p.p.m. of the applied field. If we were now to double the field and frequency, the separation expressed in microtesla or hertz would also double according to Eq. (7.15), but it would still be just 1.39 p.p.m.

Chemical shift measurments are, of course, formally based on the resonance position of the bare hydrogen nucleus (the proton) as the primary standard; for this there is no shielding and hence $\sigma = 0$. Since this is a quite impracticable standard, it is necessary to choose some reference substance as a secondary standard and to measure the resonance positions of other hydrogen nuclei from this in parts per million; the substance now almost universally selected for hydrogen resonances is tetramethyl silane, $Si(CH_3)_4$, usually abbreviated to TMS. Since TMS is immiscible with water, in aqueous solutions the salt $(CH_3)_3SiCH_2CH_2CH_2SO_3Na$ is used; its methyl protons resonate at exactly the same position as those of TMS. Both of these substances are also excellent references for ^{13}C spectra.

Tetramethyl silane has several advantages over other substances which have been used as standards:

1. Its resonance is sharp and intense since all 12 hydrogen nuclei are equivalent (i.e. have the same chemical environment) and hence absorb at exactly the same position.

2. Its resonance position is to high field of almost all other hydrogen resonances in organic molecules (that is σ_{TMS} is large) and hence can be easily recognized.

3. It is a low boiling point liquid (b.p. 27 °C) so can be readily removed from most samples after use.

Thus if about five per cent of TMS is added to a sample and the complete n.m.r. spectrum produced, the sharp, high-field resonance of the TMS is easily recognized and can be used as a standard from which to calibrate the spectrum and to measure the chemical shift positions of other molecular groupings. Conventionally n.m.r. spectra are displayed with the field increasing from the left, which places the TMS resonance to the extreme right. Two measurement scales have been used for chemical shifts, both in parts per million. An old scale, the τ scale, arbitrarily put the resonance of TMS at 10 p.p.m., with scale numbers decreasing to the left. This is no longer in use and, by international agreement, has been replaced with the so-called δ scale which sets the TMS resonance at 0.0 p.p.m. for both ^{1}H and ^{13}C spectra. Since the TMS resonance is at the extreme right the δ scale numbers increase from right to left, i.e. in the direction of weaker shielding. If conversion between the two scales is needed, e.g. when spectra in the older literature are studied, the relationship is simply $\delta = 10 - \tau$.

Figure 7.11 indicates the approximate resonance positions of a few molecules and molecular groupings on the δ scale. In studying this figure, we must not lose sight of the fact that the resonance positions shown are due only to the *hydrogen* nucleus in the molecule or group concerned.

As one would expect, the position of the hydrogen resonance depends upon the atom to which it is directly attached—the top row of the figure shows this for the series CH_4, NH_3, PH_3, SiH_4, and H_2 between $\delta = 0$ and $\delta = 4.2$. Far more important, however, is the fact that, when the hydrogen is attached directly to a particular atom, e.g. carbon, its resonance position depends markedly on the nature of the *other* substituents to that atom. Most of the remaining data in the figure relate to a CH group with a variety of substituents; we may note that, in the

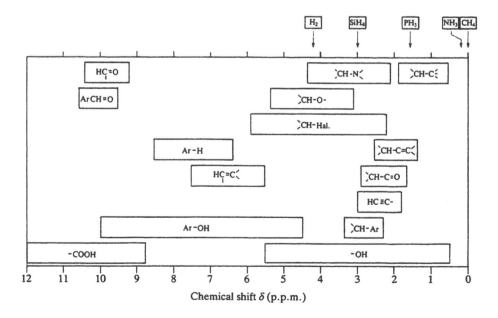

Figure 7.11 Showing the δ-scale of chemical shift values and the approximate proton chemical shifts of some simple molecules and groups.

series CH—C (at around 1 p.p.m.), CH—N (2–4 p.p.m.), and CH—O (3–5 p.p.m.), it is evident that the increasing electronegativity of the substituent progressively withdraws electrons from the neighbourhood of the hydrogen, thus reducing the shielding of the nucleus and allowing it to resonate at lower field (and higher δ number). A double- or triple-bonded substituent has a similar but smaller effect. We have already mentioned the strong deshielding effect produced by electron circulation in an aromatic ring and we note here that the resonance of a ring proton falls at around 7–8 p.p.m. The very wide range of observed resonances for OH and COOH groups occurs because the proton shielding is much affected by hydrogen bonding in these groups, so solvent, concentration, and even temperature changes can produce marked differences in their δ-values.

Figure 7.11 shows only a very small part of the huge mass of n.m.r. data available—much more is to be found in the very detailed tables and charts given in specialized texts, such as that by Williams and Fleming listed in the bibliography at the end of the chapter. Suffice it here to say that, combining detailed knowledge of chemical shift positions with the fact that the area of each resonance in the spectrum is proportional to the number of hydrogen nuclei contributing to that resonance, we see that n.m.r. is an excellent technique for both quantitative and qualitative group analysis in organic chemistry. After the following sections, where we discuss the phenomenon of energy coupling between nuclei, we shall see that the usefulness of n.m.r. extends further to the determination of the structure and configuration of molecules.

7.2.2 The Coupling Constant

Suppose that two hydrogen nuclei in different parts of a solid, e.g. a crystal lattice, are sufficiently close together in space that they exert an appreciable magnetic effect on each other—in n.m.r. terms 'appreciable' means 0.01 μT or more. We show such a case in Fig. 7.12 for two nuclei labelled A and X. Here we ignore the spin direction of X since this is immaterial for the moment, but we show the two possible directions of the z component of A's spin, either in the field direction (conventionally 'up') or opposed to that direction ('down'). The lines of force originating from A, considered as a simple bar magnet, are seen to *oppose* the applied field at X when A is up and to reinforce it when A is down. Remembering (Sec. 7.1.3) that the two A directions are virtually equally likely, we see that nucleus X will find itself in an applied field $B_0 + B_A$ or in $B_0 - B_A$ with equal probability, where B_A is the field at X due to nucleus A. Clearly this will result in the X nuclei of a sample showing two different resonance positions in the spectrum—the X signal will be a *doublet*—separated, on a tesla scale, by $2B_A$. If we now include X's spin in the argument, it is obvious that the effect of X on A is precisely the same as that of A on X, so the A signal will be split into an identical doublet.

The extent of this direct spin-spin interaction, or *coupling*, is large for hydrogen nuclei, being of the order 10^{-4} T when the nuclei are 0.1 nm apart, and is of practical use in solid-state n.m.r. spectra in measuring with some precision the interatomic distances in compounds.

When liquid or gaseous samples are considered, however, this direct spin-coupling mechanism is found not to apply because random molecular motions within the sample (molecular tumbling) continually change the orientation of the molecules within the applied field. For example, if we imagine the AX 'molecule' of Fig. 7.12(a) to be rotated through 90° in the plane of the paper (remember that the nuclear spins will *not* rotate with the molecule, since they are locked into the field direction), nucleus X will sit vertically above or below A and will thus be in a position where the field due to A *reinforces* the applied field B_0—this is just the opposite situation from that previously considered. It can be shown by proper integration of the effect of A on X at all possible orientations that continued rotation of the molecule averages the coupling exactly to zero—all situations in which A reinforces the field at X are just balanced by all those where it diminishes that field.

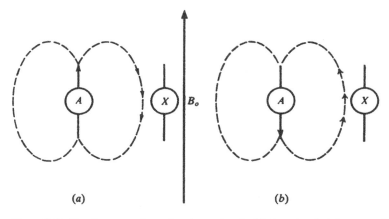

Figure 7.12 The direct coupling of nuclear spins. In (a) the spin of A decreases the net magnetic field at X, while in (b) the field at X is increased.

Nonetheless, the n.m.r. spectra of liquids do show the phenomonen of coupling but the effect is very much smaller (by a factor of 10^2–10^4 than that observed for direct coupling in solids); clearly a different mechanism is involved. Consider first Fig. 7.13(a) which shows two hydrogen nuclei joined by a pair of bonding electrons, as in the hydrogen molecule. To a first approximation we can assume that each electron 'belongs' to a particular nucleus, so we associate electron (a) with nucleus A and electron (x) with nucleus X. Plainly the most stable state energetically is that in which the electronic magnetic dipole is opposed to that of its own nucleus, but theories of chemical bonding tell us that the electrons, which occupy the same orbital, will have their spins opposed also, and so we see that the most stable state will be that in which the nucleus–electron–electron–nucleus spins alternate as shown in the figure. Consequently, the spins of A and X will preferentially be paired. This is similar to the situation

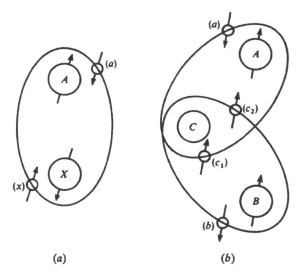

Figure 7.13 The coupling of nuclear spins via bonding electrons for (a) directly bonded atoms and (b) atoms bound to a third atom having no spin.

found for the 'across-space' effect, but is several orders of magnitude smaller; however, it represents coupling between the nuclei.

Note carefully that the above argument is not intended to be rigorous; we are not saying that the spins of A and X are immutably fixed in the paired state, simply that the paired configurations ↑↓ and ↓↑ are very slightly lower in energy than the parallel configurations ↑↑ and ↓↓. The energy difference is so minute (some 10^{-32} J) that both configurations are equally likely to occur. The effect which this has on the spectrum is shown by Fig. 7.14. At Fig. 7.14(a) we show the four energy levels of the AX system, each corresponding to one of the four possible spin combinations, ↑↓, ↑↑, ↓↓ and ↓↑; a convenient notation for these spin states is also indicated on the diagram, whereby each 'up' spin is designated α and each 'down' spin β. In this part of the diagram we imagine there to be no coupling between A and X; hence the separation between levels $\alpha\alpha$ and $\alpha\beta$, corresponding to a spin change of X only, is just the resonance energy of X, and this is clearly identical with the separation between levels $\beta\alpha$ and $\beta\beta$. Similarly, the separations $\alpha\alpha$–$\beta\alpha$ and $\alpha\beta$–$\beta\beta$ are identical and give the resonance energy of A. Transitions between the various levels give rise to two spectral lines, one at each of the chemical shift positions of A and of X, desinated ν_A and ν_X.

When coupling occurs we know that the levels $\alpha\alpha$ and $\beta\beta$ are destabilized, and that $\alpha\beta$ and $\beta\alpha$ are stabilized; this is shown in Fig. 7.14(b), where the amount of the energy change is symbolized as $J/4$—the convenience of this will become clear shortly. Now the two possible X transitions have different energies: there is $\alpha\alpha \leftrightarrow \alpha\beta$ which is smaller than in Fig. 7.14(a) by $\frac{1}{2}J$, and $\beta\alpha \leftrightarrow \beta\beta$, higher by $\frac{1}{2}J$; hence one spectral line appears $\frac{1}{2}J$ above ν_X and another $\frac{1}{2}J$ below, the separation between these two lines being just J (hence the convenience of defining the stabilization and destabilization as $J/4$ above). Similarly, the A transitions will be split into an identical doublet. The 'legs' of the doublet will be the same intensity since (1) all spin states are

Figure 7.14 To illustrate the effect of coupling between two nuclei. At (a) coupling is ignored, at (b) the coupling is taken to be small compared with the chemical shift, and at (c) the coupling is large compared with the chemical shift.

virtually equally likely and (2) detailed calculation shows that the probabilities of transitions occurring between any levels are identical. Thus we arrive at an AX spectrum: a pair of $1 : 1$ doublets, each pair separated by the coupling constant J and the midpoints of the doublets separated by the chemical shift difference Δ. Since J is observed to be a field-independent quantity, it is conveniently expressed in hertz rather than in parts per million like δ.

At this point we should emphasize that Fig. 7.14 is by no means drawn to scale. If A and X are both hydrogen nuclei, then in Figure 7.14(a) the transitions labelled X and A represent the resonance energies of such nuclei in, say, a 2.5 T field, i.e. some 100 MHz. The separation between $\alpha\beta$ and $\beta\alpha$, on the other hand, represents the chemical shift difference Δ between ν_A and ν_X, perhaps a few hundred hertz. Thus if the diagram were to scale, the distance between $\alpha\alpha$ and $\alpha\beta$ should be some million times greater than that between $\alpha\beta$ and $\beta\alpha$. Turning to Fig. 7.14(b), the quantity $J/4$ is a few hertz only, so the displacements of the various energies have been much exaggerated.

Up to now we have taken the chemical shift difference of A and X (some 100 Hz) to be large compared with the coupling constant J (a few hertz). In Fig. 7.14(c) we show what happens when the chemical shift difference decreases, i.e. when the levels $\alpha\beta$ and $\beta\alpha$ are assumed to be closer together than shown. It is a well-known fact in quantum mechanics that energy levels of the same symmetry, as are $\alpha\beta$ and $\beta\alpha$, tend to repel each other if they become close, while those of different symmetry, like $\alpha\alpha$ and $\beta\beta$, are not affected. If both $\alpha\beta$ and $\beta\alpha$ are repelled by an amount Q, as shown on the diagram, then clearly both X transitions are decreased in energy by Q, while both A transitions are increased by the same amount. The result is to push the spectral lines apart so that the chemical shift positions, ν_A and ν_X, are no longer at the midpoints of their respective doublets. Actually in the figure we have relabelled the nuclei A and B instead of A and X: this is conventional notation in n.m.r. spectroscopy, where letters close together in the alphabet are used to indicate nuclei whose chemical shift difference is small, and vice versa.

One further effect of a small chemical shift difference is illustrated in Fig. 7.14(c); this is that, although the populations of the levels are unaffected by the perturbation Q, the relative transition probabilities are very much affected, and detailed calculations show that the result is for the centre lines of the AB spectrum to gain intensity at the expense of the outer. In the limit when the chemical shift difference becomes zero (A_2 system instead of AB, since A and B become identical), the centre lines coalesce into one and the outer lines have vanishing intensity. Thus although coupling between the A_2 nuclei certainly exists, its effect is not observable in a spectrum. A typical AB spectrum is illustrated in Fig. 7.15 where it is evident that Δ (measured in hertz) is about ten times J.

Let us now return to Fig. 7.13(b). This shows the coupling of hydrogen nuclei which are each attached to a third non-spinning nucleus, such as carbon; an example of this situation is the methylene fragment $>CH_2$. Now the chain of reasoning runs as follows: the spins of A and (a) are paired, as are those of (a) and (c_1), since both of the latter occupy the same orbital; (c_1) and (c_2) have parallel spins, however, since they occupy degenerate orbitals in the same atom (cf. Hund's principle, Sec. 5.3.1); finally, (b) and (c_2) are paired and nucleus B has its spin paired with electron (b). We see that the lowest energy state, according to this electron path, is that wherein the spins of A and B are *parallel*, not paired. In this situation the coupling constant, J_{AB}, is defined to be negative, whereas it is defined positive for the previous case shown in Fig. 7.13(a).

Similar arguments to the above indicate that J is again positive for coupling via three bonds (for example H — C — C — H) and negative via four, etc., provided this type of electron path is the predominant contributor to the coupling; in some systems, particularly unsaturated ones (i.e. containing multiple bonds), other electron paths may become important. However, the magnitude of the coupling constant attenuates rapidly with increasing number of bonds; thus for H_2 the coupling (measured indirectly from the spectrum of HD) is some 240 Hz, for the

Figure 7.15 Part of the n.m.r. spectrum of ethyl cinnamate, $C_6H_5CH\!=\!CH.COOC_2H_5$, showing the typical *AB* pattern of the olefinic hydrogens. The resonances in the centre are due to the phenyl group; the ethyl resonance is off the scale to the right. (*Thanks are due to Mr A. Taylor of the University of York for this spectrum.*)

H—C—H fragment (e.g. in methane) it is some 12 Hz, for H—C—C—H (e.g. in ethanol, CH_3CH_2OH) it is 7 Hz and for H—C—C—C—H it is on the present limit of measurement, 0.5 Hz or less. However, in unsaturated molecules, in which electrons occupy orbitals extending over more than two nuclei and are thus more mobile, the couplings are somewhat larger and have been observed over more than four bonds.

It is not usually easy to determine the sign of a coupling constant experimentally, but where such data exist they are often in good agreement with the alternation theory outlined above. Further, calculations can be made predicting the magnitudes of certain couplings by applying quantum mechanical methods to the very diagrammatic approach of Fig. 7.13; these, too, are in generally good agreement with experimental values.

7.2.3 Coupling between Several Nuclei

So far we have considered the effect of coupling between two hydrogen nuclei only, but such nuclei often occur in molecules as groups, particularly CH_3 and CH_2 groups. We turn now to consider coupling between groups, using the ethyl fragment, CH_3CH_2, as our example.

In the ethyl fragment the three hydrogens of the methyl (CH_3) group have the same chemical shift since all the CH bonds are identical and the shielding at each of the nuclei is the same. Such nuclei are called *chemically equivalent*. In the same way the two nuclei of the methylene (—CH_2—) group are chemically equivalent but their chemical shift is, of course, different from that of the methyl nuclei. Further, there is some freedom of rotation of the methyl group in this fragment and hence the interaction between the methyl and methylene nuclei is averaged to the same value—the coupling constant is the same between any methyl and any methylene hydrogen. The nuclei in the methyl group (or the methylene group) are said to be *magnetically equivalent* as well as chemically equivalent. This property affords a considerable simplification of the overall spectrum because the *couplings within a group of magnetically and chemically equivalent nuclei do not affect the spectrum* and can be ignored, as in the case of A_2 considered earlier. In the present example, this means we can neglect coupling between the methyl hydrogens themselves, or between the two methylene hydrogens, and need consider only the coupling between a methyl and methylene nucleus. Since all the latter couplings are equal, as explained above, the system has just one J value to be considered. If the chemical shift between methyl and methylene is large, then we have a system A_3X_2 (where A is a CH_3 hydrogen nucleus and X a CH_2 nucleus) with a coupling constant J_{AX}.

When considering the spectrum resulting from an A_3X_2 system it is quite possible to construct an energy-level diagram analogous to that of Fig. 7.14 for the AX case. Such a diagram is complex, however, and it is much simpler to use another approach, which may be called the 'family tree' method; in this, coupling between grouped nuclei is considered stepwise.

Thus in Fig. 7.16(a) we start by imagining the A_3 and X_2 groups to be uncoupled, and hence as giving rise to one line each of intensity 3 and 2 units, respectively. When we let the A_3 group couple to just *one* of the X_2 pair, the A_3 line is split into a doublet, separation J_{AX}, each line having intensity $1\frac{1}{2}$ units. If now each leg of this doublet is considered to couple with the second X nucleus, each will be split into a doublet, separation J_{AX}, intensity $\frac{3}{4}$ unit. However, the inner line of each doublet will overlap, because the coupling is identical for both, so the overall spectrum will have the appearance of a *triplet*, intensity of each line $\frac{3}{4}$, $1\frac{1}{2}$, $\frac{3}{4}$ (i.e. a $1:2:1$ triplet) with the coupling J_{AX} appearing twice in the spectrum.

The splitting of X_2 by coupling with A_3 proceeds similarly. Coupling to one A yields a $1:1$ doublet, to the second a $\frac{1}{2}:1:\frac{1}{2}$ triplet, and to the third a $\frac{1}{4}:\frac{3}{4}:\frac{3}{4}:\frac{1}{4}$ quartet (i.e. a $1:3:3:1$ quartet) with the coupling constant J repeated three times.

This argument is very readily generalized: a group of p equivalent nuclei splits a neighbouring group into $p+1$ lines with intensities given by the pth line of Pascal's triangle:

$$
\begin{array}{cccccccccc}
p=1 & & & & 1 & & 1 & & & \\
p=2 & & & 1 & & 2 & & 1 & & \\
p=3 & & 1 & & 3 & & 3 & & 1 & \\
p=4 & 1 & & 4 & & 6 & & 4 & & 1 \\
\end{array}
$$

where each entry is obtained by summing the two numbers to its right and left in the line immediately above. Thus the two nuclei of the X_2 group split the A_3 resonance into $2+1=3$ lines, while the X_2 resonance is itself split into $3+1=4$ lines by the A_3 nuclei.

The family tree method thus predicts a quartet and triplet structure for the $-CH_2CH_3$ spectrum, the former having a total intensity of two units, the latter of three. Comparison with the spectrum of CH_3CH_2OH at the foot of Fig. 7.16 shows that this prediction is amply justified. It must be stressed, however, that if the chemical shift separation of the CH_3 and CH_2 resonances were to be decreased (forming an A_3B_2 system), e.g. by lowering the operating field of the instrument or by changing the nature of the substituent on the CH_3CH_2- group, then the family tree method becomes a successively poorer approximation to the observed spectrum. Already in the A_3X_2 spectrum of Fig. 7.16 the inner lines are seen to have a slightly enhanced intensity with respect to the outer (cf. the AB pattern of Fig. 7.14), even though the chemical shift separation is some 20 times the coupling constant. At smaller shifts not only is the intensity distortion increased but additional lines begin to appear, quite inexplicable with the

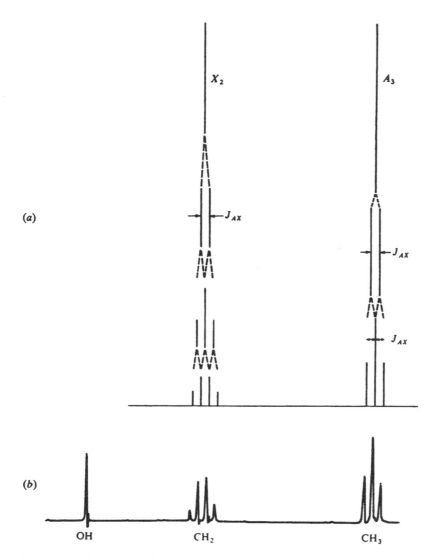

(a)

(b)

OH   CH₂   CH₃

Figure 7.16 (a) A theoretical A_3X_2 spectrum and (b) the actual spectrum of ethanol, CH_3CH_2OH.

family tree method. Under these conditions accurate fit between theory and experiment is only possible by detailed calculation of the energy levels and transition probabilities for the system.

Some other typical coupling patterns are shown in Fig. 7.17. In Fig. 7.17(a) we see an A_2X_2 spectrum given by the substance cyanhydrin, $CNCH_2CH_2OH$. The methylene groups have different chemical shifts and so it is to be expected that coupling between them will split each resonance into a 1 : 2 : 1 triplet; this is to be observed in the spectrum. Other chemical structures giving rise to similar spectral patterns are 1,2-disubstituted benzenes, or five-membered unsaturated heterocyclic systems such as furan, although in such molecules the spectra are rather more complicated owing to additional couplings which arise between the nuclei. In Fig. 7.17(b) we show an A_6X spectrum arising from an isopropyl group, $(CH_3)_2CH$ —. In this all six methyl hydrogens are equivalent and hence they split the lone methylene hydrogen into a septet, while they themselves are split into a doublet by the single nucleus. Such a pattern is easily recognizable even if the two outer lines of the septet are too weak to be observed, and is very characteristic.

Finally, in Fig. 7.17(c) we show the spectrum produced when three different nuclei couple together—an *AMX* system if all the chemical shifts are large. Within this system there are three different coupling constants: J_{AM}, J_{AX}, and J_{MX}, so that, for instance, the *A* resonance is split into a doublet of spacing J_{AM} and then each line of the doublet is further split into a doublet by J_{AX}. Thus each resonance gives rise to a symmetrical 1 : 1 : 1 : 1 quartet as shown in the figure. This pattern might arise from the three ring hydrogens of a monosubstituted furan or

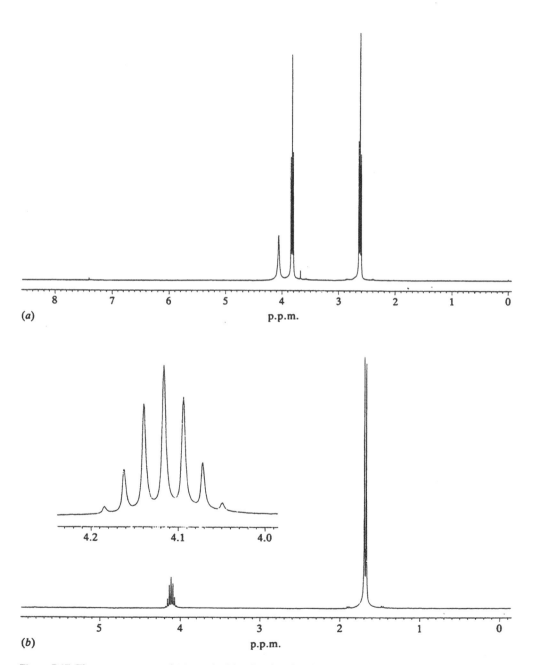

Figure 7.17 The n.m.r. spectra of (*a*) cyanhydrin, $CN.CH_2CH_2OH$; (*b*) 2-iodo-propane, $(CH_3)_2CHI$

(c)

p.p.m.

Figure 7.17 (c) Part of the n.m.r. spectrum of 2-furanoic acid, OCH═CH.CH═C.COOH. Signals from the ring hydrogens only are shown, the resonance of the COOH being off-scale to the left. (*Thanks are due to Mr A. Taylor of the University of York for this spectrum.*)

similar molecule (in this example we show the spectrum of an α-furan) or it might be from a vinyl group CH_2═CH—, where again all three hydrogen nuclei have a different chemical shift.

These coupling patterns, like that for the CH_3CH_2— fragment, are considerably complicated if the chemical shift between coupled nuclei is small. Usually, however, the additional fine structure produced and the intensity perturbations do not prevent recognition of the overall pattern, particularly when some experience has been gained from studying actual spectra. The tremendous analytical value of such patterns is obvious since their recognition immediately gives information about the chemical groupings present in the molecule under examination. We discuss this aspect of n.m.r. spectra more fully in the next section.

7.2.4 Chemical Analysis by N.M.R. Techniques

In the preceding sections we have built up a picture of the application of n.m.r. to constitutional and structural studies. Thus the observation of the τ values of lines in a spectrum (or of the centres of multiplets if coupling is occurring) immediately indicates, with very little ambiguity, the types of hydrogen-containing groups within the molecule, while the relative intensities of the lines yield directly the proportions in which these groups occur.

Further, the multiplet structure of each group in the spectrum gives information on the number of hydrogen nuclei coupled to that group and in this way shows which groups are near neighbours in the molecule. Thus groups such as CH_3CH_2, —CH_2CH_2—, $(CH_3)_2CH$—, etc., can be instantly recognized from the n.m.r. spectrum.

As an example of the use and limitations of n.m.r. spectroscopy in analysis, consider the spectrum shown in Fig. 7.18. Resonances are centred at $\delta = 8.1$, 7.5, 4.4, and 1.3, the position of the former two and the coupling pattern of the latter two suggesting that they arise from a phenyl group (C_6H_5—) and an ethyl group (C_2H_5—), respectively. The integral trace on the spectrum shows these resonances to have relative intensities of $5:2:3$ (the two phenyl

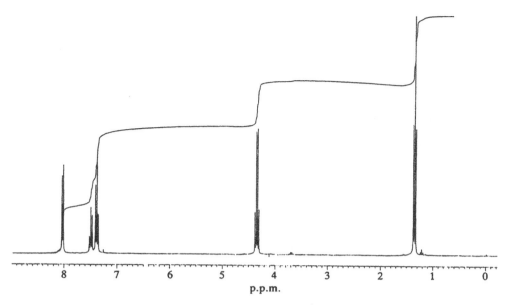

Figure 7.18 The n.m.r. spectrum of ethylbenzoate, $C_6H_5CO.OCH_2CH_3$, to illustrate the use of n.m.r. as an analytical technique. (*Thanks are due to Mr A. Taylor of the University of York for this spectrum.*)

resonances here being summed) so we know the phenyl and ethyl to be present in a 1 : 1 ratio. If we also know that the molecular formula of the substance is $C_9H_{10}O_2$, we can rapidly deduce that the resonances at $\delta = 8.1$ and 7.5 are due to hydrogen nuclei, respectively ortho and meta/para to a carbonyl group, while those at 4.4 and 1.3 are consistent with the grouping $CH_3CH_2.O.CO$ —. The molecule is thus ethyl benzoate, $C_6H_5CO.O.CH_2CH_3$.

Note that the n.m.r. spectrum, while allowing us to deduce directly the presence of phenyl and ethyl groups, does not indicate the *presence* of groups not containing magnetic nuclei, in this case O and CO. However, once these groups are known to be present, either by determination of the molecular formula or by observation of the infra-red spectrum, then the n.m.r. spectrum does indicate their *position* in the molecule. Thus the coupling pattern shows clearly that the CH_3 and CH_2 groups are directly bound and not, for example, joined via O or CO ($CH_3.O.CH_2$— or $CH_3.CO.CH_2$—) since in these latter configurations the coupling constant would be immeasurably small.

This very simple example serves to indicate the method of approach when using n.m.r. for analytical purposes. Of course, only in the simplest cases is a complete structural determination possible from the n.m.r. spectrum alone, but when taken in conjunction with other techniques, in particular infra-red spectroscopy, a great deal of useful structural information can usually be obtained about an unknown molecule.

7.2.5 Exchange Phenomena

The student may have been puzzled by one aspect of the alcohol spectra shown in Figs 7.9, 7.16, and 7.17; in these three spectra the resonance of the —OH hydrogen is shown as a single line whereas we might now expect it to be coupled with neighbouring CH_3, CH_2, or CH nuclei and hence have multiplet structure—quartet, triplet, or doublet, respectively. The reason it does not so couple is attributable to the fact that the hydroxyl hydrogen is readily exchanged with other hydrogen nuclei or ions in its vicinity; when this happens, the replacement hydrogen does not necessarily have the same spin direction as that being displaced and, if the exchange occurs sufficiently rapidly, the neighbouring CH nucleus experiences a 'coupling field' which is averaged to zero. Thus there is no net coupling between the OH and the neighbouring groups.

If the exchange is prevented by rigorous drying of the alcohol samples, coherent coupling appears and the OH resonance has the expected multiplet structure. It can be shown theoretically that for a coupling constant of J Hz the coupled nucleus must exchange more rapidly than $J/2\pi$ times per second for the multiplet to collapse to a singlet. For the alcohols considered above $J \approx 6$ Hz, and an exchange rate of only about once per second is sufficient to destroy coherent coupling. Of course, the transition from coupling to no coupling is not abrupt; at exchange rates rather lower than $J/2\pi$ the lines of the OH multiplet begin to broaden, at $J/2\pi$ they are so broad that all trace of line splitting is obliterated, and at higher rates of exchange the broad line sharpens until the single very sharp resonance of the illustrated spectra is seen.

An obvious application of this effect is to the study of hydrogen exchange kinetics; exchange rates frequently vary with temperature or concentration and such variations can be followed very precisely by observation of the change in line shape of n.m.r. signals.

As well as possibly destroying coupling between nuclei, exchange processes can also affect the observed chemical shifts and widths of n.m.r. resonances. Imagine a system which undergoes chemical exchange but where we can, experimentally, change the *rate* at which the exchange occurs—the most obvious ways of doing this systematically are by adding a catalyst or changing the temperature. We have already seen an example of the former—the exchange of the hydroxyl proton of methanol when H^+ is added. A simple example of the latter is to be found in the molecule dimethylnitrosamine:

$$O{=}N{-}N\begin{array}{c} CH_3 \\ CH_3 \end{array}$$

At room temperature this molecule is flat, so the two methyl groups are in different chemical surroundings—one *cis* (or *syn*) to the oxygen and the other *trans* (or *anti*)—and have different chemical shifts. At temperatures above about 200 °C, however, rotation about the N—N bond occurs rather freely and the positions of the methyl groups are exchanged frequently; as the temperature is lowered rotation becomes increasingly hindered and slower until at room temperature the molecule is essentially fixed in the planar form.

Suppose, then, we have some such system with two different hydrogen groups, A and B, exchanging at different rates depending on the conditions. It is convenient to discuss their n.m.r. resonances in terms of the ratio R, defined as follows:

$$R = \frac{\text{chemical shift difference of } A \text{ and } B \text{ without exchange}}{\text{rate of exchange}} = \frac{\Delta}{k}$$

where we use Δ for the difference in chemical shift and k as the rate constant for the exchange 'reaction'. Provided we measure Δ in hertz the ratio R is dimensionless, since k also has units of s^{-1}.

In Fig. 7.19 we plot spectra calculated for various values of R. When the exchange rate is small or zero, R is large, and the spectrum is exactly as might be expected—two separate peaks, one at the chemical shift position of A, the other at B—and the chemically shifted resonances are individually well defined. This situation is shown in Fig. 7.19(a), where we have chosen the chemical shift to be six times the exchange rate ($R = 6$). Equally, when the exchange is very rapid (small R), we would expect, and see, just one resonance line; physically this is because A and B become indistinguishable—each spends 50 per cent of its time on both sites and the average chemical shift becomes well defined as the mean of ν_A and ν_B. This is drawn at Fig. 7.19(f), for $R = 0.1$.

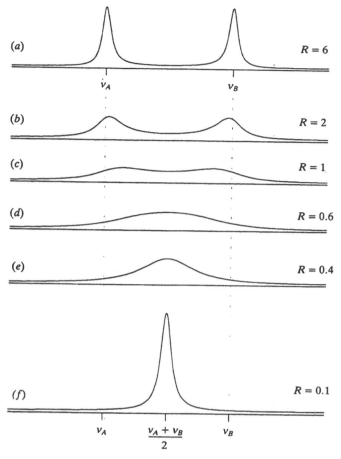

Figure 7.19 Calculated spectra for two exchanging chemical groups, A and B, showing the spectra as the rate of exchange increases from (a) to (f).

At intermediate exchange rates two things happen. Firstly, the separation between the peaks representing the two groups decreases; this is again because each group spends some time— although less than 50 per cent—at the other's site, and so takes on some of its characteristics. As the exchange rate increases—Fig. 7.19(b) to (e)—the peaks move closer until finally coalescing at high rates, as in Fig. 7.19(f).

The second change is that the n.m.r. resonance become broader. This can be explained in terms of the Heisenberg uncertainty principle (Chapter 1, Eq. (1.10)). Recall that this relates the width of a spectral line to the lifetime of the state under consideration:

$$\delta\nu \approx \frac{1}{2\pi\delta t}$$

Physically this means that, when the exchange rate is slow, A and B spend a long time on their 'own' sites, so δt is large, $\delta\nu$ is small, and the lines are sharp; conversely, as exchange increases, the effective lifetime in each state is shorter, $\delta\nu$ increases, and the lines become broader. This is evident in Fig. 7.19(b) to (e). As the rate of exchange increases still more to the point at which A and B become indistinguishable, there is just one sharp line at the mean of their chemical shifts, as in Fig. 7.19(f).

All simple exchange systems of this type behave similarly and their n.m.r. spectra depend only on the parameter R. Hence comparison of their spectra with calculated patterns like those in Fig. 7.19 yield—within limits—a value of R for the system under observation. Thus, if the chemical shift difference, Δ, for no exchange can be measured independently (e.g. by lowering the temperature to freeze exchange or by removing the catalyst), the rate constant of the exchange can be found from $k = \Delta/R$. The limits are fairly severe, however. For R much greater than 10, spectra change very little—because the 'exchange linewidth' becomes less than the natural line-width of the resonance, governed by normal relaxation processes; similarly, there is very little change for R much less than 0.1. Therefore, it is only R values between about 10 and 0.1 which can be determined with accuracy. The limits of Δ are, for protons, perhaps 1000 Hz down to about 1 Hz. So the maximum k value which can realistically be determined is 1000/0.1, and the minimum is 1/10, which limits the exchange rates to between about 10^4 and $0.1 \, \text{s}^{-1}$.

Such 'reactions' are slow in usual chemical terms, but are normal for hindered rotations about bonds, or the bending and twisting motions of large polymer molecules. Hence it is in the study of such conformational changes that the band shapes of n.m.r. spectra have given much useful information. Equally, exchange effects resulting from proton transfers, such as hydrogen bonding and tautomerism, often fall into this range and have been much studied by n.m.r. techniques.

7.2.6 Simplification of Complex Spectra

Frequently when studying all but the simplest compounds, the n.m.r. spectrum is a complicated mass of overlapping lines—groups showing two or more different coupled splittings with their patterns distorted by small chemical shifts, for example. There are several techniques now available to simplify such spectra and to convert them to more readily recognizable and analysable patterns.

Firstly, we remember that chemical shift differences vary with the strength of the applied field, whereas coupling constants do not. Thus if faced with a spectrum obtained with an applied field of 2.35 T (a 100 MHz spectrum) which shows a distorted pattern because a chemical shift difference is similar in magnitude to a coupling, it is relatively simple to re-examine the sample in an instrument operating at a higher field. Fields of 5.1 T (220 MHz) are now common, and some instruments have fields of up to 15 T (640 MHz), the latter using powerful 'superconducting' magnets. 220 MHz spectra yield chemical shifts twice as large as those at 100 MHz, and usually offer considerable simplification.

Another method by which chemical shift separations may be increased is by adding various materials to the sample solution. It has long been known that a change of solvent or of concentration could have a small effect on chemical shift positions, the magnitude of the effect depending on the extent of solute–solvent interactions, but usually not being more than 1 p.p.m. However, it has been discovered that some complexes of rare earth metals, particularly of europium and praseodymium, have exceptionally marked effects. For example, Fig. 7.20 shows the 100 MHz spectrum of 4-*epi*-trichodermol, both pure in solution (Fig. 7.20(*a*)) and with 5 mg and 10 mg of a europium agent added (Fig. 7.20(*b*) and (*c*), respectively); dashed lines connect the bases of some resonances to show how they move under the influence of the 'chemical shifting agent'.

The actual reagent used here is tris(1,1,1,2,2,3,3-heptafluoro-7,7-dimethyl-4,6-octanedione) europium III (or Eu(fod)$_3$ for short); as befits its impressive name, this is a large molecule and, when it is attached to the OH group of 4-*epi*-trichodermol, it not only perturbs the resonance of the OH but its influence spreads over the whole molecule, decreasing inversely with distance— the further away a particular hydrogen nucleus is from Eu(fod)$_3$, the less it is shifted. Thus the OH resonance moves most rapidly, and, of the two hydrogen atoms labelled (a) and (b) on the

Figure 7.20 The 100 MHz n.m.r. spectrum of 4-*epi*-trichodermol: (*a*) a solution of the pure compound; (*b*) solution plus 5 mg Eu(fod)₃; (*c*) solution plus 10 mg Eu(fod)₃. The arrow on the formula indicates where the Eu(fod)₃ is attached. (*Thanks are due to Dr J. R. Hanson and Mr P. Dew of the University of Sussex for assistance in obtaining this spectrum.*)

ecular picture, (b), the nearer to the shift reagent, moves more rapidly than (a); this is seen in the spectrum by the collapse of their coupled doublets into one line as the chemical shifts of (a) and (b) coincide—the AB system becomes A_2. Equally, of the three CH_3 groups labelled (d), (e), and (f) on the molecular picture, (d) and (e) show markedly greater shift than (f).

In this particular example most of the individual resonances are well resolved and distinct in the spectrum of the pure compound, but the use of the shifting agent helps in the assignment of certain resonances to particular nuclei. In other cases, however, where the original spectrum is badly overlapped, a shift reagent may offer considerable simplifications. In this context it should be mentioned that, while europium reagents shift resonances to low field (high δ) as seen here, praseodymium shift reagents normally cause shifts in the opposite direction.

A third approach towards simplification concentrates on the coupling constants. If coupling can be destroyed between nuclei an obvious simplification follows; thus in the spectra of alcohols discussed earlier, the OH signal is seen to be a single line rather than a multiplet due to the kinetic exchange of the hydrogen nucleus, and similarly the neighbouring proton group is decoupled from the OH and exhibits a simple spectrum.

This decoupling can be deliberately carried out in other cases by rapid exchange of the *spin* of the nucleus, rather than exchanging the nucleus itself. Consider two coupled nuclei, A and X, giving the typical AX spectrum of Fig. 7.14(*b*); bathing the sample in radiation of frequency ν_A allows the spectrum of nucleus A to be observed, but it is quite simple *simultaneously* to apply a strong radiofrequency field at frequency ν_X. This causes nucleus X to undergo rapid transitions between its two spin states and, although these changes are not directly observed by the spectrometer (since that is tuned to frequency ν_A), the reversals do result in a decoupling of the X spin from that of A; consequently the original doublet at ν_A collapses into a single line. Since two separate radiofrequencies are applied to the sample, this technique is known as 'double resonance'.

Double resonance has further uses besides simplification of spectra. For instance, the technique may be used to locate a particular resonance. Thus if the chemical shift of nucleus X places its resonance so that it is overlapped and lost within other spectral lines, it may be found by double irradiation of a nucleus coupled to X; the decoupling results in X's signal collapsing from a multiplet to a single sharp line, a change which is relatively easy to observe. Again, in a weak sample where some signals may be lost in the background noise, decoupling may result in a multiplet changing to a single line which shows up well above the noise. This application will be considered further in particular connection with ^{13}C spectra in Sec. 7.3.2. A final application of double-resonance techniques involves the application of a *weak* secondary radiofrequency, rather than the strong radiation required for decoupling. Under these conditions only partial decoupling may occur and, in some circumstances, *extra* multiplet splitting may be observed. These experiments, known graphically as 'spin tickling' experiments, lead to measurements of fundamental n.m.r. constants, particularly the relative signs of coupling constants.

7.3 NUCLEAR MAGNETIC RESONANCE SPECTROSCOPY: NUCLEI OTHER THAN HYDROGEN

7.3.1 Nuclei with spin $\frac{1}{2}$

In general, any nucleus with a spin of $\frac{1}{2}$ will give rise to n.m.r. signals provided the appropriate magnetic field and radiofrequency are applied. There are many such nuclei, of which probably ^{13}C is the most important for the general chemist; formerly this nucleus was difficult to study since it gives rise to extremely weak signals and, even after concentration of the isotope considerably above its naturally occurring 1 per cent level, satisfactory n.m.r. spectra were not easy to obtain. The introduction during the last few years of Fourier transform methods, however (see Sec. 7.4.2), has caused a big upswing in the study of this nucleus. Other spin $\frac{1}{2}$ nuclei, frequently studied in the past, include ^{19}F and ^{31}P, both being the only isotopes of their particular elements, ^{29}Si (4.7 per cent) and various metals such as ^{117}Sn, ^{119}Sn, ^{195}Pt, ^{205}Tl, or ^{207}Pb.

For all these nuclei the phenomena of chemical shift and spin-spin coupling are observed in the spectra and generally both are considerably larger than their hydrogen counterparts. Thus, while a range of some 15 p.p.m. contains virtually all the known hydrogen chemical shifts, values for phosphorus-containing compounds span some 400 p.p.m., fluorine some 600 p.p.m., and a few metals, notably ^{205}Tl and ^{207}Pb, range over 14 000 p.p.m. or more. This behaviour can be traced to the greater number and increased mobility of the extranuclear electrons, leading to greater variation in diamagnetic shielding.

The increase in spin-spin coupling constants is probably attributable to the same cause although the increase is not so marked as in the case of chemical shifts. Thus the coupling between two directly bonded phosphorus nuclei is some 600 Hz, while it may be as high as 1400 Hz for phosphorus bonded to fluorine. These figures, however, are only some 3–6 times larger than the corresponding direct H—H coupling of 240 Hz. A further general point is that couplings tend to attenuate less rapidly with an increase in the number of bonds separating the coupled nuclei than do those of hydrogen.

Two practical advantages follow from the large chemical shifts of these nuclei. Firstly, less precise instrumentation is required in order to obtain data useful for structural determinations since the tolerances with which chemical shifts must be measured are correspondingly larger. Secondly, the spectra obtained are simpler to analyse in that they have less of the complications which arise in hydrogen spectra when the chemical shifts are comparable in magnitude with coupling constants. The spectra are thus of the $A_aM_mX_x$ . . . type, rather than $A_aB_bC_c$ . . . .

In the following section we give a very brief discussion of ^{13}C n.m.r., concentrating on areas where there are differences with proton spectroscopy, and in Sec. 8.3.3 there is an even shorter discussion of some recent studies in ^{31}P spectroscopy. The interested reader is referred to books in the bibliography for more detailed coverage of the n.m.r. of these and other nuclei.

7.3.2 ^{13}C N.M.R. Spectroscopy

The principles governing ^{13}C n.m.r. are exactly similar to those of ^{1}H spectroscopy, although the scale of observed shifts and couplings is, as has been stated, greater for the former. Chemical shifts in ^{13}C work are very conveniently measured (in p.p.m.) from tetramethyl silane (TMS) as reference so that the same sample can be used to study both nuclei; the δ scale, where TMS is arbitrarily assigned the value $\delta = 0.0$, is invariably used. On this scale ^{13}C shifts range from 0 to about 250 p.p.m. and Fig. 7.21 shows the approximate regions of the spectrum in which resonances of carbon nuclei in different chemical surroundings are found. Also, as in proton spectra, the precise chemical shift of a nucleus depends on the atom or atoms attached to it, and there are correlations with the electronegativity of substituents. In ^{13}C considerable success has been achieved in assigning observed resonances to specific atoms within molecules by using additivity rules for the combined shift effect of several substituents.

Because of the greater range of chemical shifts ^{13}C spectra nearly always contain a separate and distinct resonance for each chemically shifted nucleus in the molecule—very little overlap of resonances occurs. This obviously simplifies spectra-structure correlation, but we shall see later that the *intensities* of ^{13}C resonances usually cannot be so simply correlated with the number of atoms contributing to each resonance.

A further simplification in ^{13}C spectra is apparent when one considers spin-spin coupling—with normal samples one can completely ignore direct coupling between neighbouring ^{13}C

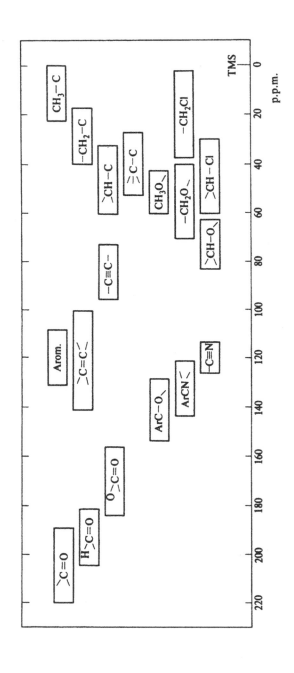

Figure 7.21 The approximate chemical shift ranges of some ^{13}C resonances.

nuclei. This is not to say that such coupling does not exist—it most certainly does, and two ^{13}C nuclei bound directly together in a molecule, or separated from each other by two or three bonds, will give rise to the *AX*-type spectrum mentioned in Sec. 7.2.2 for ^1H spectra. However, remembering that ^{13}C is present in a normal, non-enriched sample, only to the extent of about one per cent natural abundance, we can see that the probability of two such nuclei being found within a few bonds of each other is very remote. Indeed, if there are fewer than 100 carbon atoms in each molecule of sample, then there will be, on average, only one ^{13}C nucleus per *molecule*.

Coupling between ^{13}C and neighbouring *protons*, on the other hand, shows up clearly in ^{13}C n.m.r. spectra, where it can be both a help and a hindrance. It is helpful in assigning resonances to particular nuclei in a molecule. For example, a $\geqslant$C—H group in a molecule will show a doublet structure in the ^{13}C spectrum, with a J_{CH} coupling constant of about 125 Hz; a double-bonded carbon, as in $>$C$=$C—H, will show a coupling of some 170 Hz, and a triply bonded one, C$\equiv$C—H, of some 250 Hz. These values are large, and thus easily seen and distinguished in a spectrum. Substituents often affect the values considerably, but again useful empirical correlations exist; equally long-distance coupling, over two or three bonds, is well characterized. As in ^1H spectra, the *pattern* of the coupling is also very diagnostic. Thus a C—H fragment yields a doublet in the ^{13}C spectrum, CH_2 gives a triplet pattern, and CH_3 a quartet. Figure 7.22 shows just such a spectrum for CH_3C*OOH; here it is the starred carbon which is recorded, and its coupling to the three distant protons by some 10 Hz gives a clear quartet pattern.

^{13}C coupling is a hindrance in that it 'spreads out' the intensity of an already weak resonance, thus making observation more difficult. This can be overcome very elegantly by an extension of the double-resonance technique mentioned in Sec. 7.2.6. There we saw that selective radiation at a particular chemical shift position in a ^1H spectrum destroys coupling to neighbouring nuclei and thus simplifies spectra. In ^{13}C work it is common to decouple *all* the

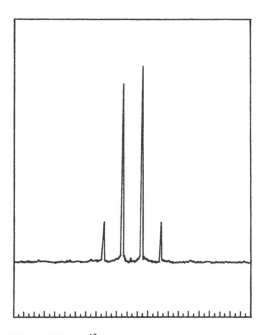

Figure 7.22 The ^{13}C n.m.r. spectrum of acetic acid, enriched in ^{13}C at the carbonyl carbon, CH_3*COOH. (*Thanks are due to Mr A. Taylor of the University of York for this spectrum.*)

protons in the molecule from the carbons. To do this requires irradiation of the sample at the ^{1}H frequency (100 MHz, for example), while observing the spectrum at the ^{13}C frequency (25.4 MHz in the same field). If the 100 MHz radiation is strong and sufficiently 'wide band' to cover all the proton frequencies in the sample—i.e. some 100 p.p.m., or 1000 Hz wide—then the protons will change their spins sufficiently often to be effectively decoupled from the carbons. This technique, often called *noise decoupling*, simplifies and intensifies the ^{13}C spectrum by collapsing the spin-spin multiplets to single lines. This results in the loss of the pattern recognition advantages mentioned in the previous paragraph, and it is often highly informative to compare the noise-decoupled spectrum with the non-decoupled one, thus finding how many hydrogens are attached to each ^{13}C.

Noise decoupling also leads to an intensity bonus due to a phenomenon known as the nuclear overhauser effect, or NOE. Briefly, we already know that the intensity of an n.m.r. signal is closely dependent on the (small) difference in population of excited and ground spin states, and that once sufficient nuclei have become excited, no more energy can be absorbed from the exciting radiation until relaxation has, at least partially, restored equilibrium conditions. In the presence of noise-decoupling radiation it turns out that other relaxation processes, which do not themselves absorb or emit radiation, are induced, so that equilibrium conditions are attained more rapidly. This effect, which in ^{13}C spectra typically gives rise to an approximate doubling of the signal intensity, is also known as *spin pumping*.

Relative intensity measurements of ^{13}C peaks often give apparently anomalous results. In Fig. 7.23 we sketch the ^{13}C spectrum of ethyl acetate as an example. The assignment of the four resonances to the four carbon nuclei in the molecule is very straightforward; the resonance at $\delta = 175$ is obviously in the C=O region (cf. Fig. 7.21), and the others are due to the saturated CH groups. In general, the resonance of CH$_2$ groups attached to electronegative atoms (like oxygen) occurs between $\delta = 40$ and $\delta = 70$, which establishes C$_3$; and finally the proximity of the carboyxl group (O.C=O) to C$_1$ shifts its resonance to higher δ from C$_4$. The intensities, however, are all 'wrong', since we would expect four resonances of equal intensity. The reason for this is to be found in the differing relaxation times of the four ^{13}C nuclei, as we discuss below.

Figure 7.23 Representation of the ^{13}C spectrum of ethyl acetate.

When observing n.m.r. spectra of low-intensity samples, like ^{13}C, by Fourier transform methods, the computer averaging technique (see Sec. 1.9) is invariably used. This requires repeated observation and accumulation of the spectrum, the accumulated real signals becoming enhanced, while the noise components essentially cancel and disappear. However, this technique can only be really effective if sufficient time is given between excitation pulses to allow the sample to return to equilibrium; otherwise maximum energy absorption from successive pulses will not occur. Strictly an infinite time should be allowed to elapse, but effectively a period of some four or five relaxation times is sufficient for virtually complete equilibration. If, as often happens in ^{13}C spectra, the relaxation times for nuclei in different surroundings are quite different, then four or five times the *longest* relaxation time should be allowed. For protons relaxation times are usually fairly short, and a delay of two or three seconds between pulses is adequate. For ^{13}C, however, relaxation times are in general rather longer, and often very much longer (1–100 second is typical), and a delay of four or five times 100 seconds between pulses would obviously make the collection of data very lengthy. Normally a delay of some 10 seconds is chosen, which is sufficient for nuclei with a short relaxation time to decay and so give rise to their 'proper' intensity in a spectrum, but those with a longer relaxation time will be somewhat smaller than they should be, and those with a very long relaxation period may well be very small indeed. A very rough rule is that the intensity of a particular signal is inversely proportional to the relaxation time of the nucleus concerned.

To explain differences in relaxation times we need to understand a little about the relaxation process itself. In order to give energy to their surroundings, excited nuclei require a magnetic (or electric) fluctuation to occur at approximately their precessional frequency—in this way resonance interaction with the surroundings can be set up. The most common source of local fluctuating magnetic fields is other spinning nuclei and, since ^{13}C nuclei are very isolated from each other, it is essentially only hydrogen nuclei attached to a particular ^{13}C nucleus, or perhaps two or three bonds away, which provide the relaxation mechanism. These ^1H nuclei offer a fluctuating magnetic field by virtue of the random movements of the molecule and, because we require these movements to occur at about 10^7–10^8 Hz (so as to resonate with the ^{13}C), we are restricted to fairly rapid tumbling of molecules in solution. Molecular rotations and vibrations are too fast (10^{10}–10^{13} Hz, approximately) and so do not contribute.

We can explain the intensity 'anomalies' of Fig 7.23, then, as follows. The C_2 carbon has no attached hydrogens and so has a relatively long relaxation time (typically about 30 seconds), thus giving rise to a line of low intensity. The CH_2 has, of course, two hydrogens and so can relax relatively easily (typical relaxation time of some 2–4 seconds), to given an intense line. The two CH_3 groups, although having three hydrogens, undergo rotation about the C—CH_3 bond; this motion is sufficiently fast to destroy resonance relaxation. Typical relaxation times for 'free' CH_3 groups are 10–20 seconds (although any steric hindrance to this rotation lowers the time), and the resultant low intensity of their ^{13}C peaks is often used to distinguish them from CH or CH_2 groups in a spectrum.

The intensity anomaly can be overcome either by spending *very* much longer collecting a spectrum or by adding a solution of a paramagnetic ion, such as iron or chromium. The 'free' electrons in such ions can be very efficient relaxers, and a ^{13}C spectrum usually exhibits a near-normal intensity pattern after the addition of such a reagent. At times, however, and particularly if the reagent forms a chemical complex with the molecule of interest, relaxation at some sites becomes too efficient, and the n.m.r. signal becomes too broad to be observed.

Observation of spin-lattice relaxation times, therefore, can sometimes be of assistance in assigning complex spectra—the low intensity of CH_3 group spectra and of carbon atoms not attached to hydrogens, for example. In addition, however, T_1 values are beginning to give insights into various aspects of molecular motion such as hindered rotation or movements of parts of a large molecule, e.g. side chains in polymers. Although we have illustrated this section with reference to ^{13}C spectra, relaxation is a phenomenon observed with all spinning nuclei, and relaxation times will become just as useful for other nuclei.

7.3.3 Nuclei with Spin Greater than $\frac{1}{2}$

We saw at the beginning of this chapter that the application of a magnetic field to any nucleus with spin I causes the spin vector to become oriented in any one of $2I + 1$ possible directions, each associated with a slightly different energy level. Thus Fig. 7.1 shows the situation for a nucleus with a spin of 1, for example ^{14}N. Since, in a given field, the spacing between the energy levels of a particular nucleus are all identical, transitions induced between any neighbouring levels will result in the emission or absorption of energy at the same frequency. Thus only one resonance line will appear for each nucleus.

In principle, then, *any* spinning nucleus will give rise to a single resonance line when the appropriate field and frequency are applied; the position of the resonance for a given nucleus will vary with its chemical surroundings and it will be split into a multiplet by interaction with other spinning nuclei—in other words, the phenomena of chemical shift and spin-spin coupling will be observed. In practice, however, the spectra of nuclei with spin greater than $\frac{1}{2}$ are usually intrinsically weak and not easy to observe and they are not much used, in themselves, for analytical or structural studies. However, elements such as ^{14}N (spin $I = 1$) and the halogens, chlorine, bromine (both of spin $= \frac{3}{2}$), and iodine (spin $= \frac{5}{2}$), occur widely in chemistry, and we should consider what effect the presence of these nuclei might have, by virtue of spin-spin coupling, on the n.m.r. spectrum of neighbouring *hydrogen* nuclei.

Let us take the ^{14}N—H group as an example: we would expect the coupling here to be reasonably large because the two nuclei are directly bonded (J, in fact, is observed to be some 50 Hz for this group). Now the unit spin of the nitrogen nucleus can take up one of three orientations in an applied field (cf. Fig. 7.1), which we may represent as ↑, →, and ↓. Comparing with Fig. 7.12 we see that the spin direction ↑ will reduce the field at the hydrogen nucleus, ↓ will reinforce it to the same extent, while it is plain that → will leave it unaltered. We expect the hydrogen resonance to be split into a *triplet*, therefore, and since all three spin orientations of the nitrogen are equally likely, each line of the triplet will have the same intensity. The formation of this 1 : 1 : 1 triplet is shown schematically in Fig. 7.24(a). In Fig. 7.24(b) we reproduce the spectrum of acidified methylamine, CH_3NH_2, to show the triplet structure of the hydrogen resonance. The solution is acidified merely to suppress the otherwise rapid exchange of the hydrogens attached to nitrogen, an exchange which would destroy the N—H coupling. One result of this is that the molecule becomes converted to the methylammonium *ion*, $CH_3NH_3^+$, and so the methyl resonance is split into a 1 : 3 : 3 : 1 quartet by spin coupling with the —NH_3^+ hydrogen nuclei; the fact that this quartet is sharply defined implies that the exchange rate is small. In spite of this we note that the NH triplet consists of very broad resonance lines; the broadening is due to quadrupole relaxation which we shall discuss shortly.

We can easily generalize the above discussion to other nuclei. A nucleus with spin I can take up one of $2I + 1$ equally likely spin orientations in an applied field; of these, some will reinforce and some reduce the field experienced by a neighbouring nucleus, so the resonance of the neighbour will be split into a multiplet of $2I + 1$ lines with equal intensity. Thus a single chlorine nucleus would tend to produce a quartet structure in its neighbours (spin $= \frac{3}{2}$), while a single iodine nucleus ($I = \frac{5}{2}$) would produce sextets. In practice, however, these splittings are not observed because of quadrupole relaxation which causes rapid transitions between the spin states (see the next section).

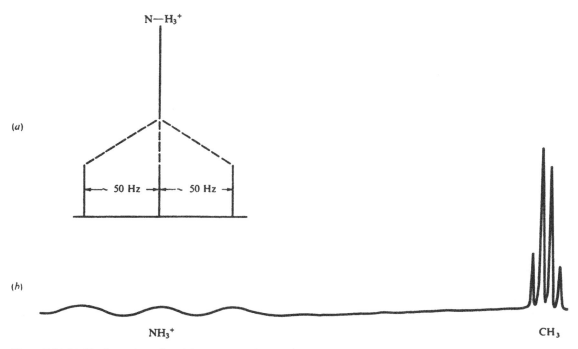

Figure 7.24 (*a*) The formation of a triplet resonance by spin-spin coupling of a hydrogen nucleus and a directly bonded ^{14}N nucleus; (*b*) the spectrum of acidified methylamine, CH_3NH_2.

7.3.4 Quadrupole Effects

In addition to the magnetic moment discussed throughout this chapter, all nuclei with a spin ≥ 1 also possess an *electric quadrupole moment*, which arises because the nuclei are not spherical. Such nuclei, in fact, are shaped either like a symmetrical egg or like a tangerine (elongated or flattened at the poles, respectively). Even if the charge density within the nucleus is constant, the distorted shape gives rise to a charge distribution which is non-spherical; the electric quadrupole moment is a measure of the departure from sphericity, being positive for egg-shaped and negative for tangerine-shaped nuclei. For spherical nuclei (i.e. spin $\frac{1}{2}$ or 0) the electric quadrupole moment is zero.

The electric moment interacts strongly with an applied electric field and, if such a field has a pronounced gradient at the nucleus, the nuclear moment will tend to lie in the field direction; further, if the gradient changes direction, the nuclear moment will try to follow this change. Thus, consider the pyramidal molecule of ammonia, NH_3. This has the three hydrogen nuclei arranged at the corners of the base of a pyramid, the nitrogen nucleus being above and equidistant from them. The presence of three positively charged nuclei on one side of the nitrogen nucleus produces a strong electric field gradient at that nucleus which tends to orient its quadrupole moment axis perpendicularly to the plane containing the hydrogen nuclei. If, due to molecular tumbling, the molecule rotates as a whole, the nitrogen nucleus tends to follow this rotation. So strong is the coupling between nuclear electric moment and field gradient that it is sufficient to prevent the coherent alignment of the nuclear *magnetic* moment in an applied magnetic field—in other words the quadrupole electric moment supplies a mechanism by which the spin orientation may be relaxed. Here, then, we have another—and usually very efficient—relaxation mechanism (cf. Sec. 7.1.5) called *quadrupole relaxation*.

If, on the other hand, we consider the ammonium ion, NH_4^+, which consists of a nitrogen nucleus at the centre of a regular tetrahedron of hydrogen nuclei, there is no field gradient at the nitrogen nucleus. Thus its spin can be oriented by an applied magnetic field quite independently of the orientation of the hydrogen nuclei—the quadrupole relaxation is extremely weak in this case. We see, then, that the efficiency of quadrupole relaxation will depend upon the symmetry of the surroundings. Univalent atoms, such as the halogens, will always be situated in a field gradient which will experience efficient relaxation; the relaxation of polyvalent atoms, such as nitrogen, will vary from molecule to molecule.

Quadrupole relaxation has two effects in n.m.r. spectroscopy. The first of these, the broadening of the n.m.r. signal from the nucleus possessing a quadrupole moment, is very similar to the broadening caused by fast relaxation (cf. Sec. 7.1.5). When quadrupole relaxation occurs the lifetime of a particular spin state may be as short as 10^{-4} s or less. This, as we saw in Sec. 7.1.5, is comparable to the relaxation time of nuclei in solids, for which we calculated a line width of some 1000 Hz. Such a 'line' would be so broad that it would be impossible to detect with a normal high-resolution n.m.r. spectrometer, since it would be indistinguishable from a very gentle wandering of the background noise.

For the second effect of quadrupole relaxation we recall the discussion of double resonance in Sec. 7.2.6, wherein an experimental method for 'stirring' or relaxing coupled spins was described. In quadrupole relaxation we have another method for destroying coherent coupling between nuclei. If the relaxation is highly efficient as, for example, in halogen nuclei, the 'stirring' is so rapid that coupling is completely destroyed and the resonances of neighbouring nuclei remain sharp; for less efficient relaxation, as in many nitrogen-containing molecules, the coupling is only partially destroyed and multiplet but broad resonances result from neighbouring nuclei. An example of this has already been seen in the very broad triplet formed by the $—NH_3^+$ group of the methylammonium ion of Fig. 7.24. In routine n.m.r. spectroscopy, the only commonly occurring quadrupolar nucleus whose effect is noticeable in spectra is, in fact, that of ^{14}N: here the effect is usually to broaden the resonance of neighbouring nuclei, sometimes with the appearance of multiplet structure. Often the broadening is so great that the hydrogen resonances disappear completely into the background noise.

A single crystal of a solid substance has a well-ordered and regular structure, apart from possible lattice defects, and any quadrupolar nucleus contained in a crystal will find itself in exactly identical surroundings as similar nuclei in other regions of the lattice. Thus all the quadrupole moments will tend to lie in the same direction, insofar as this is consistent with the Boltzmann energy distribution. Even without the application of an external magnetic field such nuclei will be able to absorb energy coherently from a beam of radiation at an appropriate frequency. These absorptions, which occur in the region 1–1000 MHz, are known as the *nuclear quadrupole resonance* (n.q.r.) spectrum or sometimes the *pure quadrupole spectrum* to emphasize the fact that an external field is not applied. N.q.r. spectroscopy essentially uses the quadrupolar nucleus as a probe to detect and estimate electric field gradients in the crystal and the data obtained are invaluable in applications of crystal field theory. The spectra are not easy to interpret, however, and will not be discussed here; the interested reader is referred to the book by Lucken listed in the bibliography.

7.4 TECHNIQUES AND INSTRUMENTATION

Basically an n.m.r. spectrometer requires components with functions similar to those already described for other spectroscopic techniques (radiofrequency source, sample holder, radiofrequency detector, recorder, etc.) but with the addition of a powerful magnet. We have seen that a resolving power of 0.5 Hz requires that the magnetic field by stabilized to within one part in 10^8

or better and, in addition, it must be the same over all parts of the sample under test. These requirements are very stringent.

Many earlier spectrometers used permanent magnets which, when carefully thermostatted, gave very precise and constant fields. They had two disadvantages, however. Firstly, the field could not be varied, so a separate crystal oscillator was required for each nucleus studied; and, secondly, the field strength was limited to about 2 T (some 90 MHz for protons) and, as we saw earlier, there are advantages in using higher fields.

Electromagnets can achieve fields up to a little over 2.5 T, although very sophisticated electronics is needed to stabilize the current to the 1 in 10^8 needed. Higher fields (the current limit is about 15 T) require the use of 'superconducting' magnets. These consist esssentially of a few turns of heavy gauge wire cooled in liquid helium; at such a low temperature the electrical resistance of the wire is negligible and so a large, steady current, once started, can pass without loss or change in the field. The obvious disadvantage of such a system is the expense and difficulty of working at low temperatures. The sample, of course, is placed in a thermally insulated cavity in the field and is not at the temperature of liquid helium. In fact, it is often desirable to record n.m.r. spectra at a variety of temperatures. This is usually achieved by passing a stream of nitrogen, pre-cooled or pre-heated to the required temperature, round the sample in its holder.

Homogeneity of the field over the sample is ensured, in an electromagnet, by very careful machining of the pole faces and by the use of secondary 'shim' coils to adjust the field gradients. Also the sample is usually spun rapidly to give effective averaging of the field 'seen' by all the nuclei.

Basically the same apparatus is used to observe spectra in the continuous wave mode (frequency domain) and in the Fourier transform mode. We deal with the former first.

7.4.1 Continuous Wave N.M.R. Spectroscopy

The usual arrangement of the components for this method is shown in Fig. 7.25 which, it should be noted, is not drawn to scale. For proton n.m.r. measurements about $0.5 \, cm^3$ of liquid sample is required, and this is held in a glass tube of 0.5 cm outside diameter and about 15 cm long, closed at one end; the liquid occupies the bottom 3 or 4 cm of the tube. For ^{13}C work about $3 \, cm^3$ of liquid is required, and the sample tube is some 1.2 cm in diameter. A trace of reference compound (such as TMS) is usually added directly to the sample. However, if the reference compound might interact with the sample, it can be held in an annulus external to the sample tube. The magnet poles are 20–30 cm in diameter, and the gap between them is only some 2–3 cm. When recording a spectrum in the field-sweep mode, the radio-frequency oscillator (the source) bathes the sample in radiation of, say, 100 MHz by means of the coil placed near the sample in a vertical plane. The magnetic field is set to 2.5 T and this is smoothly raised (swept) by means of a current produced in the sweep generator fed to auxiliary coils round the magnet poles. As each nucleus is brought to resonance it absorbs energy from the oscillator and then, when it reverts to the ground state, the emitted energy is collected by the detector coil wound round the sample, amplified, and passed to the recorder.

This method is by far the simplest technically, since it is easy to change a magnetic field smoothly, and fairly easy to maintain a precise frequency by using a carefully thermostatted crystal oscillator. The alternative frequency-sweep mode, wherein the magnetic field is maintained constant and the radiofrequency oscillator swept through a frequency range, is more difficult, but has advantages, particularly for some double-resonance experiments, and is sometimes employed.

Both of these sweep techniques have proved successful for an adequate quantity of a sample containing either hydrogen, fluorine, or phosphorus nuclei. For instance, some 20 mg of sample

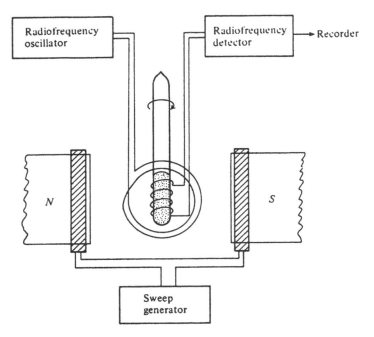

Figure 7.25 Schematic diagram of a continuous-wave n.m.r. spectrometer.

dissolved in $\frac{1}{2}$ ml of a solvent enable a good spectrum to be obtained in about 5 min; 10 mg samples take longer, perhaps 20 min.

7.4.2 Fourier Transform N.M.R. Spectroscopy

When applied to n.m.r., the FT technique is invariably used in the emission mode; we saw in Sec. 1.8 that this entails collecting the radiation simultaneously emitted by all the excited nuclei in the sample as a function of time, storing the collected information in a computer and mathematically transforming the result into the conventional frequency-domain spectrum. The apparatus required for this is thus very similar to that for continuous sweep measurements, except that the magnet sweep coils are not needed, and a computer is essential to program the pulses and carry out the Fourier transformation.

The sample must initially be excited by a pulse of radiation with the appropriate mean frequency (e.g. some 100 MHz for protons in a field of 2.5 T) but of a sufficiently wide frequency range to ensure that *all* the protons (or other nuclei) whose spectrum is to be recorded experience radiation of their particular resonance frequency and become excited. In fact, there is an inverse relationship between the length of the pulse and its frequency spread—a shorter pulse gives a wider range of frequencies, and vice versa—so the required scan width of the spectrum determines the maximum possible length of the pulse.

We saw earlier (Sec. 7.1.6) that the length of the pulse fixes the angle through which the bulk magnetic vector, **M**, is rotated. A 90° pulse rotates **M** into the xy plane and so gives maximum signal intensity. Before a second pulse can usefully be applied, however, it is necessary to wait for **M** to decay back into the z direction—a process which, theoretically, takes an infinite time, but which, in practice, can be thought of as being complete in some four or five half-lives of the decay process, i.e. some four or five times the spin-lattice relaxation time T_1. We cannot usefully collect the emitted signal during the whole of this waiting period, however, because, being an exponential decay, the signal decreases rapidly initially and soon becomes lost in the noise. The useful collection period is referred to as the acquisition time, $T_{acq.}$.

The efficient production of an n.m.r. spectrum, then, consists in carefully balancing the various factors of maximum pulse length, signal decay, and acquisition time against each other; in fact, the best signal/noise can usually be achieved by using a sequence of pulses with a length calculated from the ratio of $T_{acq.}$ to T_1, giving an optimum pulse angle less than $90°$.

7.5 ELECTRON SPIN RESONANCE SPECTROSCOPY

7.5.1 Introduction

In Chapter 6 we saw that the majority of stable molecules are held together by bonds in which electron spins are paired; in this situation there is no net electron spin, no electronic magnetic moment, and hence no interaction between the electron spins and an applied magnetic field. On the other hand, some atoms and molecules contain one or more electrons with unpaired spins, and these are the substances which are expected to show *electron spin resonance* (e.s.r.) spectroscopy; since such substances are also paramagnetic, this type of spectroscopy is often referred to as *electron paramagnetic resonance* (e.p.r.).

Substances with unpaired electrons may either arise naturally or be produced artificially. In the first class come the three simple molecules, NO, O_2, and NO_2, and the ions of transition metals and their complexes, for example Fe^{3+}, $[Fe(CN)_6]^{3-}$, etc. These substances are stable and easily studied by e.s.r. Unstable paramagnetic materials, usually called *free radicals* or *radical ions*, may be formed either as intermediates in a chemical reaction or by irradiation of a 'normal' molecule with ultra-violet or X-ray radiation or with a beam of nuclear particles. Provided the lifetimes of such radicals are greater than about 10^{-6} s they may be studied by e.s.r. methods; species short-lived at room temperature may also be studied if they are produced at a low temperature, either in the solid state or trapped in a solid 'matrix' of a host material such as a solid inert gas. This is called *matrix isolation* and increases the lifetime of the trapped radical. Obviously, while there are many and various substances to which e.s.r. can be applied, the requirement for an unpaired electron means that the technique is generally less applicable than n.m.r.

When unpaired electrons exist in a substance their spins are aligned at random in the absence of a field. When placed in a magnetic field, however, they will each have a preferred direction and, since the spin quantum number of an electron is $\frac{1}{2}$, each can be thought of as spinning either clockwise or anticlockwise about the field direction. E.s.r. spectroscopy essentially measures the energy required to reverse the spin of an unpaired electron.

Virtually all the theory which we shall need in the discussion of e.s.r. spectra has been dealt with in preceding sections of this chapter; Sec. 7.1 is particularly relevant here, since it covers such matters as the electron's magnetic moment, its interaction with an applied magnetic field to give two energy levels, the populations of those levels, the Larmor precession, and relaxation process. In fact, with the exception of the magnitudes of some quantities, the whole of Sec. 7.1 can be carried straight over into the discussion of e.s.r. spectroscopy. Parts of Sec. 7.2, such as the remarks regarding spin-spin coupling, are also relevant, as we shall see.

As in all forms of spectroscopy, four properties of the spectral lines are of importance, viz. their intensity, width, position, and multiplet structure. The first two we can deal with quite briefly; the final pair merit separate sections.

The *intensity* of an e.s.r. absorption is proportional to the concentration of the free radical or paramagnetic material present. Thus we have immediately a technique for estimating the amount of free radical present; the method is extraordinarily sensitive, in favourable cases some 10^{-13} mol of free radical being detectable.

The *width* of an e.s.r. resonance depends, as in the case of n.m.r., on the relaxation time of the spin state under study. Of the two possible relaxation processes, the spin-spin interaction is usually very efficient, unless the sample is extremely dilute, and gives a relaxation time of $10^{-6}–10^{-8}$ s; the spin-lattice relaxation is efficient at room temperature (some 10^{-6} s) but becomes progressively less so at reduced temperatures, often becoming several minutes at the temperature of liquid nitrogen. For most samples, then, we could choose 10^{-7} s as a typical relaxation time and, using this in the Heisenberg uncertainty relation of Eq. (1.11), we calculate a frequency uncertainty (line width) of $(2\pi\,\delta t)^{-1} \approx 1$ MHz. A shorter relaxation time will increase this width, and 10 MHz is not uncommon. Clearly this is a *much* wider spectral line than in the case of n.m.r., where we found a normal line width for a liquid to be some 0.1 Hz.

The wider e.s.r. lines have advantages and disadvantages. On the credit side, the homogeneity of the applied magnetic field is far less critical and where, for n.m.r., it is essential to use a magnetic field homogeneous to 1 in 10^8 over the sample, for e.s.r. a figure of 1 in 10^5 is adequate; this represents a considerable easing of manufacturing tolerances. On the debit side, however, broad-lined e.s.r. spectra mask any effects equivalent to the n.m.r. chemical shift. In addition, a broad line is more difficult to observe and measure than a sharp one and, for this reason, e.s.r. spectrometers nearly always operate in the derivative mode (see Chapter 1, Sec. 1.4 and Fig. 1.11).

7.5.2 The Position of E.S.R. Absorptions; the g Factor

We know from Sec. 7.1.2 that the spin energy levels of an electron are separated in an applied magnetic field, B_0, by an amount:

$$\frac{\Delta E}{h} = \frac{g\beta B_0}{h} \quad \text{Hz} \tag{7.16}$$

where β is the Bohr magneton (9.273×10^{-24} J T^{-1}) and g the Landé splitting factor. A resonance absorption will thus occur at a frequency $v = \Delta E/h$ Hz. From Eq. (7.16) we see that the position of absorption varies directly with the applied field and, since different e.s.r. spectrometers operate at different fields, it is far more convenient to refer to the absorption in terms of its observed g value. Thus, rearranging Eq. (7.16) we have:

$$g = \frac{\Delta E}{\beta B_0} = \frac{hv}{\beta B_0} \tag{7.17}$$

and if, for example, resonance were observed at 0.34 MHz in a field of 9506.690 T, it would be reported as resonance at a g value of 2.0023. This very precise figure is the g factor for a free electron (rather than the slightly approximate value of two given by putting $L = 0$ in Eq. (5.28)), but it is a remarkable fact that virtually all free radicals and some ionic crystals have a g factor which varies only some ± 0.003 from this value. The reason for this is essentially that in free radicals the electron can move about more or less freely over an orbital encompassing the whole molecule (as we shall see in the next section) and it is not confined to a localized orbital between just two of the atoms in the molecule. In this sense it behaves in very much the same way as an electron in free space, having $L = 0$.

Some ionic crystals, on the other hand, have very different g factors, values between about 0.2 and 8.0 having been reported. The difference here is that the unpaired electron is contributed by, and 'belongs' to, a particular atom in the lattice, usually a transition metal ion. Thus the electron is localized in a particular orbital about the atom, and the orbital angular momentum

(*L* value) couples coherently with the spin angular momentum, giving rise to a *g* value consistent with Eq. (5.28).

Nonetheless, many ionic crystals show a *g* factor very close to the free electron value of 2; this may come about in two ways:

1. The ion contributing the electron may exist in an *S* state (that is $L = 0$). For example, the ground state of Fe^{3+}, in which five *d* electrons are unpaired (that is $S = \frac{5}{2}$, $2S + 1 = 6$), has zero orbital momentum. Thus $L = 0$, $J = S + L = S$, and the term symbol is $^6S_{5/2}$ (cf. Chapter 5). Since $J = S$, $g = 2$ (Eq. (5.28)).
2. The electric fields set up by all the ions in a crystal may be sufficiently strong to *uncouple* the electron's orbital momentum from its spin momentum—i.e. coherent Russell–Saunders coupling breaks down and, on the application of a magnetic field, the electron spin vector precesses *independently* about the field direction. Thus the value of *L* is immaterial and the *g* factor reverts to two. On the other hand, if the internal crystal field is weak or if the paramagnetic electron is well shielded from the field (e.g. as in a rare earth metal, where the relevant electron orbit is buried deep within outer electron shells), *L* and *S* couple to produce a resultant *J* which itself precesses about the applied magnetic field, and *g* is given by Eq. (5.28). Intermediate cases also occur where *L* and *S* are only partly uncoupled, the residual orbital contribution to the energy giving rise to a *g* value not easily predictable theoretically.

7.5.3 The Hyperfine Structure of E.S.R. Absorptions (Electron–Nucleus Coupling)

In e.s.r. spectra we must distinguish between two kinds of multiplet structure; there is *fine structure*, which occurs only in crystals containing more than one unpaired electron spin, and *hyperfine structure*—a smaller effect—which arises when an unpaired electron can get close to a nucleus with non-zero spin. We shall deal with the latter in this section and leave until Sec. 7.5.5 our discussion of fine structure.

It is the interaction of the magnetic moment of the unpaired electron with those of any surrounding nuclei having non-zero spin which gives rise to hyperfine structure. This coupling is exactly analogous to the coupling between nuclear spins discussed in Sec. 7.2.2 earlier—indeed, the bulk of that discussion and Fig. 7.12 apply exactly to the electron–nucleus coupling if we label the nucleus as *A* and the electron *X*. For simplicity we shall limit our treatment here to nuclei of spin $\frac{1}{2}$ which will split electron resonances into doublets. If more than one nucleus can interact with a given electron, the result can be predicted using the 'family tree' method discussed in Sec. 7.2.3 and Fig. 7.16.

Electron–nucleus coupling constants are very much bigger than those for nucleus–nucleus interactions, firstly because an electron can approach a nucleus more closely than can another nucleus and secondly because the electron's magnetic dipole is some 1000 times larger than that of a nucleus. Thus in the hydrogen atom the electron resonance shows two equal lines with a separation of about 0.05 Tesla. We have seen that a bulk field of 0.34 T is equivalent to a frequency of about 9500 MHz, so 0.05 T is some 1400 MHz, which is an enormous coupling compared to the largest nucleus–nucleus coupling of about 2 kHz.

For most organic molecules with an unpaired spin the coupling constants are of order 10^{-4}–10^{-3} tesla (2–20 MHz); these are smaller than in the hydrogen atom because an unpaired electron in a molecule is never confined to just one nucleus and seldom even to one bond, but can move over several bonds with relative ease, spending only part of its time at any one nucleus. Another way to look at this is in terms of *electron density*. In the hydrogen atom the electron density of the unpaired electron is 1—the electron is in the atom and nowhere else. To a good approximation the magnitude of the coupling constant is directly related to the electron density, and we can write:

$$A = R\rho \tag{7.18}$$

where A is the coupling constant, ρ the electron density, and R the intrinsic coupling for unit density. For hydrogen, $\rho = 1$ and A, we have seen, is 0.05 T; thus $R = 0.05$ T. Figure 7.26 shows, at the bottom, the e.s.r. spectrum of the methyl radical, $\cdot CH_3$. Remember that, since this is a derivative spectrum, the actual centre of each peak should be measured at the point where the slope is zero, i.e. where the downward-sloping line crosses the x axis. However, the height of the maximum of each point above the x axis is a good approximation to the intensity of the normal, non-derivative, line in the spectrum, and it is clear that Fig. 7.26 arises from a $1:3:3:1$ quartet. Intuitively we would expect the electron to spend an equal amount of time on each hydrogen (the electron density at each should be identical), and so to couple equally to all three; this would result in the electron's resonance being split into a $1:3:3:1$ quartet, as shown by the family tree in the top part of the figure. Clearly the $\cdot CH_3$ spectrum below corresponds exactly to this pattern. The separation between the lines—the coupling—is found to be about 2.3 millitesla (mT), and so we can use Eq. (7.18) to calculate $\rho = A/R = 0.0023/0.05 = 0.046$. Thus we have the electron density at each hydrogen to be 0.046, or 4.6 per cent, and so we know

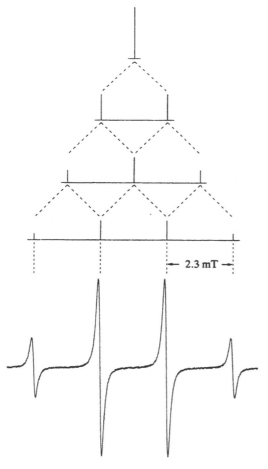

Figure 7.26 The e.s.r. spectrum of the methyl radical, $\cdot CH_3$, and the 'family tree' of couplings which produce it.

that the electron spends nearly 5 per cent of its time at each hydrogen and the remaining $100 - 3 \times 4.6 \approx 86$ per cent on the carbon.

A slightly more complex example is shown in Fig. 7.27, which illustrates the e.s.r. spectrum of the benzene radical, $\cdot C_6H_6$. Here we see a septet of lines, indicating equal coupling to all six hydrogens; the coupling constant is $0.38\,mT$, and we calculate $\rho = 0.38 \times 10^{-3}/ 0.05 \approx 7.6 \times 10^{-3}$. There are six hydrogens with this electron density, making a total of $6 \times 7.6 \times 10^{-3} = 0.046$, or nearly 5 per cent in total. Clearly the remaining 95 per cent of the electron's time is spent equally on the six carbon atoms, making nearly 16 per cent on each. Note that Figs 7.26 and 7.27 are drawn to a very different horizontal scale.

If the electron can couple to more than one set of nuclei, the pattern becomes more complicated, but can be rationalized by using the family tree again. The spectrum of the naphthalene radical is shown in Fig. 7.28. This complex pattern can be easily understood once it is realized that, because of the symmetry of the molecule, there are only two different sorts of hydrogen—the positions are numbered in the conventional sense in the molecular picture drawn in the diagram. A moment's thought will make it clear that hydrogens 1, 4, 5, and 8 are an equivalent set, as are hydrogens 2, 3, 6, and 7. The electron couples to the latter group with $A = 1.83 \times 10^{-4}\,T$ and, since there are four hydrogens, the pattern will be a $1:4:6:4:1$ pentet. One such pentet is picked out on the left of the diagram. Because of the other group of four hydrogens ($A = 4.95 \times 10^{-4}\,T$), this pentet will itself be split into a $1:4:6:4:1$ pentet; finding these in the spectrum is left as an exercise for the student, but the centre of one other pentet has been picked out on the diagram to help. The assignment of the $1.83 \times 10^{-4}\,T$ coupling to the 2, 3, 6, and 7 hydrogens is not made arbitrarily, of course. It can be checked by substituting one or more of the hydrogens with a non-spinning nucleus (e.g. a methyl group), or with a deuterium atom (which has a smaller coupling constant to the electron because of its different magnetic moment), and noting how the spectrum changes.

This use of e.s.r. techniques allows us to build up a qualitative picture of the electron distribution within a molecule which may help, for instance, in understanding chemical reactions—a positively charged reactant will plainly tend to attack that part of a molecule where the electron density is greatest, and vice versa. However, the observed results are not always in good quantitative agreement with the predictions of the molecular orbital theory, particularly for situations where simple molecular orbital theory predicts zero electron density at some points within a molecule—often coupling is observed to occur with nuclei at these points. For instance,

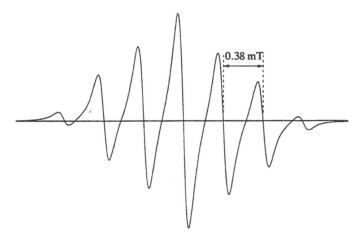

Figure 7.27 The e.s.r. spectrum of the benzene radical, $\cdot C_6H_6$. (*Thanks are due to Dr A. Whitwood of the University of York for this spectrum.*)

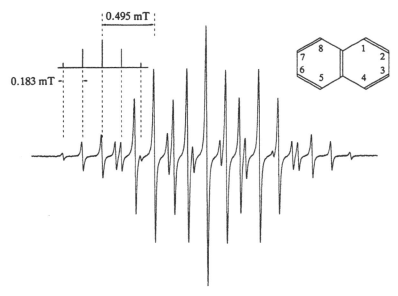

Figure 7.28 The e.s.r. spectrum of the naphthalene radical, showing the assignment of some of the lines. (*Thanks are due to Mr R. D. Farley and Prof. B. C. Gilbert of the University of York for this spectrum.*)

the methyl radical already mentioned is a planar molecule, with the carbon and three hydrogen nuclei in the same plane (Fig. 7.29(a)); the unpaired electron is considered to be held in a *p* orbital with lobes above and below this plane and, since the plane is a region of zero electron density for the orbital, there can be no electron density at the hydrogen nuclei arising directly from the unpaired spin. Thus $\rho = 0$ and so, since A is observed to be non-zero, it would seem that Eq. (7.18) does not apply.

A better account of the observed coupling can be given by following an argument very similar to that given in Sec. 7.2.2 and Fig. 7.13(a) for indirect nucleus–nucleus coupling. Thus in Fig. 7.29(b) the unpaired electron, with spin-up, say, can be considered to polarize the other carbon electrons preferentially into the spin-up situation (Hund's rule), so that the other electron in each C—H bond will have a spin-down orientation (Pauli's principle), and this will tend to align the hydrogen nuclear spins in the up direction. Hence the unpaired spin and the nuclear spin will tend to lie in the same direction and coupling exists. We can still apply Eq. (7.18) provided we agree to take ρ as a measure of the *spin* density of the unpaired electron (i.e. its influence at a remote point) rather than its *physical* density.

By the same argument it is clear that negative coupling—*negative spin density*—can occur. If we add a further carbon atom into the chain, the situation is as shown in Fig. 7.29(c); here the

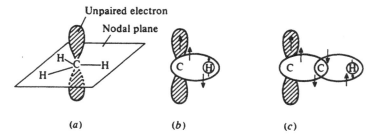

Figure 7.29 (a) The ·CH$_3$ radical showing the nodal plane of the *p* orbital containing the unpaired electron; (b) positive and (c) negative coupling between the unpaired electron and a neighbouring nuclear spin.

two electrons on the new carbon atom will tend to have parallel spins (Hund's rule again), and so the unpaired electron–nucleus configuration will be in the opposite sense to that in Fig. 7.29(*b*).

These 'configuration interaction' effects can be allowed for quantitatively in a refined version of the molecular orbital theory, and then excellent agreement is observed with experimental results. It seems that there is now ample justification for using e.s.r. spectroscopy in the accurate measurement of electron spin densities at various points within a molecule.

7.5.4 Double Resonance in E.S.R.

The concept of double resonance in e.s.r. spectroscopy exactly parallels that in n.m.r.—observation of a spectrum at one frequency while simultaneously irradiating at another. In e.s.r. there are two possibilities—the second frequency may be either at *nuclear* or at *electron* resonance frequencies—and these are called *endor* (electron–nuclear double resonance) and *eldor* (electron–electron double resonance), respectively. We shall deal only with the former; it is technically the simpler and is used more often than *eldor*.

Consider the simple case of a single unpaired electron interacting with a nucleus of spin $\frac{1}{2}$— an organic radical where the electron couples with a hydrogen nucleus is an obvious example. In an applied field both the electron and the nucleus will occupy one of two different energy states (spin 'up' or 'down'), and so there is a combined total of four energy states available. We may build up the energy-level diagram for the system in the following way (Fig. 7.30). The electron energy is split by the applied field into two widely separated states—if the field is 0.34 T then the separation is some 9500 MHz—as shown on the extreme left. Remember (and cf. Sec. 7.1.2) that the 'down' spin of an electron is more stable than the 'up' spin, because of the electron's negative charge. In the same field the hydrogen energy is split by some 13 MHz only, so we can initially simply add or subtract the nuclear energy, depending upon whether the nuclear spin is down or up, arriving at the four levels labelled ↓↑, ↓↓, ↑↑, and ↑↓ in the figure. Here we show the electron

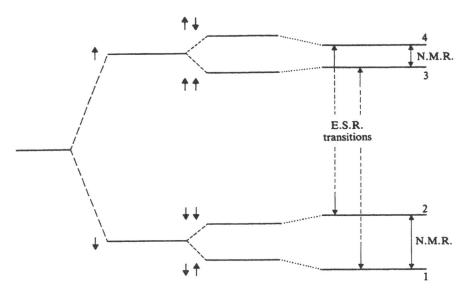

Figure 7.30 An energy-level diagram illustrating coupling between electron and nuclear spins.

spin first, and remember again that the nuclear spin 'up' is stabilized in a field. This gives the enegies in increasing order as shown.

The arguments so far, however, have ignored electron–nucleus coupling—we have simply superimposed the nuclear energy values on to those of the electron and have not included the possibility of mutal interaction of the energy states. When we do include this, we remember that states with paired spins ($\downarrow\uparrow$ and $\uparrow\downarrow$) tend to be stabilized, or lowered in energy, while parallel spins ($\uparrow\uparrow$ and $\downarrow\downarrow$) are raised. The dotted lines of Fig. 7.30 indicate the effects of these interactions, yielding the set of energy levels on the right of the figure. The reader should note the similarity between this figure and Fig. 7.14, which describes the energy levels for two coupled *nuclei*. The only difference is that of scale—in Fig. 7.14 the nuclear frequencies are all the same order of magnitude; here the electron and nuclear energy-level spacings differ by a factor of 1000.

Transitions in the e.s.r. region (those which change the electron's spin only) are shown in Fig. 7.30 and, as expected, give rise to a doublet due to coupling with H. Similarly, the n.m.r. transitions, involving the hydrogen spin only, also show two resonances. It is not particularly easy to observe the n.m.r. resonance of a radical directly—many factors are against this, such as the small concentration of radical giving a weak signal; broadening due to the efficient relaxation by the 'free' electron; stability problems with the radical, etc. We can, however, see the *effect* of applying n.m.r. frequency radiation while looking at the e.s.r. spectrum.

Imagine bathing the sample in fairly intense e.s.r. radiation so as to saturate, say, the $1 \rightarrow 3$ transition. This means that there will be no net e.s.r. signal observed, because the populations of levels 1 and 3 are equal. If we now also apply radiofrequency (n.m.r.) radiation to the sample, while still observing at the e.s.r. frequency, we can imagine sweeping the n.m.r. radiation slowly upwards. At some point we shall induce the $3 \rightarrow 4$ transition, thus raising some molecules to level 4 and so decreasing the population of level 3. As soon as this happens, e.s.r. radiation will be absorbed once more, because there is now 'room' in level 3 for the transition $1 \rightarrow 3$ to occur. So sweeping through the *n.m.r.* resonance causes a signal in the *e.s.r.* spectrum. If we continue to raise the n.m.r. frequency until resonance with the $1 \rightarrow 2$ transition occurs, we shall again disturb the relative populations, and another e.s.r. signal will be seen.

Initially, then, the endor spectrum of this system does not look very exciting—two lines separated by the coupling constant, which is exactly the same information as can be gained from the e.s.r. spectrum alone. There are, however, two reasons why endor techniques are useful.

Electron spin resonances are often very broad and it is not always possible to resolve spectra sufficiently well to see splitting due to nuclear coupling. Endor spectra, on the other hand, depend on the width of the *nuclear* resonance, because it is only while passing through the latter that an e.s.r. signal is observed. Although the nuclear signal is often considerably broadened by relaxation *via* the free electron, it is nonetheless much sharper than e.s.r., so nuclear coupling is often very much easier to observe, and the coupling constants themselves are more accurately evaluated, from the sharper spectral lines given by endor.

Additionally, if the e.s.r. coupling is to a nucleus with a spin greater than $\frac{1}{2}$, perhaps to nitrogen ($I = 1$) in an organic molecule or to a metal ion in a crystal, the e.s.r. signal is split into $I + 1$ lines. If resolved, this splitting gives a complex spectrum; if unresolved, it gives a very broad signal. The endor spectrum is much simpler; whatever the spin of the coupled nucleus, its resonance is split only into a doublet by interaction with the single electron.

We should point out that the above description of endor does not represent the way in which such experiments are actually carried out. We have discussed fixed energy levels and varying applied frequencies. In fact it is technically simpler to keep the applied e.s.r. frequency constant and to vary the energy levels by varying the magnetic field until resonance occurs. It is simpler to think of the process in the way we have done above, however, and none of the principles are altered.

7.5.5 The Fine Structure in E.S.R. Spectra (Electron–Electron Coupling)

We have seen (Sec. 7.5.3) that hyperfine structure occurs as a result of interaction between nuclear spins and the unpaired electron. Fine structure, on the other hand, occurs only for substances with more than one unpaired electron and is due to electron–electron coupling. Such conditions only normally arise in crystals, and particularly those containing complex ions.

Consider the case of a crystal which contains molecules or ions with two parallel, rather than paired, electron spins, resulting in a total spin of 1; this is, of course, a *triplet* state, since $S = 1$ and $2S + 1 = 3$. Molecular triplet states are often unstable, reverting to the singlet state with paired spins. For example, in the case of naphthalene irradiated with ultra-violet light, individual molecules undergo excitation to the triplet state which decays quite rapidly to the ground state. Here, however, by cooling the crystal to low temperatures, or by diluting the naphthalene in a solid, inert lattice, the triplet state can be maintained and examined by e.s.r. techniques. On the other hand, transition metal ions often exist quite stably in a triplet state and are easily studied at room temperature by e.s.r.

The two electrons per molecule or per ion forming the triplet state can be treated for most e.s.r. purposes as a single particle of spin 1. Thus the angular momentum vector corresponding to $S = 1$ is given by:

$$\mathbf{S} = \sqrt{S(S+1)}\,\frac{h}{2\pi} = \sqrt{2}\ \text{units}$$

and, in the absence of a field, the vectors orient themselves randomly. When a magnetic field is applied, or if an internal field exists within the crystal, the vectors can take up one of three directions only—essentially with, across, or against the field direction, as shown in Fig. 7.31(a); the components of the angular momentum in the field direction are, of course, $S_z = +1$, 0, or -1. In the field the $S_z = +1$ state is raised in energy, the $S_z = -1$ lowered, and the $S_z = 0$ unaffected, as shown in the centre of Fig. 7.31(b). (Note that the $+1$ state is *raised* in energy, whereas (cf. Fig. 7.1(c)) when dealing with nuclei of spin 1 the $+1$ state is lowered; this reflects, of course, the opposite sign of nuclear and electron charges.) Under the selection rule, $\Delta S_z = \pm 1$ only, two transitions are allowed, but both would have identical energy, and hence give rise to just one spectral line. In fact *two* lines of different energy are invariably observed, and we must now consider why it is that the energy levels are split unsymmetrically; there are several reasons.

Firstly, remembering that each $S = 1$ state is, in fact, made up of two electrons with parallel spins, we know that each spin produces a small magnetic field in the vicinity of its partner. This effect is identical with that of dipolar spin-spin coupling described in Fig. 7.12 for two *nuclei*, except that it is in the opposite direction, since nuclei and electrons have opposite charges—thus

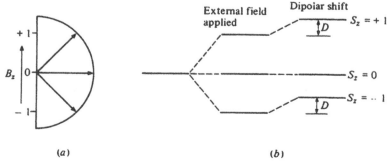

(a)   (b)

Figure 7.31 (a) The allowed orientations of two parallel electron spins in a magnetic field, B_z. (b) The splitting of the energy into three levels, and the dipolar shift, D, raising the $S_z = \pm 1$ states.

here an up spin increases its neighbour's field, while a down spin decreases it. Hence for our two electrons, *both* spins in the $S_z = +1$ state feel an applied field rather greater than the external field, B_z, and this state is *raised* in energy; for $S_z = -1$ both feel a smaller field, which again *raises* their energy (remember that increasing the applied field *raises* the $S_z = +1$ energy, but *lowers* the $S_z = -1$). In the case of $S_z = 0$, the dipolar field is *across* the main applied field so neither adds nor subtracts from it. The $S_z = 0$ energy is, therefore, quite unaffected by dipolar coupling.

The net effect of dipolar interaction is, then, to raise *both* the $S_z = +1$ and $S_z = -1$ states with respect to $S_z = 0$, as shown on the right of Fig. 7.31(b). Here we show the dipolar shift, D, to be smaller than the splitting caused by the main applied field; for small fields, however, it is clear that both $S_z = +1$ and $S_z = -1$ may lie above $S_z = 0$ in energy, since the magnitude of the dipolar shift is quite independent of the applied field.

Secondly, there may be spin-orbit coupling, whereby the electrons' orbital angular momenta and spin momenta are combined. Since the net spins of $S_z = +1$ and $S_z = -1$ are oppositely aligned, such coupling may again disturb the energy-level pattern, but the $S_z = 0$ state will be unaffected. The spin-orbit effect may be in the same sense or opposed to the dipolar interaction.

Finally, if there is a strong internal electric field within the crystal (and there will be, unless the substance is highly symmetrical), this will result in further perturbations to the energy levels of $S_z = +1$ and $S_z = -1$, but not of $S_z = 0$.

The net result of these three energy perturbations is usually considered under the one heading of the *zero-field* (or *crystal-field*) effect, since they produce an energy-level shift of the $S_z = +1$ and -1 states with respect to $S_z = 0$, even in the absence of an external field. This situation is shown in Fig. 7.32(a), where the $S_z = \pm 1$ states are assumed to lie above the $S_z = 0$; the opposite shift may also arise, but this does not alter the following discussion of the spectrum. When a steadily increasing magnetic field is applied to the crystal, as in Fig. 7.32(b), the $S_z = \pm 1$ levels diverge and, for a given radiation frequency, it is clear that there will be transitions at two different applied fields; thus the spectrum will consist of two fine structure lines.

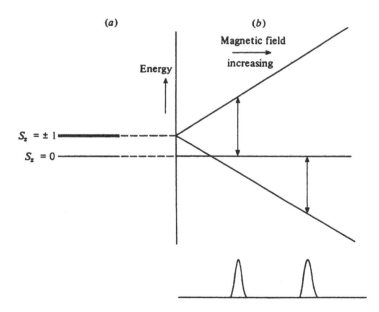

Figure 7.32 (*a*) The zero-field splitting of a triplet state. (*b*) The effect of an applied magnetic field and the allowed transitions between levels.

The zero-field splitting is usually a large effect; for example, the observed separation between the $S_z = 0$ and the $S_z = \pm 1$ levels in the naphthalene triplet state is equivalent to about 3000 MHz, which is to be compared with the energy separation of some 8000 MHz for an externally applied field of 0.3 T. Some transition metal ions have much larger zero-field splittings. Thus in a field-sweep spectrum, as indicated at the foot of Fig. 7.32(b), the fine structure separation is large, some 0.1–0.2 T or more, compared with the line width of 10^{-3}–10^{-4} T.

Extension of the above arguments shows that fine structure splitting will occur similarly for crystals containing molecules or ions with more than two parallel spins. In general, if there are n parallel spins, there will be n equally spaced resonances in the e.s.r. spectrum.

7.5.6 Techniques of E.S.R. Spectroscopy

We have seen that e.s.r. spectroscopy is usually carried out in the microwave region; thus the description of microwave apparatus for measuring molecular rotations (Chapter 2, Sec. 2.5) applies equally to e.s.r., with the addition of a magnet to produce the required field. Many commercial spectrometers now use a klystron source operating at a fixed frequency of 9500 MHz with an electromagnet at about 0.34 T equipped with subsidiary coils so that the field can be swept through and spectra recorded.

Additionally, coils are usually fitted to supply an oscillating magnetic field of strength a few microtesla and frequency some hundreds of cycles per second; this has two purposes. Firstly, it provides a modulator (or 'chopper') frequency to which the detector-amplifier system may be tuned in order to improve the signal-to-noise ratio. Secondly, this arrangement can be used to display the spectrum in the derivative mode, as is usual for e.s.r. In Fig. 7.33(a) we show the

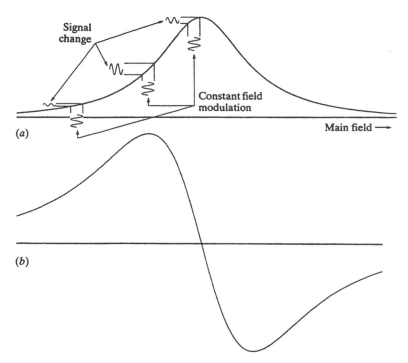

Figure 7.33 (a) An absorption spectrum showing the constant amplitude of the field modulation and consequent signal change leading to (b), the corresponding derivative spectrum.

(constant) amplitude of the modulation field as the main field is swept over the absorption spectrum. In regions where the absorption spectrum is varying slowly the signal change monitored over each cycle of the modulation is small, whereas when the spectrum changes rapidly, the signal change is large. Thus the signal change measures the *slope* of the spectrum and this may be recorded directly as the derivative curve, as in Fig. 7.33(*b*). The latter has two main advantages for e.s.r. spectroscopy: (1) the point of maximum absorption is difficult to measure accurately with a broad absorption curve, but is shown with much greater precision as the intersection of two lines in the derivative mode; (2) it is also more accurate to estimate the intensity of a derivative signal than that of the corresponding broad absorption.

BIBLIOGRAPHY

Abraham, R. J., J. Fisher, and P. Loftus: *Proton and Carbon-13 N.M.R. Spectroscopy: An Integrated Approach*, Heyden & Son, 1978.

Akitt, J. W.: *N.M.R. and Chemistry: An Introduction to N.M.R. Spectroscopy*, 2nd ed., Chapman & Hall, 1983.

Carrington, A., and A. D. McLachlan: *Introduction to Magnetic Resonance*, Chapman & Hall, 1979.

Derome, A. E.: *Modern N.M.R. Techniques for Chemical Research*, Pergamon Press, 1987.

Ebsworth, E. A. V., D. W. H. Rankin, and S. Cradock: *Structural Methods in Inorganic Chemistry*, 2nd ed., Blackwell, 1991.

Farrar, T. C., and E. D. Becker: *Pulse and Fourier Transform N.M.R.: Introduction to Theory and Methods*, Academic Press, 1971.

Gerson, F.: *High Resolution E.S.R. Spectroscopy*, John Wiley, 1971.

Harris, R. K.: *Nuclear Magnetic Resonance, A Physiochemical Review*, 2nd ed., Longman, 1986.

Kalinowski, H.-O., S. Berger, and S. Braun: *Carbon-13 N.M.R. Spectroscopy*, John Wiley, 1988.

Kevan, L., and R. N. Schwartz (eds): *Time Domain Electron Spin Resonance*, John Wiley, 1979.

Kemp, W.: *N.M.R. in Chemistry: A Multinuclear Approach*, Macmillan Press, 1986.

Lucken, E. A. C.: *Nuclear Quadrupole Coupling Constants*, Academic Press, 1969.

Martin, M. L., G. J. Martin, and J. J. Delpuech: *Practical N.M.R. Spectroscopy*, Heyden & Son, 1986.

Wertz, J. E., and J. R. Boulton: *Electron Spin Resonance*: *Elementary Theory and Practical Applications*, Chapman & Hall, 1986.

Williams, D. H., and I. Fleming: *Spectroscopic Methods in Organic Chemistry*, 4th ed., McGraw-Hill, 1989.

PROBLEMS

(Useful constants: $h = 6.626 \times 10^{-34}$ J s; $\beta_N = 5.051 \times 10^{-27}$ J T^{-1}.)

7.1 (*a*) A particular n.m.r. instrument operates at 30.256 MHz; what magnetic fields are required to bring ^{1}H and ^{13}C nuclei to resonance at this frequency?

(*b*) Calculate the resonance frequency of the ^{1}H nucleus when subjected to field strengths of 2.5 and 5.25 tesla. (Use data from Table 7.1.)

7.2 The four lines from an *AX* spectrum are observed at $\delta = 5.8, 5.7, 1.1, 1.0$ (measured from TMS with an instrument operating at 100 MHz). What are the chemical shift positions (in δ) of the *A* and *X* nuclei, and the coupling constant (in hertz) between them?

7.3 Vinyl fluoride exhibits the following approximate coupling constants:

$$J_{H'H''} = 5 \text{ Hz} \qquad J_{H'F} = 85 \text{ Hz}$$
$$J_{H'H'''} = 13 \text{ Hz} \qquad J_{H''F} = 50 \text{ Hz}$$
$$J_{H''H'''} = -3 \text{ Hz} \qquad J_{H'''F} = 20 \text{ Hz}$$

Sketch the n.m.r. spectrum of the fluorine nucleus, assuming that the chemical shift differences between the hydrogen nuclei are all large compared with their couplings.

7.4 A fictitious non-spinning nucleus B forms pentavalent compounds; one such is $H_4B\!-\!CH_3$, where the BH_4 group forms a square pyramid with B at the apex, and the methyl group is free to rotate about the $B\!-\!C$ bond. Sketch the H n.m.r. spectrum, assuming that all chemical shifts are large compared with the couplings.

7.5 How do n.m.r. spectra change when a nucleus exchanges between two non-equivalent sites within a molecule? You should describe three situations: (*a*) where the rate of exchange is rapid, (*b*) where the rate of exchange is slow, and (*c*) an intermediate condition. Describe the appearance of the n.m.r. spectrum of methanol when it is (*i*) very dry and (*ii*) has a trace of water added to it. Account for the differences between them.

7.6 (*a*) What effect does the isotopic substitution of deuterium for hydrogen have on the appearance of ^1H and ^{13}C spectra?

(*b*) Deuterated acetone (CD_3COCD_3) is often used as a solvent for proton n.m.r. spectra since it should exhibit no proton resonance. In fact, samples usually contain some residual CD_3COCD_2H; predict the proton n.m.r. spectrum of this molecule, assuming no coupling between the CD_3 and CD_2H groups and remembering that deuterium has a spin of 1.

7.7 Although the low temperature n.m.r. spectra of solids usually show broad resonance, raising the temperature often causes a sudden narrowing, even though the sample is still well below its melting point. Explain this observation.

7.8 Predict the number of lines in the e.s.r. spectrum of each of the following radicals:

(*a*) $[CF_2H]^{\cdot}$ (*b*) $[^{13}CF_2H]^{\cdot}$
(*c*) $[CF_2D]^{\cdot}$ (*d*) $[CClH_2]^{\cdot}$

(The spin of D is 1; other spins are given in Table 7.1.)

7.9 The e.s.r. spectrum of the *p*-benzosemiquinone radical:

shows a $1:4:6:4:1$ quintet with a spacing of about 0.24 mT. Using Eq. (7.18) and some of the data in Sec. 7.5.3, and assuming that the $C\!-\!H$ groups in this molecule behave similarly to those of benzene, calculate the percentage of time spent by the unpaired electron at each hydrogen, carbon, and oxygen nucleus.

7.10 (*a*) Assign the remaining lines in the e.s.r. spectrum of naphthalene (Fig. 7.28).

(*b*) Predict the e.s.r. spectrum for 1-iodo-naphthalene and for 2-iodo-naphthalene, assuming that the electron densities, and hence couplings, are unchanged from those of naphthalene itself.

SOLID STATE AND SURFACE SPECTROSCOPIES

The importance of solids and their surfaces in everyday life has led to a wealth of research into their properties and behaviour. Work in these areas is particularly demanding experimentally because surfaces can be very complex and sample sizes are often small. Thus the development of spectroscopies which can be routinely used to investigate them has until recently been hampered by the need for high sensitivity. Success has relied on a number of technological advances which have taken place over the last few years.

This chapter outlines the principles of a few of the wide range of techniques which have now been established to investigate surfaces and the solid state.

Two approaches to the investigation of surfaces are generally adopted. The first is to investigate high-area samples, such as finely divided metal particles supported on oxides like silica or rare earth oxides; these are studied in order to model 'real' metal catalysts. The resulting large surface area (of the order of 10^{19} atoms per square cm) allows large numbers of molecules to be adsorbed, resulting in the need for comparatively low sensitivity. The main disadvantage is that the surface of a dispersed metal can be highly complex and typically contains different types of crystalline facets leading to a wide variety of available adsorption sites. As a result, spectra often show overlapping contributions from several different forms of adsorbed species.

The second approach largely overcomes this problem. The faces of highly ordered single crystal samples are studied. These can be well characterized and the types of adsorption sites available determined by diffraction measurements. However, the disadvantage of this work is that sample sizes on single crystal surfaces are very small compared to bulk samples. Typically, single crystal surfaces have diameters of $\sim 1\,\mathrm{cm}$ and contain of the order of 10^{15} atoms cm^{-2}. This is equivalent to sampling nanogram quantities of bulk materials, so very high sensitivity is required. In addition, at a gas pressure of $\sim 10^{-6}$ mbar, the surface will be bombarded at a rate of $\sim 10^{14}$–10^{15} molecules cm^{-2} each second. If the gas contains traces of a reactive impurity, such as oxygen, the surface may become saturated by it and contaminated in seconds. In order to overcome this difficulty, studies of single crystal samples must be carried out under ultra-high vacuum (UHV) conditions at pressures of some 10^{-10}–10^{-11} mbar (about one-tenth of the pressures found in space within the solar system), so that there is sufficient time ($\sim$ 1 hour) to take meaningful measurements before the surface becomes contaminated. Methods which enable us to do this have taken a long time to develop.

8.1 VIBRATIONAL STUDIES OF SURFACES

Studying the vibrational motions of adsorbed species is a very powerful technique for finding their geometries and composition, and for determining the type of site occupied and the bonding to the surface.

8.1.1 Electron Energy Loss Spectroscopy (EELS)

The first method developed to probe the vibrations of adsorbates and of the surface itself was electron energy loss spectroscopy (EELS). The technique involves directing a beam of electrons with an accurately defined energy on to the sample. Most of the emerging beam consists of electrons which have been elastically scattered (i.e. have neither lost nor gained energy), together with some which are inelastically scattered (i.e. have lost or gained energy). The data are recorded as a plot of the total number of electrons against their scattered energy.

A typical EELS spectrometer is shown in Fig. 8.1; in the electron gun a white hot tungsten filament emits electrons with a wide spread of energies, which are accelerated and pass into the monochromator. This consists of a series of slits and electric fields, and the radius of the circular path which the electrons follow in the fields depends on their energy and the strength of the applied field. The monochromator allows only electrons with a narrow range of energies to pass through, the rest being screened out by colliding with the edges of the slits, as is shown schematically on the right of the figure. After scattering from the sample, the electrons pass through the analyser, operating in exactly the same way as the monochromator, and so to an electron multiplier, where they are collected and counted. The monochromator is rotatable around the point where the beam strikes the sample, so that different angles of incidence can be chosen. Equally the energy of the electrons hitting the sample can be varied by changing the electric fields in the monochromator.

The energy of the electron beam used is typically in the range 1–10 eV. The resolving power is inherently poor in comparison with infrared techniques. The best EELS analyser today can

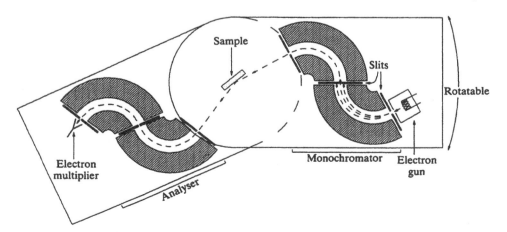

Figure 8.1 The electron energy loss spectrometer.

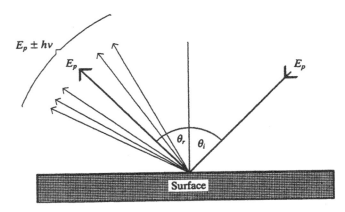

Figure 8.2 The principle of the EELS experiment where θ_i is the angle of incidence, θ_r is the angle of reflection, E_p is the energy of the incident beam (the primary energy), and $E_p \pm h\nu$ the energy of the loss electrons.

resolve about $1 \, \text{meV}$ of electron energy; this, being equivalent to about $8 \, \text{cm}^{-1}$, compares badly with typical infra-red resolutions of $1 \, \text{cm}^{-1}$ or less.

The beam is directed on to the surface at a particular angle of incidence θ_i, as shown in Fig. 8.2. Electrons not losing or gaining energy at the surface are reflected directly from it at the same angle; this, because of its similarity to reflection of a beam of light from a mirror, is called specular reflection. However, about 0.1 per cent of the electrons undergo inelastic scattering, losing or gaining a quantum of energy, and may leave the surface at any angle, including the specular. For the purposes of this discussion it is the energy change which is important, since this will be a quantum of *vibrational* energy which has been gained or lost by the adsorbed molecule.

There are several mechanisms by which the electron beam can interact with adsorbed molecules, but here we shall consider only the most important, known as dipole scattering. This is due to the long-range interaction between the electric fields associated with the incident electrons and the vibrating dipoles of the adsorbed species. The interaction causes a small change in momentum of the electron and leads to scattering close to the specular direction.

If, as is usual, the adsorbent surface is a metal, electrons in the metal surface itself (in the 'conduction band') are free to move within the surface. This means that any adsorbed fragment with oscillating bonds giving rise to a changing dipole will readily induce an equal and opposite dipole in the surface. This is illustrated in Fig. 8.3. If the molecular vibration is *parallel* to the surface (Fig. 8.3(a)), it is effectively 'cancelled out' by the induced dipole, whereas, if it is *perpendicular* (Fig. 8.3(b)), the resultant is a doubling of the dipole—the plus and minus ends of the dipole are twice as far apart in the metal–molecule combination as they would be in the

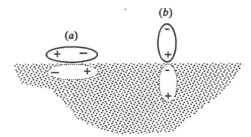

Figure 8.3 (a) Cancellation of the dipole moment of an adsorbate with the dipole parallel to the surface. (b) Enhancement of the dipole moment by a factor of 2 for an adsorbate with the dipole perpendicular to the surface.

molecule alone. Obviously vibrations at intermediate angles will have intermediate effects. This is the origin of the metal surface selection rule for dipolar scattering—namely that *only vibrational modes exhibiting dynamic dipoles perpendicular to the surface will interact with the beam* (they are said to be 'allowed'). This selection rule has proved very revealing for the analysis of adsorbate geometries.

Figure 8.4 shows the EELS spectrum of ethyne adsorbed on the close-packed Cu(111) surface. The attachment to the surface is via bond formation between the C≡C bond and the metal surface (so-called π bonding) in addition to bonds formed between the C and Cu atoms directly (σ bonding). The adsorbed molecule can be assumed to lie parallel to the surface and those vibrational modes with dipole changes perpendicular to the surface are shown in the small diagrams on the figure. Table 8.1 gives the approximate CC and CH symmetric stretching frequencies for gaseous ethyne, ethene, and ethane. If we make the reasonable assumption that the EELS bands at 1307 and 2920 cm^{-1} are, respectively, the CC and CH stretching frequencies, it is clear that the adsorbed molecule has bond characteristics between those of the C—C and C=C bonds. This represents a very strong attachment to the surface.

Table 8.1 Symmetric stretching frequencies for C—C and C—H bonds in ethyne, ethene, and ethane

	CC symmetric stretch (cm^{-1})	CH symmetric stretch (cm^{-1})
Ethyne, HC≡CH	1974	3374
Ethene, H$_2$C=CH$_2$	1620	3020
Ethane, H$_3$C—CH$_3$	990	2870

Figure 8.4 Electron energy loss spectrum of ethyne adsorbed on the Cu(111) surface. (*Reproduced with permission from B. J. Bandy, M. A. Chesters, M. E. Pemble, G. S. MacDougall and N. Sheppard, Surface Science, 139, 87, 1984.*)

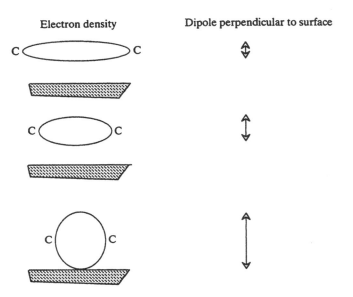

Electron density   Dipole perpendicular to surface

Figure 8.5 The oscillating dipole of the C≡C bond, with the bond axis lying parallel to the surface.

If the CH bonds were parallel to the surface, the CH stretching vibration would produce a dipole change also parallel to the surface, and this mode would consequently be difficult to observe in the EELS spectrum. The appearance of the band at $2920\,\text{cm}^{-1}$ implies that the CH bonds are probably tilted away from the surface as shown.

At first sight it might be thought that, because of the metal surface selection rule, the CC stretch would not be observed. However, if the bond is considered as a barrel-like electron distribution, when the carbon atoms oscillate the barrel becomes stretched and compressed as illustrated in Fig. 8.5, leading to variable electron transfer between the CC bond and the surface. This results in a dipole whose magnitude oscillates perpendicularly to the surface, even though the atoms move parallel to it.

The frequency positions of the vibrational features can also be used to assess the site of an adsorbate. For example, when CO binds via the carbon to the surface, the position of the CO stretching mode will indicate whether the CO is terminally bound to one metal atom (2100–$2000\,\text{cm}^{-1}$) or is bridging across two or more metal atoms (1900–$1700\,\text{cm}^{-1}$).

8.1.2 Reflection–Absorption Infra-red Spectroscopy (RAIRS)

Infra-red spectroscopy is normally carried out, as we have seen in Chapter 3, by passing a beam of IR radiation through a sample and recording which frequencies are absorbed. This technique is often referred to as 'transmission' or 'absorption' spectroscopy. Such a method is not easily applicable to surface studies (but see Sec. 8.1.3 later) because the metal itself is not transparent to IR radiation, but merely reflects it. However, in order to reach the metal surface and be reflected from it, the radiation must pass twice through any molecules adsorbed on the surface—once on the way 'in' and once on the way 'out'. Although the resulting 'path length' is minuscule it is sufficient to form an absorption spectrum of the material, and it is on this that the RAIRS technique depends.

Essentially IR radiation of the appropriate frequency interacts with vibrations of the adsorbed species, exactly analogously to ordinary IR spectroscopy. However, since the interaction is with vibrating dipoles, which induce equal and opposite dipoles in the metal surface, the metal surface selection rule applies (see Sec. 8.1.1). It is clear from this that IR radiation plane polarized *perpendicularly* (p-polarized) to the metal surface interacts much more strongly with molecular vibrations than radiation polarized parallel to the surface (s-polarized). Further, radiation striking the surface at very small incident angles (grazing incidence) will interact most strongly, partly because the path length through the absorbed species is longer, but principally because the p-polarized electric field vector of the radiation is almost perpendicular to the surface. Usually, therefore, the experiment is carried out with p-polarized IR radiation at grazing incidence.

The majority of the radiation is reflected from the surface due to the interaction of the incident radiation with the conduction electrons in the surface. The intensity of the reflected radiation is very much greater than that absorbed by the monolayer (or less) of adsorbate. Results are usually recorded, therefore, as the difference between the spectrum of the clean metal surface and that of the surface plus adsorbate. Thus if, at a particular IR frequency, the clean surface reflects an amount R of the radiation, and the surface + adsorbate reflects $R - \Delta R$ (energy having been absorbed by the adsorbed molecule), it is usual to plot $\Delta R/R$ against frequency, although sometimes absorbance plots, $\log_{10}(\Delta R/R)$, are used.

The practical difficulties involved in obtaining such high-sensitivity spectra can be illustrated by considering adsorbed carbon monoxide. When adsorbed on a surface, CO has a large dynamic dipole and is one of the most strongly infra-red absorbing molecules, but in the RAIRS experiment absorption bands associated with it often amount to only about 1 per cent $\Delta R/R$. Considering that a typical hydrocarbon molecule absorbs infra-red 10–100 times less strongly than CO, the need for high sensitivity (of the order of $1:10^5$ in terms of the signal-to-noise ratio) is clear, particularly since it is also desirable to record spectra of surfaces containing less than one monolayer of adsorbate.

At first RAIRS measurements were carried out using dispersive IR spectrometers, but only with the development of FT-IR spectrometers has the technique really been able to achieve the sort of sensitivity required to carry out routine measurements. The main advantage over EELS is that RAIRS can obtain considerably higher resolution. This is of particular importance in the study of the adsorption of polyatomic molecules; for example, a hydrocarbon molecule such as adsorbed but-2-yne may have several allowed vibrations in the CH stretching region. At EELS resolutions it will be impossible to distinguish between the modes, whereas RAIRS with its inherently high resolution will be able to separate fundamentals and overtones in the same region. This enables a more categorical assignment of the surface species to be made.

One point worth mentioning at this stage is that, as with any other type of IR spectroscopy, window materials such as NaCl, KBr, etc., must be used. Unlike EELS, where the entire spectrometer is encased in an ultra-high vacuum system, for RAIRS the beam must be passed from the FT-IR spectrometer into UHV and out again to be detected. A typical arrangement for a RAIRS system is shown in Fig. 8.6. IR radiation leaves the FT-IR spectrometer as a parallel beam. After reflection off a plane mirror, it is focused on to the sample through a KBr window which is sealed to UHV. After reflection, the beam passes out of UHV via another KBr window and is ultimately focused on to a very sensitive photoconductive detector.

Figure 8.7(a) shows a typical RAIRS spectrum, obtained when ethene was adsorbed on the Pt(111) surface at 360 K. The spectrum indicates that the molecule rearranges on adsorption to give the ethylidyne fragment, C—CH$_3$; comparison with the data in Table 8.1 shows that the RAIRS band at 2884 cm^{-1} is very close to the symmetric CH stretch of the methyl group, and the band at 1118 indicates that the C—C bond is close to being a single bond. The frequency of the symmetric deformation vibration of pure ethane (the opening and closing of the 'methyl umbrella') is at 1380 cm^{-1}, which correlates well with the strong RAIRS band at 1339 cm^{-1}. The molecular motions corresponding to these three bands are depicted schematically in Fig. 8.7(c). Figure 8.7(b) is discussed in the following section.

Figure 8.6 A typical reflection–absorption infra-red spectrometer (RAIRS).

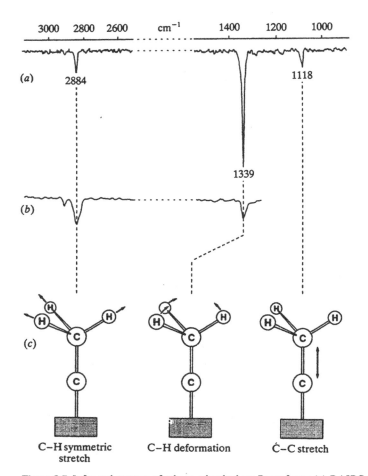

Figure 8.7 Infra-red spectra of ethene adsorbed on Pt surfaces: (*a*) RAIRS of ethylidyne on Pt(111), (*b*) transmission infra-red spectrum of ethene adsorbed on platinum supported on silica, and (*c*) the modes observed for ethylidyne. (*(a) and (b) reproduced with permission from M. A. Chesters, C. De la Cruz, P. Gardner, E. M. McCash, P. Pudney, G. Shahid and N. Sheppard, J. Chem. Soc. Faraday Trans., **86**(15), 2757–2763, 1990.*)

8.1.3 Transmission IR of Supported Metals

Transmission IR studies of oxide-supported metals constituted the first vibrational measurements of chemisorbed adsorbates. Earlier work centred on physical adsorption on the oxides of light elements such as silica, aluminium oxide, and zeolites. Very small particles, with high surface areas, could be produced, which made the IR techniques favourable because of the relatively high concentration of adsorbates obtainable. In addition, the small oxide particles gave weak scattering of the incident radiation because the particle diameters were much less than the radiation wavelength. The oxides could easily be pressed into porous discs and mounted in a cell with suitable windows through which IR radiation could be passed.

The use of FT-IR techniques made it possible to carry out transmission studies on larger particles formed by heavier metal oxides, which have higher refractive indices and therefore scatter more of the incident radiation. As the sensitivity of spectrometers and detectors improved it became possible to record spectra of adsorbates on oxide-supported metals, such as, for example, finely divided platinum on silica. The oxide support inhibits the sintering of the separate metal particles at the high temperatures used for surface cleaning and also provides windows between the opaque metal particles.

The major disadvantage of the use of oxide-supported samples for IR studies is that there are regions where strong absorptions by the oxides themselves make observations impossible. For example, in silica the entire transmission range below $1300 \, \text{cm}^{-1}$ is blacked out by the oxide, while for alumina, blackout occurs below $1100 \, \text{cm}^{-1}$. However, many adsorbate vibrations of interest fall in the higher wavenumber ranges and can be studied.

Figure 8.7(b) shows the IR transmission spectrum of ethene adsorbed on platinum supported on silica. Comparison of this spectrum with the RAIRS spectrum of the same species in Fig. 8.7(a)—discussed at the end of the previous section—shows that the transmission spectrum has largely identical features but a much better signal-to-noise ratio. It also contains an additional band at $2922 \, \text{cm}^{-1}$ and a small feature at $1496 \, \text{cm}^{-1}$, due to ethene adsorbed in the di-σ form, where the double bond has opened to give two C—Pt bonds. This transmission spectrum, however, cannot show either of the CC stretching modes, since absorption by the silica support masks everything below about $1300 \, \text{cm}^{-1}$.

8.1.4 Raman Spectroscopy

At first glance it might appear that, because the Raman scattering process is inherently very weak and single crystal surface sample sizes are very small, the application of Raman spectroscopy to adsorbates would be rather unproductive. However, recent technological advances are beginning to make it possible to observe Raman signals from molecules adsorbed on single crystal surfaces. The experimental details are essentially the same as those described in Chapter 4—a laser beam is shone on to the sample and the scattered radiation is collected and recorded as a function of scattering angle and frequency. The difference for surface work is that, as always, the sample must be in UHV, and so the radiation is passed into and out of the evacuated sample chamber through quartz or glass windows.

It will be recalled from Chapter 4 that the intensity of Raman scattering depends both on the *polarizability* variations of vibrating bonds (rather than their dipole) and on the symmetry of the vibration with respect to the radiation beam. This is equally true of surface studies and it turns out, again, that molecular motions *perpendicular* to the surface give the most intense scattering. However, other vibrations do give weak scattering, and observation of the relative scattering intensities of various vibrational modes can be used to decide the type of motion occurring and its orientation with respect to the surface—all useful in the determination of surface structures.

The Raman spectrum of a submonolayer of deuterated benzene, C_6D_6, adsorbed on Ag (111) is shown in Fig. 8.8. The relatively strong band at $945\,cm^{-1}$ can be assigned to the symmetric C—C stretching vibration (the so-called 'ring breathing frequency') of the molecule, where all six C—D groups simultaneously move together and apart in the plane of the ring. Observation of this vibration, at exactly the same frequency as in the Raman spectrum of the pure liquid, indicates that the ring probably lies parallel to the metal surface, a bond being formed by the interaction of the π-electron cloud with the surface. Any other bonding would destroy the symmetry of the ring and would probably modify this vibrational mode. The much weaker band at about $493\,cm^{-1}$ is assigned to the symmetrical out-of-plane bending of the deuterium atoms—all moving up and down simultaneously—whose frequency in the pure liquid is $503\,cm^{-1}$.

Surface Raman spectra have also been obtained for high surface area materials, such as oxides or supported metals. The sensitivity requirements are far less than for Raman measurements on single crystals (analogous to the comparison between RAIRS and transmission IR), but there are other problems which make these difficult samples to study. Pulsed techniques have been used to overcome the problem of fluorescence, which has a longer timescale than Raman scattering, but experiments on highly coloured samples (like supported metals) are hampered by local heating by the laser beam (see Chapter 4, Sec. 4.6).

In addition, the vibrational modes of adsorbates on the roughened surfaces of the coinage metals, copper, silver, and gold, are sometimes observed to have about one million times the intensity that would be predicted by comparison with their Raman spectra in the gas phase. This startling observation has led to the development of surface enhanced Raman spectroscopy (SERS) and has revealed a vast body of data on adsorbate vibrations on these few metals, particularly silver. In part this strong enhancement results from conduction-electron resonances of the metal particles that are excited in the visible region, and does not apply to other metals.

Figure 8.8 Raman spectra of benzene-d_6 adsorbed on the (111) surface of silver. (*Reproduced with permission from A. Campion, Chapter 8 in* Vibrational Spectroscopy of Molecules on Surfaces, *ed. J. T. Yates Jr and T. E. Madey, Plenum Press, 1987.*)

8.1.5 Inelastic Helium Scattering

From the previous sections, it is clear that, although many techniques have been applied to studying the vibrations of surfaces and adsorbates, they tend to produce spectra which result from only those vibrations which have dipoles (be they permanent or induced) perpendicular to the surface. In this final section on the vibrational spectroscopies of surfaces, we will look at a type of spectroscopy developed recently, which mainly detects modes parallel to the surface. The technique is helium atom scattering and has been used as a diffractive probe for many years. It is expensive to set up and maintain and, as such, it is unlikely that its use will become routine.

Experiments are carried out by firing a pulsed, supersonic beam of helium atoms at a sample and recording the arrival times of the scattered atoms (i.e. time-of-flight measurement using a mass spectrometer for detection). Like the incident electrons in EELS, the helium atoms can either be elastically or inelastically scattered. As the energy of the incident beam is known, the time difference between the arrival of the elastically scattered beam and the inelastically scattered atoms is used to find the energy loss or gain. In the helium scattering case, the elastic scattering process can be described in terms of billiard ball type collisions, because both the helium and the surface atoms can be pictured as hard spheres.

A minority of the helium atoms undergo inelastic collisions where a quantum of vibrational energy is lost or gained due to adsorbate vibrations or to vibrations of the surface atoms themselves (surface phonon modes which are localized motions of the surface atoms). Phonon modes which can be detected have large amplitudes at the surface, but the amplitude decays away into the bulk. The adsorbate vibrations which are detected are usually those parallel to the surface.

Supersonic helium beams are produced by expanding helium at high pressure (usually several atmospheres) through a small nozzle (~ 10–40 μm). Supersonic beams are highly mono-energetic because the distance between interatomic collisions is very short under the high-pressure conditions in the source, and each atom undergoes many collisions, leading to a 'sharing out' of energy. The beam energies are typically in the range 10–100 meV (~ 80–800 cm^{-1}), which matches the energies of the low-frequency modes of the adsorbates.

The helium atom scattering spectrum of carbon monoxide adsorbed on the Pt(111) surface is shown in Fig. 8.9(a). There are two very low frequency vibrational modes evident in this spectrum; the first has a frequency of 48 cm^{-1} (and an overtone at just twice this, i.e. 96 cm^{-1}) and the second, much weaker and only discernible in the enlarged, $\times 50$, spectrum, at 129 cm^{-1}. The central, very intense, peak is due to helium atoms elastically scattered from the surface. To the left of this the scattering is recorded of helium atoms with *less* energy than the original beam, and to the right those with greater energy. As always with vibrational modes (see Chapter 3, Sec. 3.1.3) the ground state is more populated, and there is a greater chance that a colliding helium atom will give energy to a molecule rather than take it from a previously excited molecule, so the left-hand side of the spectrum is slightly more intense.

The nature of the vibrational modes concerned is unusual. In the gaseous state CO molecules are free to rotate and to translate (i.e. to move bodily throughout the gas); when bound to the metal surface actual rotation and translation are prevented, but are converted to the type of vibrational motion found here—often referred to as 'frustrated rotation' and 'frustrated translation'. Figure 8.9(b) and (c) shows, schematically, the frustrated motions for CO.

In Figure 8.9(b) we show, at (i), the rotating molecule and at (ii) we see the molecule bound to the surface, but still attempting to rotate—clearly the motion tends to bend the metal–C–O fragment into a zig-zag shape. Bond-restoring forces will then cause the fragment to undergo vibration as shown at (iii), where we depict, much exaggerated, the two extremes of the vibration. In a similar way, Fig. 8.9(c) shows, at (i), the translating molecule and then, at (ii) and (iii), the frustrated translation and the resulting pendulum-like swinging of the C and O atoms about

(a)

(b)

(c)

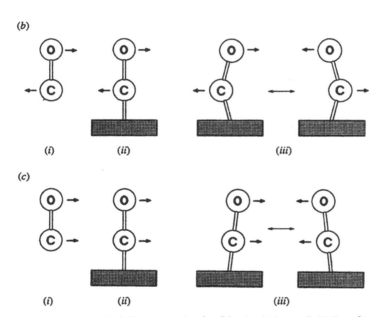

Figure 8.9 (a) Inelastic helium scattering for CO adsorbed on a Pt(111) surface at 300 K. The feature observed at 129 cm^{-1} is due to the hindered rotation which is illustrated in (b), while those at 48 and 96 cm^{-1} are due to the hindered translation mode shown in (c) and its first overtone. ((a) *is reproduced with permission from A. Lahee, P. Toennies and Ch. Wöll*, Surface Science, **177**, 371, 1986.)

the metal. This latter motion has the lower frequency, and so is assigned to the band at 48 cm^{-1}, the zig-zag vibration being at 129 cm^{-1}.

8.2 ELECTRONIC SPECTROSCOPIES OF SURFACES

The earliest techniques usefully applied to the investigation of surface and near-surface properties were those probing their electronic behaviour. The application of electronic spectroscopies to surfaces has provided a wealth of valuable information and has considerably advanced our understanding of surfaces and adsorbates. There are many such techniques available but we will concentrate on the most common.

The term 'near-surface properties' is used here because electron spectroscopies probe the first few atomic layers of the surface as well as one or more adsorbate layers, unlike the vibrational spectroscopies described in the previous section, which probe only the adsorbate vibrations. This arises because electrons used as probes can travel through solids. However, an important limitation for all electron spectroscopies of surfaces is that electrons can only travel short distances without losing energy via interactions with the bulk atoms.

Electrons with kinetic energies in the range 20–1000 eV can travel only some 5–15 Å, i.e. a few atomic layers through solids. This means that electrons 'escaping' from a solid surface must originate from near the surface, and so will carry information about surface atoms and those in the few layers just below the surface only.

8.2.1 Photoelectron Spectroscopy (PES)—X-Ray (XPES) and Ultra-violet (UPES)

We have already discussed photoelectron spectroscopy briefly in both Chapter 5 (Sec. 5.5) and Chapter 6 (Sec. 6.5). There we found that illuminating a sample with a beam of high-energy radiation results in the emission of electrons whose energies are related to the composition and structure of the sample.

The same techniques can be applied to the study of surfaces and surface-adsorbed species. When the radiation used is in the ultra-violet region the technique is referred to as UPES (ultra-violet photoelectron spectroscopy) and the electrons emitted are the valence electrons of atoms on or close to the surface; if X-radiation is used, the technique is called XPES and this is sufficiently energetic to cause core electrons to be emitted; further, since the X-rays penetrate into the sample, the emitted electrons may arise from deeper in the bulk material than in the case of UPES.

In Chapter 5, Sec. 5.5, we saw that the energy of an incident photon, $h\nu$, is related to the kinetic energy (E_{KE}) of ejected electrons as follows (see Eq. (5.26)):

$$h\nu = E_{KE} + BE$$

where BE is the binding energy (the ionization potential) of the level from which the electron was emitted.

For both XPES and UPES the kinetic energy of the ejected electrons is measured using a hemispherical analyser such as that shown earlier in Fig. 5.13. Experiments on surfaces are carried out using similar equipment, with the added necessity of using UHV conditions.

The XPES spectrum of niobium deposited on a silicon surface is shown in Fig. 8.10. The major peaks below 500 eV can all be quite precisely correlated with the ionization energies of electrons from the various orbitals of niobium or silicon, as has been done on the figure. The peak just above 500 eV arises from the $1s$ orbital of oxygen contaminating the surface. The sloping background is due to secondary electron emissions within the bulk of the sample. Clearly the technique is invaluable for qualitative analysis—the electrons being emitted come from deep within the atom concerned, well shielded from any effects due to chemical binding, and any atomic species near or on the surface can be instantly recognized from its ionization potential. Quantitative analysis is also possible, by observing the relative intensities of electron emission. The theory of such measurements is difficult, however, and it is usual to rely on comparison spectra obtained from samples of known composition.

UPES spectra are similar but, since the method excites outer (valence) electrons, chemical bonding affects the emission energies (see Chapter 6, Sec. 6.5). This chemical shift information can, in surface studies, give valuable insights into the structure of adsorbed molecules.

8.2.2 Auger Electron Spectroscopy (AES)

X-ray photoelectron spectroscopy (see the preceding section) and Auger electron spectroscopy are very closely related; in the former we look at electrons directly ejected from the inner core of

Figure 8.10 The XPES spectrum of niobium deposited on a silicon surface. (*Thanks are due to Dr C. Walker of the Universities of Leeds and York for recording this spectrum.*)

surface atoms, while in the latter it is electrons emitted during secondary processes, *after* the emission of the first electron, which are studied. Another minor difference is that a high-energy electron beam (> 1 keV) is normally used in Auger work, rather than an X-ray beam, to eject the first electron. The equipment for detecting emitted electrons and analysing their energies is very similar to that for PES spectroscopy described earlier.

To understand the Auger process we must imagine an electron to be ejected from a core level, E_x, of a surface atom in the ground state, as shown in Fig. 8.11(*a*). The ion formed, with a vacancy or 'hole' in E_x, is shown as the initial excited state in Fig. 8.11(*b*). The hole in E_x can be filled by an electron dropping down from a higher level, E_y. This releases considerable energy $(E_x - E_y)$, part of which can be absorbed by an electron in a higher orbital, E_z, which will be ejected with kinetic energy representing the excess of $E_x - E_y$ over E_z. The latter is known as the Auger energy, E_{Auger}, as shown in Fig. 8.11(*c*). The resulting ion, with 'holes' in both E_y and E_z, is shown in Fig. 8.11(*d*); note that, although this is referred to there as the final state, it is nonetheless higher in energy than the original ground state, and yet more electron rearrangements will subsequently occur, perhaps involving electron capture, or perhaps emission of successively less-energetic Auger electrons, often referred to as a cascade process, before the ground state is regained.

Because energy is conserved, it is clear that the energy released by electron E_y dropping down, $(E_x - E_y)$, must be equal to the energy taken up by the ejection of electron E_z with kinetic energy E_{Auger}, so we can write:

$$E_{\text{Auger}} + E_z = E_x - E_y$$

or

$$E_{\text{Auger}} = E_x - E_y - E_z \qquad (8.1)$$

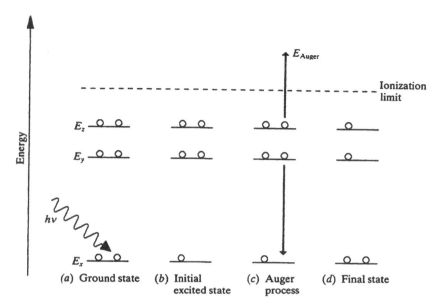

Figure 8.11 The Auger process.

If, as here, we think of E_x, E_y, and E_z as the orbital energies of electrons in the ground state of the atom, this equation is not strictly accurate—the Auger process deals with electrons in ionized states, which will have slightly different energies. This need not concern us here, however; the important point is that the observed Auger energy is directly related to the binding energies of electrons in surface atoms, and so is characteristic of them. Auger processes are more likely in the lighter elements, where the core electron energy is not very different from the energies of the outer electrons (and see X-ray fluoresence in the next section). AES is, therefore, widely used to detect light element contaminants on metals.

Any chemical bonding to a metal surface, such as by an adsorbed molecule, will necessarily change the energies of electrons in outer orbitals of the surface atoms, which means that Auger spectroscopy gives information regarding the type of bonding which occurs. Interpretation is not easy, however, and will not be dealt with here. Equally an Auger spectrum can give quantitative information, but, like photoelectron work discussed earlier, usually it is best interpreted by observation and comparison with samples of known composition.

Figure 8.12 shows the AES spectrum of silicon. This spectrum, like most AES spectra, is displayed as a *differential* (see Chapter 1, Sec. 1.4). The reason is that, if it were drawn as a direct plot of the number of electrons emitted versus energy, the cascade process whereby successively less energetic electrons are emitted as each excited atomic state undergoes further decay gives a heavily sloping background. Since the slope is relatively constant, however, differentiation removes it and emphasizes the sharp AES features.

The principal peak at 92 eV in the figure arises from the following process (X-ray notation for the levels is given in brackets): initial excitation removes an electron from the 2s level (L); subsequently an electron drops down to fill this level from the 3s (M), while an Auger electron is ejected from the 3p (M). Using X-ray notation, the whole process would be referred to as an *LMM* transition.

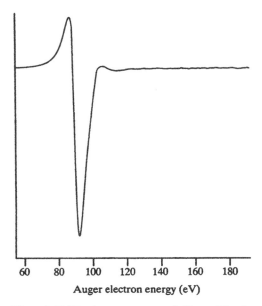

Figure 8.12 The Auger spectrum of silicon. (*Thanks are due to Mr N. J. Curson of the University of Cambridge for obtaining this spectrum.*)

8.2.3 X-Ray Fluorescence (XRF)

In heavy elements the core electrons are very firmly held by the high charge of the nucleus while the outer electrons, being shielded from the nucleus by the inner orbitals, have much the same binding energies as in light atoms. Once the initial excited state (Figure 8.11(b)) is reached, the electron dropping down releases much more energy from heavy atoms than from light ones and this makes the Auger process inefficient—rather than interacting with outer electrons, the energy released appears directly in the form of secondary X-rays emitted by the sample.

The high-energy beam required to eject inner core electrons from heavy elements is usually obtained from hard X-rays and, as the re-emitted energy is also in the X-ray region, the whole process is known as X-ray fluorescence. It is entirely complementary to Auger spectroscopy, the former process dominating in the lighter elements, the latter in the heavier.

There are two further important differences between Auger and X-ray fluorescence spectroscopies. Firstly, the hard X-rays required to excite core electrons in heavy atoms can readily penetrate into the metal; thus the technique gives information about the *bulk* sample, not just the surface atoms as in Auger work. Secondly, Auger spectroscopy depends on the energies of bonding electrons, while X-ray fluoresence involves core electrons only, the energies of which are unaffected by bonding. Thus X-ray fluorescence is an excellent 'fingerprint' technique for recognizing elements, since their emission frequencies are relatively independent of the chemical state of the element. Together this means that X-ray fluorescence is much used for the rapid analysis of bulk metal samples—impurities or alloying elements can be immediately recognized by their emission energies and the amount of each is accurately proportional to the emission intensities.

8.3 NUCLEAR MAGNETIC RESONANCE

Nuclear magnetic resonance is an extremely valuable analytical technique and the desire to develop it as a probe for solids and surfaces has been intense among researchers. There are, however, major difficulties in such applications, mainly due to the very broad resonances which normally occur and to the inherent lack of sensitivity of the technique.

8.3.1 Broad-Line N.M.R.

There are several factors contributing to the breadth of the resonance lines in solid-state n.m.r. spectra. We consider each of them briefly below, together with some discussion of how each problem may be overcome.

Dipolar coupling   Every magnetic nucleus in a sample produces its own local field which, to a greater or less extent, is 'felt' by its neighbours; this is known as dipolar coupling. In Fig. 8.13(a) we show a single nucleus, with magnetic moment μ, placed in an external field, B_0, applied along the z axis; we remember (cf. Chapter 7, Sec. 7.1.4) that the field causes the nucleus to precess with a constant frequency, the Larmor frequency. The nuclear magnetic moment can be considered to be resolved into two components, one along the z axis and the other in the xy plane; these are labelled μ_z and μ_{xy} in the figure. Clearly the z component remains static in magnitude and direction during the precession, while μ_{xy} rotates in the xy plane, although its magnitude remains constant. As we shall see, both these magnetic fields—the static and the rotating—can interact with neighbouring nuclei, but in what immediately follows we shall consider the static field only.

The magnitude of the effect of one nuclear magnet on another depends markedly on the distance between them and their relative orientation to each other. We illustrate this in Fig. 8.13(b) where we show the static z components of two nuclei, labelled μ_{1z} and μ_{2z}. The field which nucleus 1 produces at nucleus 2 can be shown to be proportional to $1/r^3$, where r is the distance between the nuclei, and to $(3\cos^2\theta - 1)$, where θ is the angle which a line joining the nuclei makes with magnet 1, i.e. the z direction. Obviously nucleus 1 experiences a field from nucleus 2 in exactly the same way.

A normal n.m.r. sample has a very large number of nuclear spins, and the effect of dipolar coupling at any one nucleus is the summation of the effects of all the surrounding nuclei. In practice, because of the $1/r^3$ term, we need consider only the near neighbours. We must remember that (for spin $\frac{1}{2}$ nuclei) each nucleus can be aligned either *with* the main applied

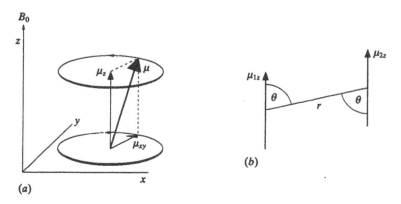

Figure 8.13 (a) The origin of the dipolar coupling phenomenon. (b) The dipolar coupling effect in solids.

field or *against* it, and so its effect on neighbouring nuclei can either increase the effective field which they feel or decrease it.

In gas and liquid phases dipolar coupling can be ignored since molecular tumbling ensures that θ undergoes rapid changes; simple calculus shows that the average value of $(3\cos^2\theta - 1)$ is zero, so there is no dipolar coupling. Such tumbling does not occur in solids, however. Overall, of course, we have seen (see Chapter 7, Sec. 7.1.3) that in an applied field there are almost exactly as many spins up as there are spins down, and we might feel that this should also average the dipolar coupling to zero. For this to occur, however, it would be necessary that there should be equal numbers of up and down spins in *every* microregion of the sample. This is clearly unreasonable: in some places an up spin will perhaps be totally surrounded by down spins, in others totally surrounded by up spins, and in yet others (and more likely) surrounded by both up and down spins, the proportion of each varying from place to place. Thus some nuclei will feel a large total field in one direction from its neighbours, others will feel a large field in the opposite direction, and some will feel a smaller field or even no field at all. Since the resonance frequency of each nucleus depends on the total magnetic field which it experiences, it is clear that dipolar coupling will result in a wide resonance peak made up of all the different individual resonances of each nucleus. Because of the statistical distribution of nuclear spins the shape of each broad resonance is Gaussian.

This is the major problem in solid-state n.m.r. for at least two reasons. Firstly, broad resonances are inherently difficult to observe and measure—they can easily become totally lost in the noise background of a spectrum. Secondly, the average dipolar coupling between protons is about 0.2 mT, which means that the resultant line broadening will prevent the direct observation of chemical shifts and coupling constants, which are smaller by factors of 10–100. It is normally possible, however, to treat an observed peak as a set of overlapped Gaussian curves which can often be resolved into their components by mathematical methods, nowadays left to the computer.

Although molecular tumbling cannot average the $(3\cos^2\theta - 1)$ term to zero, it is possible with some samples to use a mechanical method of making this term disappear. For $(3\cos^2\theta - 1)$ to be equal to zero we must have $\cos\theta = 1/\sqrt{3}$, or $\theta = 54.7°$. It turns out that spinning the sample very rapidly about an axis at 54.7° to the main applied field can be a sufficiently close approximation to molecular tumbling to remove dipolar coupling; 54.7° is usually called the *magic angle*, and the process is known as magic angle spinning (MAS) or magic angle rotation (MAR). Unfortunately, the spinning rate required depends on the width of the broad resonance, and hence on the dipolar coupling; mechanically it is not yet possible to spin samples at rates greater than 10 kHz and it is rare for dipolar broadening to be narrower than this. Magic angle spinning is usually much more effective in removing the effects of chemical shift anisotropy (see below).

Chemical shift anisotropy　As we saw in Chapter 7, Sec. 7.2.1, the chemical shift in n.m.r. arises because of electron circulation induced in the molecule by the applied field. This circulation sets up its own small magnetic field which shifts the resonance of neighbouring nuclei. In gases and liquids, molecular movements ensure that each molecule experiences the applied field from all possible directions over a short time and the net electron circulation is effectively a summation of all these possibilities. Thus the chemical shift is *isotropic* and has a value for the molecular fragment as a whole.

In solids each molecule has a fixed orientation in the applied field and the electron circulation may depend markedly on that orientation. A very simple example is a molecule containing an aromatic ring, as illustrated in Fig. 8.14. If the plane of the ring is perpendicular to the applied field, as on the left of the figure, the induced circulation of the π-ring will be very much greater than if it is parallel to the field (on the right) or at any other orientation. This *chemical*

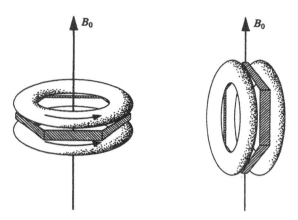

Figure 8.14 The effect on an aromatic ring of orientation in an applied field.

shift anisotropy again results in a broadening of the resonance—molecules with different orientations experience a different shielding and so have a different resonance frequency. Theory shows that the $(3\cos^2\theta - 1)$ term is important here, too. Fortunately, the broadening produced is usually not as large as that for dipolar coupling, and magic angle spinning of some 5–7 kHz can often produce narrower resonances.

Spin-spin coupling Generally the magnitude of spin-spin coupling between nuclei is independent of the phase of the sample. With solid samples, however, splitting due to such coupling is not usually resolved, because of the width of the lines, but simply leads to some additional broadening. For this reason it is normal to remove the effects of spin-spin coupling by double-resonance techniques (see Chapter 7, Sec. 7.2.6) in order to produce narrower lines.

Relaxation times In Chapter 7, Sec. 7.1.5, we found that the width of a resonance line is very dependent on the relaxation time of excited spins, which, in turn, depends on the efficiency of interactions between them (spin-spin or transverse relaxation) or between them and the surroundings (spin-lattice or longitudinal relaxation). Slow relaxation means long lifetimes in a particular state and so, from Heisenberg's uncertainty principle, sharp lines. Efficient relaxation means short relaxation times and broad lines.

In the solid state, relaxation can, under some circumstances, be extremely efficient via the spin-spin mechanism involving the *rotating* component of the magnetic moment, which we introduced in Fig. 8.13(a) above. If two hydrogen nuclei are neighbours and their precessional frequencies are similar (i.e. they are not too different in chemical shift), their rotating magnetic components will be close in frequency; this means that coherent interaction can readily occur and they can exchange spin energy very efficiently with each other. This mechanism is known as dipolar relaxation and in a solid with many such nuclei held at a constant orientation to each other the exchange of energy can occur very rapidly, giving rise to very efficient relaxation and hence broad spectral lines.

The efficiency of this mechanism obviously decreases as the nuclei become further apart. This means that for nuclei such as ^{13}C which, with an abundance of only 1 per cent, are very unlikely to be near neighbours in a sample, the dipolar relaxation process is not a factor in line broadening. Neighbouring *hydrogen* nuclei cannot, of course, contribute to the dipolar relaxation of ^{13}C nuclei, because they have a very different precessional frequency, and coherent interaction between the rotating magnetic dipoles of ^1H and ^{13}C does not occur.

The long relaxation time of ^{13}C in solid samples brings with it a further disadvantage, however. We found in Chapter 7, Sec. 7.3.2, and in Fig. 7.22 of that section, that the multi-pulse technique normally used for the observation of ^{13}C spectra will only give good signal intensity if the time between pulses is long compared to the relaxation time—too short a delay means that the ^{13}C spins have not had time to return to the ground state ready to be re-excited. The very long relaxation times in solids means that unacceptably long delays must be left between pulses in order to produce sufficient signal strength.

A subtle method of overcoming this problem, called *cross-polarization*, has been devised, which essentially involves using the abundant proton nuclei as a 'sink' into which the excess ^{13}C energy can be poured. It relies on the fact that the precessional frequency of a nucleus about a given axis is proportional to the magnitude of the applied field. In the same applied field protons precess at very nearly four times the rate of ^{13}C nuclei; thus if we can apply a fourfold field strength to the ^{13}C nuclei, they will precess at the same frequency as the protons and can then exchange energy with them. This is achieved by a sequence of 90° pulses. The first pulse is at the proton frequency and is immediately followed by a second pulse at the ^{13}C frequency, but of four times the amplitude of the first; a technique known as spin-locking enables both types of nuclei to precess with the same frequency about their respective fields. The matched frequencies of the rotating components ensure efficient relaxation of the ^{13}C, whose n.m.r. signal is observed. Because ^{13}C relaxation is efficient, further pulses at the ^{13}C frequency can be applied and the summed spectrum collected. Of course, the protons undergo relaxation as well but, since they are far more abundant than the ^{13}C, many pulses at the ^{13}C frequency can be accumulated before it is necessary to apply another pulse at the proton frequency.

Figure 8.15 demonstrates the application of the various techniques described above to the 75 MHz ^{13}C spectrum of poly-methylmethacrylate ($[-CH_2.C(CH_3),CO(CH_3)O-]_n$, which is the basis of many plastics, e.g. perspex). All the spectra illustrated were obtained using cross-polarization. Figure 8.15(a) shows the broad spectrum resulting from using a non-spinning sample and without proton decoupling—it is almost completely featureless. When proton decoupling is introduced, spectrum (b) results, which begins to show some chemical shift data. The effect of magic angle spinning is shown in spectrum (c), which should be directly compared with spectrum (a), since (c) is also taken *without* proton decoupling. Finally Fig. 8.15(d) shows the result of both magic angle spinning and proton decoupling, from which we can see the much better resolution which results. It should be realized, however, that spectrum (d) still represents a very broad-lined spectrum—at the frequency of recording these spectra (75 MHz), 1 p.p.m. is equivalent to 75 Hz and the narrowest line in spectrum (d) is about 2 p.p.m. or 150 Hz broad; this should be compared with liquid spectra which normally have bandwidths of 1 Hz or less.

8.3.2 N.M.R. of Surfaces

The application of nuclear magnetic resonance spectroscopy to surface studies encounters all the difficulties of broad-line spectra discussed above. In addition, n.m.r. is inherently an insensitive technique and normally needs relatively large samples to produce satisfactory spectra. Finally, of course, the experiments must be carried out with the sample in a high vaccum. Despite these problems some progress has been made in using n.m.r. for surface studies, and exploratory experiments with materials adsorbed in zeolites and on oxide-supported metals have been successfully carried out.

δ_C (p.p.m.)

Figure 8.15 Solid-state n.m.r. spectra of polymethyl methacrylate: (*a*) static, with no decoupling, (*b*) static with decoupling, (*c*) magic angle spinning and no decoupling, and (*d*) with both magic angle spinning and decoupling. (*Thanks are due to Prof. R. K. Harris of the University of Durham for permission to reproduce this spectrum.*)

8.3.3 N.M.R. in Medicine

We cannot leave the topic of solid-state n.m.r. without a brief mention of some of its applications to medicine, which are becoming routine. Living tissue is, of course, rich in protons—in fats, proteins, and the water which constitutes a large proportion of cells—and in phosphorus; n.m.r. studies of both these nuclei have found application in medicine. Whole plants, small animals, and parts of larger animals—e.g. a human arm, leg, or head, still attached to its living host—can be placed between the poles of a suitably designed superconducting magnet and, with the provision of surface coils to direct the radiofrequency radiation to the area of interest, spectra of relatively small regions can be obtained. Local shim coils can be arranged to give a good homogenous field only over the volume of interest, while allowing the field outside this volume to change rapidly with distance; in this way resonances from outside the volume are broad and merge with the background noise.

^{31}P, a nucleus with spin $\frac{1}{2}$ and 100 per cent natural abundance, has long been of interest in n.m.r. studies. It has a usefully wide range of chemical shifts and its coupling patterns to neighbouring hydrogens can give useful structural information. In medicine, some disorders are associated with an alteration in the type or concentration of phosphorus-containing chemicals in the body, and n.m.r. studies have been useful in diagnosis. Some ^{13}C work has also been done, although here the limitation of 1 per cent abundance is severe. N.m.r. can, however, be used to trace the destination of ^{13}C atoms from enriched samples.

The medical applications of hydrogen n.m.r. spectra have developed in a rather different way. Here the aim is not to produce a *spectrum* of living cells, but to produce a *density map*, or *image*, of protons in the tissue. The resulting picture is rather like an X-ray photograph, but of the soft tissue rather than hard bones, and it is obtained without subjecting the body to harmful X-rays. The technique is usually known as magnetic resonance imaging, or MRI. Perhaps of greatest importance is the fact that the n.m.r. image obtained can distinguish between normal and cancerous cells.

If the part of the body being studied were placed in a uniform field, the result would simply be one broad resonance signal from all the protons; instead, the body is placed in a field with a known varying gradient, and the size of the proton signal is measured as a function of distance along the field; the relaxation time of the signal at each point is also measured, since it is known that this differs significantly in normal and malignant cells. The total image is built up by scanning different positions with the field gradient and using a computer to convert the signals into a density map.

Figure 8.16 shows a set of MRIs of a human brain viewed at different levels. The images in (*a*), (*b*), and (*c*) are taken from above, and show horizontal 'slices' through the head at three different levels; (*d*) is from the side. The eyes, nose, and even part of the ears of the subject can be seen.

The images shown in Fig. 8.16 are enhanced by injecting the patient with a small (and, hopefully, relatively harmless) quantity of a gadolinium complex prior to scanning. The ligands of this complex enable the gadolinium to coordinate to water molecules, thus shortening their relaxation times and producing stronger signals for the scan.

8.4 STRUCTURAL METHODS: EXAFS AND SEXAFS

Information on the detailed structure of solids and of the near-surface region, in terms of interatomic distances and angles, is available from an extension of X-ray absorption and Auger techniques. If X-ray absorption is studied, the information obtained mainly concerns the bulk material, whereas the Auger electrons provide information on the surface and near-surface region. The two techniques are known as extended X-ray absorption fine structure (EXAFS) and surface extended X-ray absorption fine structure (SEXAFS), respectively.

8.4.1 Extended X-ray Absorption Fine Structure (EXAFS)

In order to understand the basis of this technique, it is necessary to think of electrons in terms of their wave properties, rather than as particles. When an X-ray photon is absorbed and a core electron emitted, the outgoing photoelectron can be thought of as a spherical wave, as illustrated in Fig. 8.17(*a*). This shows the situation where an X-ray photon displaces an electron from the core levels of the central atom (shown shaded in the figure); the emitted electron behaves as an expanding wave which, when it reaches neighbouring atoms, will be partially reflected or 'back-scattered' (Fig. 8.17(*b*)). The scattered wave interferes with the original electron wave; if it is in phase with the original wave, constructive interference occurs and the absorption of X-rays is enhanced, while if it is out of phase, there is destructive interference and X-ray absorption is reduced. The net result is that the detector 'sees' a variation in X-ray absorption close to a transition (the so-called 'absorption edge'). Obviously, this variation or oscillation of signal will depend on the number, nature, and distance of the surrounding atoms from which the wave is back-scattered.

Figure 8.18(*a*) shows the oscillations in X-ray absorption which result from bombarding a palladium foil with X-rays. For EXAFS, the total absorption is not important; all the important

Figure 8.16 MRIs of the human brain: (*a*), (*b*) and (*c*) from above and (*d*) in profile. (*Thanks are due to Dr D. Kean of the Royal Hallamshire Hospital, Sheffield, and the University of Sheffield, for these images.*)

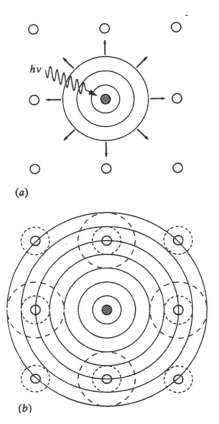

(a)

(b)

Figure 8.17 The principles of EXAFS: (*a*) a photon ejects an electron as an expanding wave; (*b*) the wave reflects from surrounding atoms to cause interference.

information can be gleaned from the relative sizes and energies of the oscillations themselves. The absorption background can therefore be allowed for and removed by computer, as shown in Fig. 8.18(*b*). The oscillations in this figure are caused by the superposition of electron waves scattering from atoms at different distances from the original emitting atom; the phase of the electron wave depends on the distance of the scattering atom from the original atom, while the number of atoms at each different distance affects its amplitude. Fourier transforming Fig. 8.18(*b*) will result in separating out these components and Fig. 8.18(*c*) results. The interatomic distances can be found from the separation of the peaks. For palladium, the nearest neighbours are 2.75 Å away, while the next nearest neighbours are at 3.89 Å, etc., as shown.

Although EXAFS has proved an extremely sensitive probe of structure, it requires a high-intensity, tunable X-ray source, such as a synchrotron. It cannot therefore be described as a routine laboratory tool, but it has already yielded much structural data on simple molecules and salts, and is being applied to more complex systems.

8.4.2 Surface Extended X-Ray Absorption Fine Structure (SEXAFS)

If, instead of observing the X-ray absorption of a solid sample, the intensity of the Auger electrons is observed, the resulting diffracted waves carry information about the surface and near-surface regions rather than the bulk material. This, as we have seen earlier, is because electrons cannot travel far through the bulk material, and can only be detected if they escape

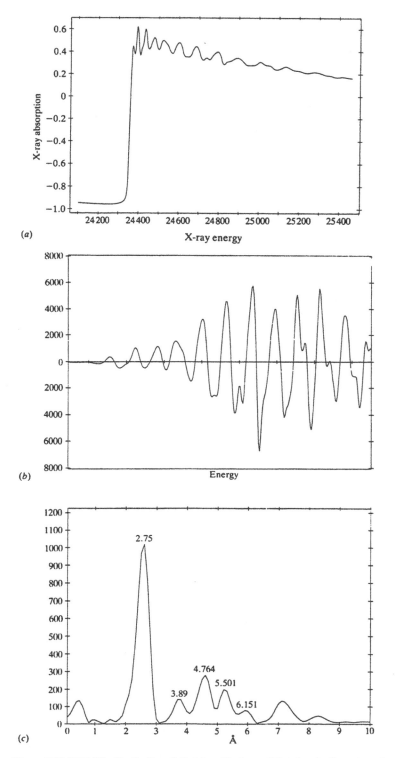

Figure 8.18 EXAFS of palladium foil: (*a*) oscillations of the signal adjacent to the adsorption edge, (*b*) data after background subtraction, and (*c*) Fourier-transformed data showing peaks from which the interatomic distances are obtained. (*Thanks are due to Dr B. Cowie of SERC Daresbury Laboratory for obtaining these data.*)

from the surface and near-surface regions. The technique is closely similar to EXAFS, however, and also suffers from the requirement for a synchroton source.

BIBLIOGRAPHY

Clark, R. J. H., and R. E. Hester (eds): *Spectroscopy of Surfaces, Advances in Spectroscopy*, Vol. 16, John Wiley, 1988.
Ibach, H., and D. L. Mills: *Electron Energy Loss Spectroscopy and Surface Vibrations*, Academic Press, 1982.
Prutton, M.: *Surface Physics*, 3rd ed., Oxford Physics Series, Oxford University Press, 1993.
Thompson, T., M. D. Baker, A. Christie, and J. F. Tyson: *Auger Electron Spectroscopy*, John Wiley, 1985.
Woodruff, D. P., and T. A. Delchar: *Modern Techniques of Surface Science*, Cambridge University Press, 1988.
Yates, J. T., and T. E. Madey (eds): *Vibrational Spectroscopy of Molecules on Surfaces*, Plenum Publishing Corporation, 1987.
Zangwill, A.: *Physics at Surfaces*, Cambridge University Press, 1988.

PROBLEMS

8.1 Water is known to adsorb on a number of transition metal surfaces, via the oxygen atom, as shown below:

    (*a*) Considering the vibrational modes of the free molecule (illustrated in Fig. 3.9), sketch the expected vibrations for the adsorbed molecule. Which of the modes would you expect to be able to observe using EELS?

    (*b*) There are three rotations for the 'free' molecule. Given that the axes for these rotations are parallel and perpendicular to the metal surface, sketch modes which arise for the adsorbed molecule, indicating for each the axis of rotation.

    (*c*) Which of the techniques described in this chapter would you use to detect these modes?

8.2 How many peaks would you expect to see in the AES spectrum of the following elements:
    (*a*) Hydrogen   (*b*) Carbon   (*c*) Sodium

8.3 Three modes are observed in the RAIRS spectrum of ethyne adsorbed on Cu(111): at 2920, 1307 and 920 cm^{-1}. These modes are assigned in this chapter. What experiment could you perform in order to prove which peaks are due to C—C modes and which to C—H modes? Outline your reasons.

8.4 Oxygen can be bound molecularly to a number of single crystal surfaces.
    (*a*) Sketch the vibrational modes which would be observed by RAIRS and EELS, given that the molecule may be adsorbed with the molecular axis (*i*) parallel to the surface or (*ii*) perpendicular to the surface.

    (*b*) How could isotopic substitution by ^{18}O of one of the oxygen atoms in the molecule be used to distinguish between the the two types of adsorption?

    (*c*) What effect would substitution of both ^{16}O atoms by ^{18}O atoms have on the spectra?

8.5 When methanol is adsorbed on an oxidized copper surface, it is deprotonated to produced a methoxy (CH_3—O) species. EELS works revealed that the species adsorbs via the oxygen atom, but, because of its poor resolution, was unable to determine whether the CO bond was perpendicular to the surface (cf. ethylidyne, shown in Fig. 8.7(*c*)) or tilted. By applying the metal surface selection rule to the C—H stretching vibrations which you would expect for the two species, suggest how RAIRS (with its high resolution) would be able to show that the methoxy species has its CO bond perpendicular to the surface.

8.6 Which technique for investigating electronic properties would you use to find the oxidation state of an adsorbed oxygen species? What information would you expect to obtain?

MÖSSBAUER SPECTROSCOPY

Mössbauer spectroscopy, named after its discoverer who received a Nobel prize in 1961 for his work, is concerned with transitions between energy levels within the nuclei of atoms. About a third of the known elements, principally the heavier ones, when formed by the radioactive decay of an isotope of the same or a different element, are initially produced in an excited nuclear state; after a very short delay, of the order of microseconds, the excited nucleus reverts to the ground state and emits energy of a very high frequency, usually in the γ-ray region of the spectrum. It is the study of this γ-ray emission and subsequent reabsorption which constitutes Mössbauer or γ-ray spectroscopy.

9.1 PRINCIPLES OF MÖSSBAUER SPECTROSCOPY

In this section we outline the essentials of γ-ray spectroscopy using as our example the iron nucleus—one of the most thoroughly studied in this respect. The isotope ^{57}Fe is conveniently produced by the decay of radioactive ^{57}Co, which is a relatively long-lived species, having a half-life of some 270 days. A simplified energy-level diagram for the process is shown in Fig. 9.1 where, following electron capture, the cobalt nucleus is seen to produce ^{57}Fe in an excited energy state, designated Fe*; Fe* very rapidly drops to the ground state Fe, the energy change involved being $\Delta E = 2.30 \times 10^{-15}$ J (per nucleus). The frequency of the emitted γ-ray is thus $v = \Delta E/h = 3.5 \times 10^{18}$ Hz. Now if a second Fe nucleus, initially in the ground state, were to be put in the emitted radiation beam, it would be expected to absorb energy from the beam and to become excited into the Fe* state. A γ-ray detector (a scintillator or a geiger counter) placed in the beam behind the absorber should show that this has happened. In fact, it was found possible to observe γ-ray absorption only if the source material were moved relative to the sample; the two main reasons for this are of fundamental importance, and we discuss them now.

Firstly, consider the width of the γ-ray emission; the half-life of the excited state Fe* is about 1.5×10^{-7} s and, replacing this in Eqs (1.10) and (1.11), we see that the Heisenberg uncertainty principle gives the energy uncertainty as $\delta E \approx 10^{-34}/\delta t \approx 10^{-27}$ J and the frequency uncertainty as $\delta v \approx 10^6$ Hz. These uncertainties are very small when compared with the energy

Figure 9.1 Simplified energy-level scheme showing the decay of radioactive ^{57}Co to excited state and then ground state ^{57}Fe.

change of 10^{-15} J and the associated frequency of 3.5×10^{18} Hz—the excited state energy level is very precisely defined. In fact, the relative line width, $\delta v/v \approx 10^{-12}$, is very much smaller than that found in any other spectroscopic technique. For instance, in n.m.r. spectra we have encountered a line width of some 0.1 Hz in 100 MHz, for which $\delta v/v \approx 10^{-8}$; in the infra-red region, $\delta v/v$ is about 10^{-5} for gases and rather larger for liquids.

The second factor is that, with some very sharp emission frequency, any effect which produces a change in the nuclear energy levels or in the radiation frequency itself will prevent resonance reabsorption. One very important effect exists in the motion of the emitting nucleus. A photon of frequency 10^{18} Hz has a relatively large momentum (given, according to the de Broglie relationship, by h/λ where λ is the wavelength) and, when this is ejected by the nucleus, the nucleus will recoil considerably in order to conserve the total momentum. A simple calculation shows that the nuclear recoil velocity is of the order of 10^2 m s^{-1}. It is well known that, when a moving body emits radiation (or sound) a stationary observer sees (or hears) a shifted frequency; this, of course, is called the Doppler effect. Specifically, the frequency shift Δv is given by:

$$\Delta v = \frac{v \upsilon}{c} \quad \text{Hz} \tag{9.1}$$

where v and c are the frequency and velocity of the emitted radiation, respectively, and υ is the relative velocity of source and observer. For our Fe* nucleus, recoiling at 10^2 m s^{-1} and emitting radiation at 3.5×10^{18} Hz:

$$\Delta v = \frac{3.5 \times 10^{18} \times 10^2}{3 \times 10^8} \approx 10^{12} \text{ Hz}$$

This shift, although small relative to the emission frequency of 10^{18} Hz, is very large indeed compared with the line width of 10^6 Hz.

It is these factors then—nuclear recoil moving the emitted frequency some millions of line widths away from the position of the absorption frequency—which earlier prevented the study of γ-ray spectroscopy. Mössbauer's main contributions—elegantly simple when seen in retrospect—were, firstly, to use solid crystal lattices as emitters, in which the emitting nucleus is firmly fixed to surrounding nuclei, and hence has a very large apparent mass within which the recoil energy can be dissipated, and further to cool both source and sample to low temperatures so that thermal motions of the lattice atoms are reduced to a minimum.

In this way it was shown, for instance, that the γ-ray emission from a piece of radioactive cobalt metal (that is Fe* nuclei in a metal lattice) could be absorbed by metallic iron. However, it was also found that if the iron sample were in any other chemical state, the different chemical surroundings of the iron nucleus produce a sufficient effect on the nuclear energy levels for absorption no longer to occur. The final requirement in our spectroscopy—we already have source, sample, and detector—is some scanning device so that we may search for the precise absorption frequency of a particular sample. The Doppler effect here becomes useful rather than a nuisance.

We have seen that a velocity of $10^2\,\mathrm{m\,s^{-1}}$ produces a huge Doppler shift; a similar calculation from Eq. (9.1) shows that a velocity of $1\,\mathrm{cm\,s^{-1}}$ ($10^{-2}\,\mathrm{m\,s^{-1}}$) produces a shift of $10^8\,\mathrm{Hz}$—which, being about 100 line widths, represents a reasonably wide scan of the spectrum. Experimentally, then, we could proceed as in Fig. 9.2(a). Here a piece of ^{57}Co is mounted on a screw thread, rotation of which gives a steady velocity drive to the source. A geiger counter mounted behind the sample will show a sudden fall in the count rate when the sample starts to absorb the γ-rays emitted by the source. A complete spectrum will have to be examined point by point, since for any one source velocity the Doppler shift is constant, and so one would need to set up perhaps a hundred different relative velocities of course and sample, varying from $+1\,\mathrm{cm\,s^{-1}}$ through 0 to $-1\,\mathrm{cm\,s^{-1}}$, to cover a reasonable range of frequencies.

A much more convenient arrangement is shown in Fig. 9.2(b) where the source metal is seen mounted on what is essentially a loudspeaker coil. An alternating current of a few cycles per

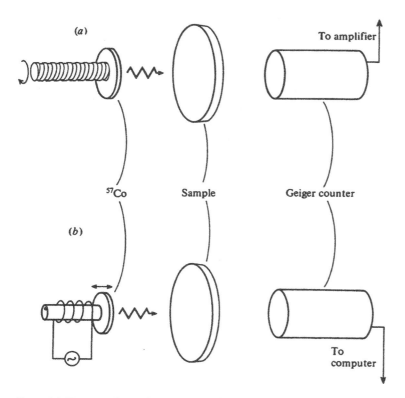

Figure 9.2 Two experimental arrangements for Mössbauer spectroscopy: (a) shows a screw-thread drive and (b) an oscillating drive for the ^{57}Co source.

second is applied to the coil so that the source oscillates back and forth. At the extremes of its motion it will have zero velocity relative to the sample, whereas at the centre it will have its maximum velocity either towards or away from the sample. The geiger counter output now needs to be fed to a multi-channel computer wherein the results from each point of the source's movement are collected and summed over each cycle. A time of several minutes to a few hours is usually sufficient to record a good spectrum. The final spectrum is displayed on counts per second versus centimetres per second scales, a fall in counts per second indicating γ-ray absorption by the sample.

In essence, then, Mössbauer spectroscopy is simple. Excited nuclei must be available from some source, which should be fairly stable (i.e. have a half-life of at least several weeks) so that its frequent replacement is unnecessary, and so that the rate of decay, and hence the intensity of γ-emission, stays sensibly constant during an experiment of several hours. The excited nuclei must decay to the ground state rapidly, emitting γ-rays in the 10^{17}–10^{20} Hz range. A Doppler shifting device, a geiger counter, and a small computer complete the apparatus. Against this apparent simplicity, however, it must be remembered that the relative velocity of source and sample must be *very* precisely controlled (an error of $0.01\,\text{cm}\,\text{s}^{-1}$ in the velocity shifts the frequency by more than one line width and could easily render an absorption undetectable) and that the source and sample should ideally be maintained at the temperature of liquid helium. However, as stated earlier, many nuclei fulfil the required conditions and have been studied by γ-ray techniques. We turn now to consider some applications.

9.2 APPLICATIONS OF MÖSSBAUER SPECTROSCOPY

9.2.1 The Chemical Shift

It has already been mentioned that a nucleus in chemical surroundings different from those of the source does not absorb at the same frequency; this effect is referred to as the chemical shift, or sometimes as the isomer shift, and an observed shift is usually reported in $\text{cm}\,\text{s}^{-1}$, conversion to hertz or megahertz being trivial but unnecessary. Such a shift is illustrated in the Mössbauer spectrum of the ferrocyanide ion, $[Fe(CN)_6]^{4-}$, in Fig. 9.3(a), where we see that the single sharp absorption peak occurs at a velocity of about $-\frac{1}{2}\,\text{mm}\,\text{s}^{-1}$ with reference to the Fe* in a ^{57}Co source as zero.

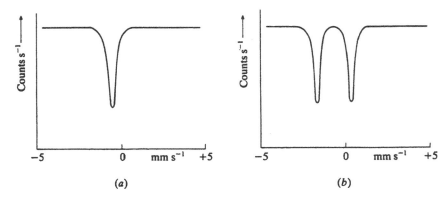

(a)   *(b)*

Figure 9.3 The Mössbauer spectra of (a) the $[Fe(CN)_6]^{4-}$ ion and (b) the $[Fe(CN)_5NO]^{2-}$ ion, showing quadrupole splitting.

Table 9.1 Chemical shift of some ^{119}Sn compounds

Valence state	Electron configuration	Chemical shift (mm s^{-1})
Sn^{4+}	$5s^0 5p^0$	0
Sn (4-covalent)	$5(sp^3)$	2.1
Sn^{2+}	$5s^2 5p^0$	3.7

The main factor affecting the magnitude of chemical shifts is the electron density at the nucleus concerned and, since p, d, etc., orbitals have zero density at their nuclei, we may be quite specific and say that it is the s orbital density which is important. Observation of shifts, then, allows measurement of relative s electron density which, in turn, gives an estimate of the bond character of atoms or ions chemically attached to the Mössbauer nucleus. Thus in a series of tin compounds, using the Mössbauer isotope ^{119}Sn, the chemical shifts given in Table 9.1 are observed; the outer electron structure of the tin atom is $5s^2 5p^2$, so in the 4+ state tin has no outer s electrons, whereas in the 2+ state it is assumed that the (higher) $5p^2$ electrons are removed, leaving two s electrons. In its four-covalent state, where the compounds are tetrahedral and the configuration is $5(sp^3)$, there is essentially only one s electron. The chemical shifts reflect these structures almost linearly. Clearly, it is a relatively simple matter to discover the valence state of an unknown tin compound from its Mössbauer spectrum.

Another important factor affects chemical shifts; nuclei in excited states usually have a different radius from those in the ground state—they may be either smaller or larger. Theoretically it may be shown that the chemical shift is given by:

$$\text{Chemical shift} = \text{const} \, (\rho_{\text{ex.}}^2 - \rho_{\text{gd.}}^2) \left(\frac{R_{\text{ex.}} - R_{\text{gd.}}}{R_{\text{gd.}}} \right) \tag{9.2}$$

where ρ^2 represents s electron density at the excited and ground state nuclei, and R is the radius of each nucleus; the constant is simply a universal proportionality factor. For any given nucleus, of course $(R_{\text{ex.}} - R_{\text{gd.}})/R_{\text{gd.}}$ is constant, and so relative electron densities can be determined as stated above. However, the *sign* of the chemical shift will depend on whether $R_{\text{ex.}}$ is larger or smaller than $R_{\text{gd.}}$. One way to discover which is the larger is to examine a sample under high pressure; here the assumption is that the compression increases the electron density near the sample nucleus, so $(\rho_{\text{ex.}}^2 - \rho_{\text{gd.}}^2)$ decreases, and then the sign of the observed chemical shift change gives the sign of $(R_{\text{ex.}} - R_{\text{gd.}})$.

Measurement of change of nuclear size on excitation clearly has great importance for theories of nuclear structure. For instance, it is an observed (but as yet unexplained) fact that in ^{129}I the excited nucleus is larger, whereas in ^{127}I the ground state nucleus has the larger radius—it is rather surprising that the addition of two neutrons should have such a profound effect on the nuclear structure. These nuclei are formed in the excited state by decay from the radioactive tellurium nuclei, ^{129}Te and ^{127}Te.

9.2.2 Quadrupole Effects

It happens that the majority of Mössbauer nuclei have a non-zero spin and, further, that most of them have half-integral rather than integral spins. The spin of the excited state is invariably different from that in the ground state (and, indeed, there is a selection rule which says that this must be so), and so it follows that either or both of the nuclear states must involve a spin greater than $\frac{1}{2}$, i.e. one or both will have a quadrupole moment (cf. Sec. 7.3.4), and this will interact with electric field gradients in the vicinity.

A fairly common situation is for the excited state nucleus to have $I = \frac{3}{2}$ (and thus a quadrupole moment) and the ground state $I = \frac{1}{2}$; this is found, for example, in ^{57}Fe, ^{119}Sn, and ^{129}Xe. For this case the four (that is $2I + 1$) possible orientations of the excited nucleus in an electric field along the vertical (z) direction are shown in Fig. 9.4(a). Since the angle which the cigar-shaped nucleus (assuming the nucleus to have a *positive* quadrupole moment) makes with the field gradient is the same in the $I_z = +\frac{1}{2}$ and the $I_z = -\frac{1}{2}$ states, these two states have the same energy in the field. Similarly, the $I_z = \pm\frac{3}{2}$ states have the same energy, although obviously this energy will be different from that of the $\pm\frac{1}{2}$ states. Thus the excited nuclear energy, shown in Fig. 9.4(b), splits into two levels, Fig. 9.4(c), when an electric field gradient exists at the nucleus. With a positive quadrupole moment the $\pm\frac{3}{2}$ states are raised in energy and the $\pm\frac{1}{2}$ states lowered, whereas if the quadrupole moment is negative (tangerine-shaped nucleus) the reverse is true. The splitting is small, with normal electric fields but, as the spectrum of the nitroprusside ion, [Fe(CN)$_5$NO]$^{2-}$, in Fig. 9.3(b) shows, it gives quite an observable effect. It should be noted that the electric field causing the splitting here is *not* externally applied, but is inherent in the structure of the ion. Thus comparing Fig. 9.3(a) with (b), we see that the Fe(CN)$_6^{4-}$ ion is sufficiently symmetrical for there to be no net field gradient at the iron nucleus, while replacement of one CN group by NO produces an internal field.

While useful data on the magnitudes of internal crystal fields can be obtained from such spectra, it should be pointed out that a simple explanation is not available for every spectrum so far observed. Thus some compounds which might be expected to possess internal electric fields do not show splitting, and some which would be expected to show a single line spectrum, show unexplained multiplet structure.

When other spin quantum numbers are involved, or when both ground and excited state nuclei have a quadrupole moment, no new principles are involved and an extrapolation of the above arguments usually yields a convincing explanation of observed spectral structure.

9.2.3 The Effect of a Magnetic Field

The non-zero spin which we have seen to be associated with either the excited or the ground state nucleus—and usually with both—will interact with a magnetic field in the manner already described in Chapter 7. Each energy state will split into $2I + 1$ separate energy levels, the spacing between them being $\Delta E = g\beta_N B_z/h$, where B_z is the magnetic field at the nucleus. In general the g values of the excited and ground states will differ, so the splitting in each state will be different; in addition, the g values may have opposite signs (as in the case of ^{57}Fe, where the excited state g is negative).

Figure 9.5(a) and (b) shows the situation for this nucleus; at (a) the energy levels are shown in the absence of a magnetic field, whereas in (b) the field is shown splitting the ground state ($I = \frac{1}{2}$, g positive) into two sub-levels and the excited state ($I = \frac{3}{2}$, g negative) into four. Following the conventions of Sec. 7.1.2, we show the $I_z = +\frac{1}{2}$ sublevel of the ground state nucleus to be lower than that of $I_z = -\frac{1}{2}$, while the sublevels of the excited state increase in energy in the order $I_z = -\frac{3}{2}, -\frac{1}{2}, +\frac{1}{2} +\frac{3}{2}$.

The selection rules for Mössbauer spectroscopy are complicated; essentially all transitions for which $\Delta I_z = 0$ or ± 1 are allowed, as shown in the diagram, but the relative transition probabilities are found to differ according to the states involved. Thus if we consider the six transitions in pairs, and write the Fe* state first, we have (1) $\frac{3}{2} \leftrightarrow \frac{1}{2}, -\frac{3}{2} \leftrightarrow -\frac{1}{2}$; (2) $\frac{1}{2} \leftrightarrow \frac{1}{2}, -\frac{1}{2} \leftrightarrow -\frac{1}{2}$; (3) $-\frac{1}{2} \leftrightarrow +\frac{1}{2}, +\frac{1}{2} \leftrightarrow -\frac{1}{2}$. Detailed calculation shows that, while the two members in each pair have the same probability, the relative probabilities of the three pairs

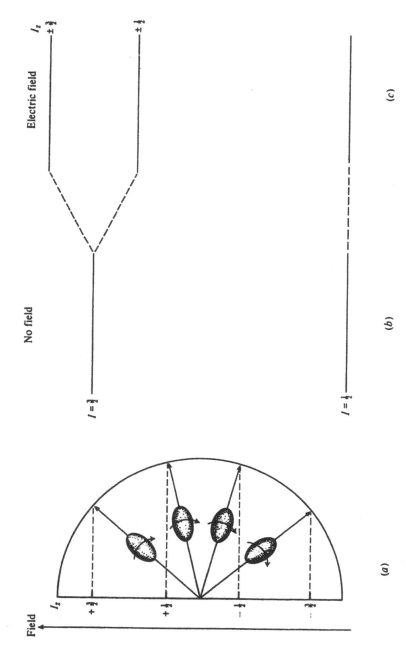

Field

I_z

$+\frac{3}{2}$

$+\frac{1}{2}$

$-\frac{1}{2}$

$-\frac{3}{2}$

(a)

No field

$I = \frac{3}{2}$

$I = \frac{1}{2}$

(b)

Electric field

I_z

$\pm \frac{3}{2}$

$\pm \frac{1}{2}$

(c)

Figure 9.4 *(a)* The four allowed directions of a nuclear spin $I = \frac{3}{2}$. *(b)* The energy levels of the excited and ground state in the absence of an electric field. *(c)* The energies in the presence of a field.

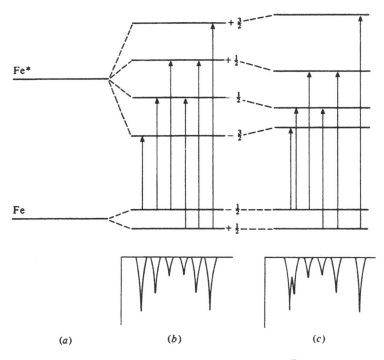

Figure 9.5 (*a*) The excited and ground state energy levels of ^{57}Fe and (*b*) the splitting produced by a magnetic field. In (*c*) the effect of applying simultaneous magnetic and electric fields is shown.

are 3 : 2 : 1 for (1) : (2) : (3). The overall spectrum will, therefore, be as shown at the foot of Fig. 9.5(*b*), and it is precisely this type of spectrum which has been observed, for instance, in the case of metallic iron.

We should note that the magnetic field necessary to cause the energy-level splitting may be applied externally, but it also often happens that *internal* effects within the sample produce sufficient field to cause observable splitting. By using an external field to 'calibrate' the spectrum, it is quite possible to estimate the magnitude of the internal fields, and in this way fields of 20–50 T have been found for various compounds of ^{57}Fe. Such fields are large compared with the 5–15 T fields created by superconducting magnets, but one should not suppose that the internal field extends uniformly through the bulk sample—the field is formed by interaction of the nucleus with its surrounding electrons and this is probably an extremely localized effect.

If an *electric* field exists within the sample as well as an internal or external magnetic field, then the quadrupolar shifts are superimposed on the magnetic splitting. In Fig. 9.4 we saw that, for a positive quadrupolar moment, the $\pm\frac{3}{2}$ states are moved upwards in energy, while the $\pm\frac{1}{2}$ states move down. This effect is shown in Fig. 9.5(*c*) and we see that, since the selection rules and transition probabilities are unchanged, a six-line spectrum is again produced, but that the lines are no longer equally spaced. A spectrum of this type is observed for iron difluoride, FeF_2.

BIBLIOGRAPHY

Goldanskii, V. I., and R. H. Herber: *Chemical Applications of Mössbauer Spectroscopy*, Academic Press, 1968.
Greenwood, N. N.: 'The Mössbauer Spectra of Chemical Compounds': *Chemistry in Britain*, **3**, 56, 1967.
Greenwood, N. N., and T. C. Gibb: *Mössbauer Spectroscopy*, Chapman & Hall, 1971.
Wertheim, G. K.: *The Mössbauer Effect*, Academic Press, 1964.

PROBLEMS

(Useful constants: $h = 6.626 \times 10^{-34}$ J s; $c = 2.998 \times 10^8$ m s^{-1}.)

9.1 A photon of wavelength λ has an equivalent momentum of h/λ. Calculate the recoil velocity of a free Mössbauer nucleus of mass 1.67×10^{-25} kg (i.e. atomic weight 100) when emitting a γ-ray of 0.1 nm wavelength. What is the Doppler shift of the γ-ray frequency to an outside observer?

9.2 A particular Mössbauer nucleus has spins $\frac{5}{2}$ and $\frac{3}{2}$ in its excited and ground states, respectively. Into how many lines will the γ-ray spectrum split if: (a) the nucleus is under the influence of an internal electric field gradient, but no magnetic field is applied; (b) there is no electric field gradient at the nucleus but an external magnetic field is applied; (c) both an internal electric field gradient and an external magnetic field are present?

GENERAL BIBLIOGRAPHY

Brand, J. C. D., and J. C. Speakman: *Molecular Structure: The Physical Approach*, 2nd ed., Edward Arnold, 1975.

Brittain, E. F., W. O. George, and C. H. Wells: *Introduction to Molecular Spectroscopy, Theory and Experiment*, Academic Press, 1970.

Chang, R.: *Basic Principles of Spectroscopy*, Krieger, 1978.

Crooks, J. E.: *The Spectrum in Chemistry*, Academic Press, 1978.

Ebsworth, E. A. V., D. W. H. Rankin, and S. Cradock: *Structural Methods in Inorganic Chemistry*, 2nd ed., Blackwell Scientific Publications, 1991.

Harris, D. C., and M. D. Bertolucci: *Symmetry and Spectroscopy*, Dover Publications, 1989 (originally Oxford University Press, 1978).

Richard, W. G., and P. R. Scott: *Structure and Spectra of Molecules*, John Wiley, 1985.

Steinfeld, J. I.: *Molecules and Radiation: An Introduction to Modern Molecular Spectroscopy*, Harper & Row, 1974.

Straughan, B. P., and S. Walker: *Spectroscopy*, Vols 1–3, Chaman & Hall, 1976.

Svanberg, S.: *Atomic and Molecular Spectroscopy*, 2nd ed., Springer-Verlag, 1992.

Wheatley, P. J.: *The Chemical Consequences of Nuclear Spin*, North-Holland, 1970.

Williams, D. H., and I. Fleming: *Spectroscopic Methods in Organic Chemistry*, 4th ed., McGraw-Hill, 1989.

Chapter 1

1.1 $2000\,\text{cm}^{-1} = 5\,\mu\text{m}$ Infrared: vibrational
     $0.15\,\text{nm} \;\; = 2 \times 10^{18}\,\text{Hz}$ X-ray: nuclear configuration
     $500\,\text{nm} \;\; = 20\,000\,\text{cm}^{-1}$ UV/visible: electronic
     $9\,\text{GHz} \;\; = 0.3\,\text{cm}^{-1}$ E.s.r.: electron spin

1.2 $10\,\mu\text{m} = 2.998 \times 10^{13}\,\text{Hz} = 1000\,\text{cm}^{-1}$
    $\Delta\varepsilon \;\; = h\nu = 1.986 \times 10^{-20}$ joules per molecule
              $= 11.96\,\text{kJ}\,\text{mol}^{-1}$
For twice the energy change, the wavelength is $5\,\mu\text{m}$.

1.3 Br_2: neither infra-red nor microwave active.
    HBr: both infra-red and microwave active.
    CS_2: infra-red, but not microwave active.

1.4 Widening the slits would reduce the resolution but increase the sensitivity.

1.5 See Sec. 1.7.2
    (*a*) Transmittance $= I/I_0 = 0.01$,
         so Absorbance $= \log_{10}(1/T) = 2$
    (*b*) Absorbance $= 2 = \epsilon c l = \epsilon \times 10^{-3} \times 1\,\text{mol}\,\text{l}^{-1}\,\text{cm}$
         Hence $\epsilon = 2 \times 10^3\,\text{l}\,\text{mol}^{-1}\,\text{cm}^{-1}$
               $= 2 \times 10^5\,\text{l}\,\text{mol}^{-1}\,\text{m}^{-1}$

1.6 (*a*) $\delta E \approx 10^{-35}\,\text{J}$; $\delta\nu \approx 0.016\,\text{Hz}$.
    (*b*) $\delta E \approx 10^{-33}\,\text{J}$; $\delta\nu \approx 1.6\,\text{Hz}$.

1.7 Energy change $4.005 \times 10^{-22}\,\text{J}$ molecule^{-1} :

$T(\text{K})$	29	290	2900
N_{upper}	368	905	990

Energy change 4.005×10^{-21} J molecule^{-1} :

T(K)	29	290	2900
N_{upper}	0	368	905

1.8 A single sharp spike at the frequency of the sine wave.

Chapter 2

2.1 HCl, CH$_3$Cl, CH$_2$Cl$_2$, and H$_2$O only.

2.2 $B = 0.35717$ cm^{-1}; $I = 7.837 \times 10^{-46}$ kg m^2; $r = 1.756 \times 10^{-10}$ m ($= 175.6$ p.p.m.); $J = 9 \to J = 10$ transition is at 7.1434 cm^{-1}; maximum intensity at $J = 17$.
 (a) 0 rotations s^{-1}.
 (b) 1.90×10^{11} rad s^{-1} = 3.03×10^{10} rotations s^{-1}.
 (c) 1.41×10^{12} rad s^{-1} = 2.25×10^{11} rotations s^{-1}.

2.3 H^{37}Cl: $B = 10.575$ cm^{-1}; D^{35}Cl: $B = 5.447$ cm^{-1}.

2.4 For HI: $B = 6.4275$ cm^{-1}; $I = 4.355 \times 10^{-47}$; $r = 1.619 \times 10^{-10}$ m; transition $J = 4 \to J = 5$.
 For DI: $B = 3.2535$ cm^{-1}; $I = 8.604 \times 10^{-47}$; $r = 1.617 \times 10^{-10}$ m; transition $J = 9 \to J = 10$;
 bond lengths are the same to within experimental error.

2.5 Maximum population is at $J = 10$. The approximate relative intensities of some other lines are as follows:

J	0	2	4	6	8	10	12	14	16	18	20
Relative intensity	1	4.9	8.2	10.7	12.2	12.8	12.4	11.3	9.7	8.0	6.2

2.6 Taking atomic masses as ^{12}C = 19.93×10^{-27} kg, ^{13}C = 21.59×10^{-27} kg, and ^{16}O = 26.56 $\times 10^{-27}$ kg, the first three lines in the spectra of ^{12}CO and of ^{13}CO are at:
 ^{12}CO: 3.8642 cm^{-1} 7.7285 cm^{-1} 11.593 cm^{-1}
 ^{13}CO: 3.6945 cm^{-1} 7.3890 cm^{-1} 11.083 cm^{-1}
so it would be necessary to distinguish lines 0.17 cm^{-1} apart (i.e. the separation between the first lines of each of the spectra). The temperature of Saturn's atmosphere could be estimated from the relative intensities of the three lines.

2.7 $\omega = (4B^3/D)^{1/2} = 2995$ cm^{-1}. Force constant = 517.6 N m^{-1}.
 (a) The calculation assumes simple harmonic motion, which is only an approximation to the truth.
 (b) D is small and so is not very accurately determined; dividing the large B^3 by the small D leads to inaccuracies.

Chapter 3

3.1 $\omega_e = 1903.98$ cm^{-1}; $\omega_e x_e = 13.96$ cm^{-1} (or $x_e = 7.332 \times 10^{-3}$); zero-point energy = 948.5 cm^{-1}; force constant, $k = 1595.0$ N m^{-1}; dissociation energy = 776.2 kJ mol^{-1} (inaccurate because cubic and quartic terms in $(v + \frac{1}{2})$ are ignored and these become important at large v—see Sec. 6.1.4).

3.2 Zero-point energy for (HCl + D$_2$) = 2937.5 cm^{-1}, while that for (DCl + HD) = 2808.5 cm^{-1}. Hence 129 cm^{-1} (or 1.542 kJ mol^{-1}) is evolved during the reaction.

3.3 The hot band has approximately 0.36 the intensity of the fundamental.

3.4 $18.795\,\text{cm}^{-1}$

3.5

Molecule	HBr	O_2	OCS	SO_2	BCl_3	HC$\equiv$CH	CH_4	CH_3I	C_6H_6
Degrees of freedom	1	1	4	3	6	7	9	9	30

3.6 $\nu_{CD} = 2422\,\text{cm}^{-1}$ (if ν_{CH} is taken as $3300\,\text{cm}^{-1}$);
$\nu_{OD} = 2619\,\text{cm}^{-1}$ (if ν_{OH} is taken as $3600\,\text{cm}^{-1}$);
$\nu_{CS} = 975\,\text{cm}^{-1}$ (if ν_{CO} is taken as $1100\,\text{cm}^{-1}$).

3.7

Molecule	$^1\text{H}^{37}\text{Cl}$	$^2\text{H}^{35}\text{Cl}$	$^2\text{H}^{37}\text{Cl}$
Estimated frequency (cm^{-1})	2988	2144	2141

3.8   The bond length increases by about 2 per cent from $v = 0$ to $v = 1$; at high J values this results in the R branch lines becoming closer together and the P branch lines further apart.

3.9   (a)   Fundamental frequency $= 2991.3\,\text{cm}^{-1}$; zero-point energy $= 1495.7\,\text{cm}^{-1}$.
(b)   Rotational constant $= 10.59\,\text{cm}^{-1}$.
(c) $P_{(3)} = 2927.8$; $P_{(2)} = 2948.9$; $P_{(1)} = 2970.1\,\text{cm}^{-1}$;
$R_{(0)} = 3012.5$; $R_{(1)} = 3033.7$; $R_{(2)} = 3054.8\,\text{cm}^{-1}$.
(d)   Your sketch should look like the spectrum at the foot of Fig. 3.6.
(e)   (i)   The R branch lines will get progressively closer and the P branch lines further apart towards the wings of the spectrum because of the breakdown of the Born-Oppenheimer approximation.
(ii)   Since HCl has two abundant isotopic forms—$H^{35}Cl$ and $H^{37}Cl$—the real spectrum consists of two sets of P and R branch lines with an approximately 3 : 1 intensity ratio and slightly different band centre and spacing.

3.10   In an alkene C$=$C group the carbon atoms carry almost identical charges (*exactly* identical if the substituents are the same on each) and so oscillation of the bond produces a very small (or zero) dipole change. Thus the infra-red absorption will be very weak (or zero). The much more polar C$=$O group produces a relatively large dipole change on oscillation and interacts strongly with infra-red radiation.

Chapter 4

4.1   (a) The ease with which the electron distribution in a molecule or atom may be distorted by the application of an electric field.
(b) There must be a change in the polarizability of the molecule during the motion.
For rotations: $\Delta J = \pm 2$
For vibrations: $\Delta v = \pm 1$; $\Delta J = \pm 2$
(c) The laser.

4.2   Raman; $B = 60.83\,\text{cm}^{-1}$, so line separation $= 243.3\,\text{cm}^{-1}$.

4.3   No, the nuclear spin affects intensities only and in this case produces a strong–weak–strong–weak alternation.

4.4 ^{14}N—^{15}N : Raman
HC≡CH : Raman
C=O : infra-red
Re—Re : Raman

4.5   Vibrational frequency is proportional to (force constant)$^{1/2}$, so the ratio of force constants is given by $(275/122)^2 = 5.08$. Re is hexavalent, so forms a single bond in the complex, but a quadruple bond in the ion (eight electrons contribute to the bond). Other factors will affect the force constant, however, and a ratio of 5 is quite reasonable.

4.6   Molecule is linear (*PR* contour of an infra-red band) and has a centre of symmetry; hence is A—B—B—A. 3374 cm^{-1} and 3287 cm^{-1} are close to the ≡CH stretching frequency (given in Table 3.4 as 3300 cm^{-1}) so molecule is acetylene, HC≡CH. Assignments:

3374 cm^{-1}:	symmetric CH stretch
3287 cm^{-1}:	asymmetric CH stretch
1973 cm^{-1}:	C≡C stretch
729 and 612 cm^{-1}:	bending vibrations

4.7   Non-linear and no centre of symmetry; hence

3756 cm^{-1} and 3652 cm^{-1} are in the region of OH stretching frequency, so molecule is H$_2$O. Assignments:

3756 cm^{-1}:	asymmetric stretch
3652 cm^{-1}:	symmetric stretch
1595 cm^{-1}:	bend

4.8   Both molecules, no centre of symmetry. N$_2$O linear (*PR* bands); hence N—N—O. NO$_2$ non-linear; hence either

Chapter 5

5.1   5331.55 cm^{-1}, 7799.29 cm^{-1}, and 9139.80 cm^{-1}. Ionization energy = 12 186.40 cm^{-1} (≡1.511 eV).

5.2 $L = 2$; $S = \frac{3}{2}$, $J = \frac{5}{2}$. Minimum number of electrons = 3; possible electron configurations include: *spp*, *ppp*, *ppd*, etc. (all non-equivalent electrons).

5.3   (a) $^1S_0$, $^3S_1$;   (b) $^1D_2$, $^1P_1$, $^1S_0$, $^3D_{3,2,1}$, $^3P_{2,1,0}$, $^3S_1$;   (c) $^1D_2$, $^3D_{3,2,1}$;   (d) $^1F_3$, $^1D_2$, $^1P_1$, $^3F_{4,3,2}$, $^3D_{3,2,1}$, $^3P_{2,1,0}$.

5.4   (a) $^1S_0$; (b) $^1D_2$, $^1S_0$, $^3P_{2,1,0}$; (c) $^1G_4$, $^1D_2$, $^1S_0$, $^3F_{4,3,2}$, $^3P_{2,1,0}$.

5.5 $^4S_1$: total spin $= \frac{3}{2}$; hence J must be half-integral; $^2D_{7/2}$: $S = \frac{1}{2}$, $L = 2$; hence J cannot be greater than $\frac{5}{2}$; $^0P_1$: $2S + 1$ cannot be zero.

5.6 $^2P_{3/2} \rightarrow ^2D_{3/2}$: 10 lines; $^2P_{3/2} \rightarrow ^2D_{5/2}$: 12 lines; $^2P_{1/2} \rightarrow ^2D_{3/2}$: 6 lines.

5.7   Terms are (in increasing energy order): $^4S < ^2D < ^2P$.

5.8   Ru: $^5F_5$; La: $^2D_{3/2}$; Nb: $^6D_{1/2}$; Ca: $^1S_0$; Eu: $^8S_{7/2}$; Gd: $^9D_2$; Ag: $^2S_{1/2}$; W: $^5D_0$; Pt: $^3D_3$.

5.9   The ground state of Tl$^+$ is $6s^2$ with term symbol $^1S_0$ and is assigned to the zero energy level. The first excited state is $6s^1 6p^1$ for which the terms, in order of energy, are: $^3P_0 < ^3P_1 < ^3P_2 < ^1P_1$; of these, the three triplet states are assigned to the energies (b) to (d), respectively. Thus the (b)–(c) separation should be *twice* the (a)–(b) separation, since the former has an upper J value of 2 while the latter has $J = 1$ in the upper state. The observed spacings are in the ratio of about 3 : 1, which shows that Russell–Saunders is a poor approximation in this case.

Chapter 6

6.1 By plotting the given data, extrapolating, and taking the area under the graph, the dissociation energy is found to be approximately $12\,100\,\text{cm}^{-1}$ or $144.7\,\text{kJ}\,\text{mol}^{-1}$.

6.2 $490.3\,\text{kJ}\,\text{mol}^{-1}$ (or $41\,000\,\text{cm}^{-1}$).

6.3 Ground state: $v_{\text{max.}} = 69$; $D_e'' = 690.0\,\text{kJ}\,\text{mol}^{-1}$
                                        $D_0'' = 680.2\,\text{kJ}\,\text{mol}^{-1}$
        Excited state: $v_{\text{max.}} = 53$; $D_e' = 581.5\,\text{kJ}\,\text{mol}^{-1}$
                                        $D_0' = 570.7\,\text{kJ}\,\text{mol}^{-1}$

6.4 Ground state: $D_0'' = 345.8\,\text{kJ}\,\text{mol}^{-1}$
        Excited state: $D_0' = 237.4\,\text{kJ}\,\text{mol}^{-1}$
The answers to Prob. 6.3 are inaccurate (in this case badly so) because Eq. (6.10) relies on the use of terms up to $(v + \frac{1}{2})^2$ only; a better treatment would involve terms in $(v + \frac{1}{2})^3$ and $(v + \frac{1}{2})^4$.

6.5 Band head is in the P branch at $p = 14$ (Eq. (6.23)), and the position of the band head is at $19\,354\,\text{cm}^{-1}$. The ground state has the larger internuclear distance.

6.6 Four peaks of equal intensity.

6.7 The single sharp line arises from the removal of an electron from a non-bonding orbital. The cluster of peaks corresponds to the removal of either a bonding or an antibonding electron. However, since the vibrational frequency of the resulting ion is *greater* than that of the neutral molecule, the force constant must be greater in the ion and the electron must have come from an antibonding orbital.

Chapter 7

7.1 (a) For H: $0.7107\,\text{T}$; for ^{13}C: $2.8266\,\text{T}$.
       (b) For $2.5\,\text{T}$: $106.44\,\text{MHz}$; for $5.25\,\text{T}$: $223.53\,\text{MHz}$.

7.2 Chemical shift positions are 5.75δ and 1.05δ; $J_{AX} = 10\,\text{Hz}$.

7.3 Eight lines of equal intensity, symmetrically disposed about the midpoint; taking the midpoint at $0.0\,\text{Hz}$, the lines are at: -77.5, -57.5, -27.5, -7.5, $+7.5$, $+27.5$, $+57.5$, and $+77.5\,\text{Hz}$. (*Note*: The J_{HH} values given are irrelevant to the ^{19}F spectrum as the chemical shifts are large.)

7.4 The BH_4 group will give rise to a $1 : 3 : 3 : 1$ quartet because of coupling to the CH_3; conversely, the CH_3 group will show a $1 : 4 : 6 : 4$ pentet, with the same separation between the lines as exhibited in the quartet.

7.5 (a) Rapid exchange gives a single sharp line at the average chemical shift.
       (b) Slow exchange gives two sharp lines, one at each chemical shift position.
       (c) Intermediate rate of exchange gives two broad, overlapping resonances, closer together than the individual chemical shifts.
       (i) Very dry methanol shows coherent coupling between the OH and CH_3 protons; the former signal is a $1 : 3 : 3 : 1$ quartet and the latter a $1 : 1$ doublet of three times the intensity of the OH.
       (ii) A trace of water facilitates OH exchange and destroys the coupling. The spectrum shows a single OH peak (more or less broadened depending on how much water has been added) and a single CH_3 peak of three times the intensity.

7.6 (a) Since D has a spin of 1, it splits the resonances of neighbouring ^{13}C and H nuclei into three lines of equal intensity.
       (b) The proton resonance of CH_2D is a $1 : 2 : 3 : 2 : 1$ pentet.

7.7   Small groups, such as CH_3 or C_6H_5, may often undergo internal rotation at temperatures not far below the bulk melting point of the material. This lengthens the relaxation time of the nuclei and so sharpens the resonance.

7.8   (a) 6; (b) 12; (c) 9; (d) 12.

7.9   At each hydrogen: 0.5% (= 2% total)
At each carbon: 10.0% (= 60% total)
At each oxygen: 19.0% (= 38% total)

7.10   (a) The assignment of the 25-line spectrum to five pentets, where the pentets have a 1 : 4 : 6 : 4 : 1 intensity ratio, is shown below:

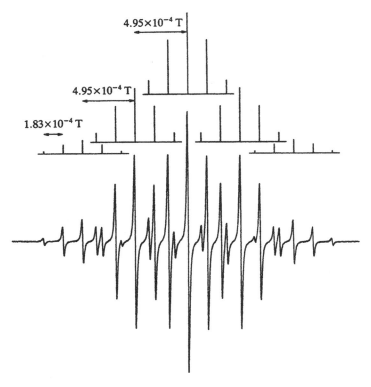

(b) In 1-iodo-naphthalene there are only three protons to coupling to the electron with $A = 1.83 \times 10^{-4}$ T. Each individual pentet in the figure will thus be reduced to a 1 : 3 : 3 : 1 quartet and the overall spectrum will be of 20 lines only. Similarly, with 2-iodo-naphthalene only three protons have a 4.95×10^{-4} coupling, so now there will be only 4 pentets in the spectrum, again making 20 lines.

Chapter 8

8.1 (*a*) Expected vibrations are:

$$v_{asym.} \qquad v_{sym.} \qquad \delta$$

Of these, the symmetric and deformation (δ) vibrations should be observed using EELS.

(*b*) Frustrated motions will be:

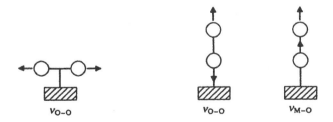

For the rocking motion, the axis is through the oxygen atom and perpendicular to the paper.

8.2 (*a*) None; (*b*) one; (*c*) two.

8.3 Observe the RAIRS spectrum of C_2D_2. Modes involving C—D will be lowered by approximately $\sqrt{2}$ in energy, while the C≡C mode will be scarcely affected.

8.4 The vibrational modes are:

$$v_{O-O} \qquad v_{O-O} \qquad v_{M-O}$$

If *both* oxygen atoms are isotopically substituted the spectrum would be unchanged except for a slight lowering in frequency due to the extra mass. If only *one* atom is substituted, the parallel-bound molecule would show only a slight frequency shift, but the perpendicularly bound species would show *two* slightly different frequencies, one due to the molecule with the ^{18}O attached to the surface and the other due to that with ^{16}O attached.

8.5 If the molecule has its C—O bond perpendicular to the surface, the symmetric stretch of the C—H bonds changes the dipole perpendicularly to the surface, but the asymmetric stretch produces a dipole change parallel to the surface, so only the former will be observed. If the bond is not perpendicular, then both modes cause a dipole change at an angle to the surface which will mean that both are likely to be seen.

8.6 UPS: chemical shift and fingerprint information; AES: fingerprint information.

Chapter 9

9.1   Recoil velocity $= 39.7\,\mathrm{m\,s^{-1}}$; Doppler shift $= 3.97 \times 10^{11}$ Hz.

9.2   (*a*) 5 lines; (*b*) 12 lines; (*c*) 12 lines.